ONE WEEK LOAN

Computer Networks and Systems
Queueing Theory and Performance Evaluation

Third Edition

Springer
*New York
Berlin
Heidelberg
Barcelona
Hong Kong
London
Milan
Paris
Singapore
Tokyo*

Thomas G. Robertazzi

Computer Networks and Systems

Queueing Theory and
Performance Evaluation

Third Edition

With 116 Figures

 Springer

Thomas G. Robertazzi
Department of Electrical Engineering
SUNY Stony Brook
Stony Brook, NY 11794-2350
USA

CIP data available.

Printed on acid-free paper.

Production managed by Jenny Wolkowicki; manufacturing supervised by Joseph Quatela.
Camera-ready copy prepared using the author's Troff files.
Printed and bound by Edwards Brothers, Inc., Ann Arbor, M.I.
Printed in the United States of America.

9 8 7 6 5 4 3 2 1

ISBN 0-387-95037-0 Springer-Verlag New York Berlin Heidelberg   SPIN 10766828

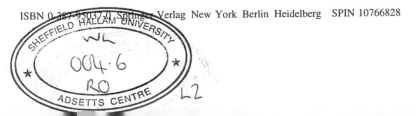

To Marsha, Rachel, and Deanna

Preface

Statistical performance evaluation has assumed an increasing amount of importance as we seek to design more and more sophisticated communication and information processing systems. The ability to predict a proposed system's performance before one constructs it is an extremely cost effective design tool.

This book is meant to be a first-year graduate level introduction to the field of statistical performance evaluation. It is intended for people who work with statistical performance evaluation including engineers, computer scientists and applied mathematicians. As such, it covers continuous time queueing theory (chapters 1-4), stochastic Petri networks (chapter 5), discrete time queueing theory (chapter 6) and recent network traffic modeling work (chapter 7). There is a short appendix at the end of the book that reviews basic probability theory. This material can be taught as a complete semester long course in performance evaluation or queueing theory. Alternatively, one may teach only chapters 2 and 6 in the first half of an introductory computer networking course, as is done at Stony Brook. The second half of the course could use a more protocol oriented text such as ones by Saadawi [SAAD] or Stallings [STAL].

What is new in the third edition of this book? In addition to the well received material of the second edition, this edition has three major new features. First of all, solutions to all of the chapter 2 through 6 problems are being published in a separate volume by Springer-Verlag. I believe this will be a great aid to students and engineers, computer scientists and applied mathematicians seeking to learn this subject. The second feature is a new chapter 7 on network traffic models that have been the subject of much interest since the publication of the second edition. Chapter 7 includes discussions of continuous and discrete time models, burstiness, self-similar traffic modeling and solution techniques. Lastly, I have added sixteen new problems to chapter 6 on discrete time queueing systems. Many of these involve switching.

I am grateful to B. L. Bodnar, J. Blake, X. Chao, J. S. Emer, M. Garrett, R. Guerin, W. Hagen, M. Haviv, H. Huang, Y. C. Jenq, M. Karol, J. F. Kurose, S.-Q. Li, A. C. Liu, J. McKenna, H. T. Mouftah, W. G. Nichols, M. Pinedo, I. Y. Wang, the IEEE, Digital Equipment Corporation and International Business Machines Corporation for allowing material, most of it previously published, to appear in this book.

My appreciation of this material has been enhanced by interaction with my students in Stony Brook's computer networks and performance evaluation courses. I am grateful to S. Rappaport for encouraging me to teach the performance evaluation course. Thanks are due for editorial assistance for this edition to Springer-Verlag's T. von Foerster and J. Mallozzi and to J. Wolkowicki for the production of this book and the solutions volume. Thanks are due to M. Gerla for reviewing the manuscript of the first edition. Thanks are also due to H. Badr, M.

Crovella, and S. Rappaport for looking over a draft of chapter 7 and providing valuable feedback. This book benefited from the drawing ability of L. Koh. Certain graphs and tables were made by J.-W. Jeng, K. Ko and J. Shor. Thanks are due to J. Eimer for typing new material for this edition. This book would not have been possible without the use of computer facilities supervised by A. Levochkin and M. Dorojevets.

Finally, I would like to dedicate this book to my wife, Marsha and my two daughters, Rachel and Deanna, who made writing this book worthwhile.

T.G.R.
Stony Brook, N.Y.

Contents

Preface   vii

Chapter 1: The Queueing Paradigm
1.1   Introduction 1
1.2   Queueing Theory 1
1.3   Queueing Models 2
1.4   Case Study I: Performance Model of a Distributed File Service
      By W.G. Nichols and J.S. Emer 6
1.5   Case Study II: Single-bus Multiprocessor Modeling
      By B.L. Bodnar and A.C. Liu 9
1.6   Case Study III: TeraNet, A Lightwave Network 11
1.7   Case Study IV: Performance Model of a Shared Medium
      Packet Switch
      By R. Guerin 13

Chapter 2: Single Queueing Systems
2.1   Introduction 19
2.2   The M/M/1 Queueing System 19
      2.2.1 The Poisson Process 20
      2.2.2 Foundations of the Poisson Process 22
      2.2.3 Poisson Distribution Mean and Variance 25
      2.2.4 The Inter-Arrival Times 27
      2.2.5 The Markov Property 28
      2.2.6 Exponential Service Times 29
      2.2.7 Foundation of the M/M/1 Queueing System 30
      2.2.8 Flows and Balancing 32
      2.2.9 The M/M/1 Queueing System in Detail 37
2.3   Little's Law 43
2.4   Reversibility and Burke's Theorem 47
      2.4.1 Introduction 47
      2.4.2 Reversibility 48
      2.4.3 Burke's Theorem 51
2.5   The State Dependent M/M/1 Queueing System 53
      2.5.1 The General Solution 53
      2.5.2 Performance Measures 55
2.6   The M/M/1/N Queueing System: The Finite Buffer Case 56
2.7   The M/M/∞ Queueing System: Infinite Number of Servers 59

2.8   The M/M/m Queueing System: m Parallel Servers with a Queue 61
2.9   The M/M/m/m Queue: A Loss System 64
2.10 Central Server CPU Model 65
2.11 Transient Solution of the M/M/1/∞ Queueing System 68
      2.11.1 The Technique 68
      2.11.2 The Solution 69
      2.11.3 Speeding Up the Computation 72
2.12 The M/G/1 Queueing System 74
      2.12.1 Introduction 74
      2.12.2 Mean Number in the Queueing System 74
      2.12.3 Why We Use Departure Instants 83
      2.12.4 Probability Distribution of the Number in the
             Queueing System 85
2.13 Priority Systems for Multiclass Traffic 89
To Look Further 91
Problems 92

Chapter 3: Networks of Queues
3.1   Introduction 101
3.2   The Product Form Solution 102
      3.2.1 Introduction 102
      3.2.2 Open Networks 102
            3.2.2.1 The Global Balance 102
            3.2.2.2 The Traffic Equations 103
            3.2.2.3 The Product Form Solution 104
      3.2.3 Local Balance 108
      3.2.4 Closed Queueing Networks 108
      3.2.5 The BCMP Generalization 111
3.3   Algebraic Topological Interpretation of the
      Product Form Solution 112
      3.3.1 Introduction 112
      3.3.2 A First Look at Building Blocks 112
      3.3.3 Building Block Circulatory Structure 118
      3.3.4 The Consistency Condition 130
3.4   Recursive Solution of Nonproduct Form Networks 132
      3.4.1 Introduction 132
      3.4.2 Recursive Examples 135
      3.4.3 Numerical Implementation of the Equations 141
3.5   Queueing Networks with Negative Customers 141
      3.5.1 Introduction 141
      3.5.2 Product Form Solutions 142
      3.5.3 The Chao/Pinedo Model 147
            3.5.3.1 Introduction 147
            3.5.3.2 The Model
                    By X. Chao and M. Pinedo 148

To Look Further 155
Problems 156

Chapter 4: Numerical Solution of Models
4.1   Introduction 163
4.2   Closed Queueing Networks: Convolution Algorithm 164
      4.2.1 Lost in the State Space 164
      4.2.2 Convolution Algorithm: Single Customer Class 166
      4.2.3 Performance Measures from Normalization Constants 170
4.3   Mean Value Analysis 197
      4.3.1 State (Load) Independent Servers 197
      4.3.2 A Closer Look at the Arrival Theorem 202
      4.3.3 State (Load) Independent Servers (Random Routing) 203
4.4   PANACEA: Approach for Large Markovian
      Queueing Networks 206
      4.4.1 Introduction 206
      4.4.2 The Product Form Solution 207
      4.4.3 Conversion to Integral Representation 208
      4.4.4 Performance Measures 211
      4.4.5 "Normal Usage" 212
      4.4.6 Some Transformations 212
      4.4.7 Asymptotic Expansions 214
      4.4.8 The Pseudonetworks 216
      4.4.9 Error Analysis 218
4.5   Norton's Equivalent for Queueing Networks 219
      4.5.1 Introduction 219
      4.5.2 Equivalence 220
4.6   Simulation of Communication Networks
      By J.F. Kurose and H.T. Mouftah 223
      4.6.1 Introduction 223
      4.6.2 The Statistical Nature of a Simulation 224
      4.6.3 Sensitivity Analysis of Simulation Results 226
      4.6.4 Speeding up a Simulation 228
To Look Further 230
Problems 232

Chapter 5: Stochastic Petri Nets
5.1   Introduction 237
5.2   Bus-oriented Multiprocessor Model 238
5.3   Toroidal MPN Lattices 243
5.4   The Dining Philosophers Problem 249
5.5   A Station-oriented CSMA/CD Protocol Model 251
5.6   The Alternating Bit Protocol 253

5.7   SPN's without Product Form Solutions 256
       5.7.1 Introduction 256
       5.7.2 Nonsafe Resource Sharing Models 257
       5.7.3 Synchronization Models 263
5.8   Conclusion 268
To Look Further 269
Problems 270

Chapter 6: Discrete Time Queueing Systems
6.1   Introduction 275
6.2   Discrete Time Queueing Systems 276
6.3   Discrete Time Arrival Processes 280
       6.3.1 The Bernoulli Process 280
       6.3.2 The Geometric Distribution 282
       6.3.3 The Binomial Distribution 284
       6.3.4 Poisson Approximation to Binomial Distribution 288
6.4   The Geom/Geom/m/N Queueing System 290
6.5   The Geom/Geom/1/N and Geom/Geom/1 Queueing Systems 294
6.6   Case Study I: Queueing on a Space Division Packet Switch 297
       6.6.1 Introduction 297
       6.6.2 Output Queueing 300
       6.6.3 Input Queueing 303
6.7   Case Study II: Queueing on a Single-buffered Banyan Network 308
       6.7.1 Introduction 308
       6.7.2 The Model Assumptions 309
       6.7.3 The Model and Solution 310
6.8   Case Study III: DQDB Erasure Station Location 312
       6.8.1 Introduction 312
       6.8.2 Optimal Location of Erasure Nodes
             By M.W. Garrett and S.-Q. Li 315
To Look Further 319
Problems 320

Chapter 7: Network Traffic Modeling
7.1   Introduction 333
7.2   Continuous Time Models 333
       7.2.1 Poisson Process (PP or M) 333
       7.2.2 Generally Modulated Poisson Process (GMPP) 334
       7.2.3 Markov Modulated Poisson Process (MMPP) 334
       7.2.4 Switched Poisson Process (SPP) 336
       7.2.5 Interrupted Poisson Process (IPP) 336
       7.2.6 Markovian Arrival Process (MAP) 337
       7.2.7 Autoregressive Moving Average Model (ARMA) 337
       7.2.8 Fluid Flow Approximation Model (FFA) 337
       7.2.9 Self-Similarity Source Model (SSS) 338
       7.2.10 Renewal Process (RP, GI) 338

    7.2.11 Semi-Markov Processes (SMP) 338
7.3   Discrete Time Models 338
    7.3.1 Deterministic Process (DP) 339
    7.3.2 Bernoulli Process (BP) 339
    7.3.3 Generally Modulated Deterministic Process (GMDP)............. 339
    7.3.4 Markov Modulated Deterministic Process (MMDP) 339
    7.3.5 Switched Deterministic Process (SDP) 340
    7.3.6 Interrupted Deterministic Process (IDP)................................... 340
    7.3.7 Discrete Time Markovian Arrival Process (DMAP) 341
    7.3.8 Discrete Renewal Process (DRP) 341
7.4   Solution Methods 341
    7.4.1 Simulation 342
    7.4.2 Linear Equation Solution 342
    7.4.3 Probability Generating Function 342
    7.4.4 Fluid Flow Approximation 342
    7.4.5 Transient Effect Models 343
7.5 Burstiness 343
    7.5.1 Ratio of Peak Rate to Mean Rate 343
    7.5.2 Coefficient of Variation of Traffic Load 344
    7.5.3 Index of Dispersion 344
    7.5.4 Spectral Characteristics 346
    7.5.5 Some Other Techniques 347
    7.5.6 Queueing Performance under Burstiness 347
7.6   Self-Similar Traffic 348
    7.6.1 Some Basics 348
    7.6.2 Self-Similarity 350
    7.6.3 The Hurst Effect 352
    7.6.4 Roots of Self-Similarity 353
    7.6.5 Detecting Self-Similarity 353
    7.6.6 Network Performance 354
To Look Further 355

Appendix: Probability Theory Review
A.1 Probability 357
A.2 Densities and Distribution Functions 358
A.3 Joint Densities and Distributions 360
A.4 Expectations 361
A.5 Convolution 362
A.6 Combinatorics 363
A.7 Some Useful Summations 363
A.8 Useful Moment-generating Function Identities 363

References 365

About the Author 403

Index 405

Chapter 1: The Queueing Paradigm

1.1 Introduction

This is a book about statistical prediction. It tells the story of how the behavior of complex electronic systems can be predicted using pencil, paper, the poetry of mathematics, and the number crunching ability of computers.

The types of prediction that we would like to make include:

→ How many terminals can one connect to a time sharing computer and still maintain a reasonable response time?

→ What percentage of calls will be blocked on the outgoing lines of a small business's telephone system? What improvement will result if extra lines are added?

→ What improvement in efficiency can one expect in adding a second processor to a computer system? Would it be better to spend the money on a second hard disc?

→ What is the best architecture for a space division packet switch?

Paradigms are fundamental models that abstract out the essential features of the system being studied. This book deals with two basic paradigms: the paradigm of the queue (chapters 1,2,3,4,6,7) and the paradigm of the Petri net (chapter 5).

1.2 Queueing Theory

The study of queueing is the study of waiting. Customers may wait on a line, planes may wait in a holding pattern, telephone calls may wait to get through an exchange, jobs may wait for the attention of the central processing unit in a computer and packets may wait in the buffer of a node in a computer network. All these examples have been studied using the mathematical theory of queues or queueing theory.

Queueing theory has its roots early in the twentieth century in the early studies of the Danish mathematician A. K. Erlang on telephone networks and in the creation of Markov models by the Russian mathematician A. A. Markov. Today it is widely used in a broad variety of applications. Moreover the theory of queueing continues to evolve.

Before proceeding to the actual study of queueing networks in chapters 2, 3, 4, 6 and 7 we will spend some time discussing models of queueing.

1.3 Queueing Models

What's in a Name?

In the operations research literature the items moving through a queueing system are often referred to as "customers" as real human waiting line may be envisioned. When a computer scientist writes about queues, he or she will often refer to the items moving through the queueing system as "jobs" since the application is often jobs circulating through a computer system. A telephone engineer, meanwhile, will speak of "calls". Finally, in computer networks, packets of data pass through the nodes and lines of such networks so a queueing model will be phrased in terms of these "packets". While "customer", "job", "call" and "packet" all are meaningful, depending on the application, we will generally use the generic term "customer" to refer to the items that move through a queueing system.

The Simplest System

The simplest queueing model involves a single queue. We can make use of a commonly used schematic representation in order to illustrate it. This is done in Figure 1.1.

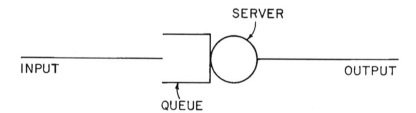

Fig. 1.1: A Single Server Queueing System

In this queueing system customers enter from the left and exit at the right. The circle represents the "server". For instance, in a supermarket check out line the server would be the employee operating the cash register. The open rectangle in Figure 1.1 represents the waiting line, or queue, that builds up behind the server.

A word about terminology is in order. The system of Figure 1.1 can be referred to as a "queueing system" or as a "queue" even though one of its components is also called the "queue". The meaning will usually be clear from the context.

In chapters 2, 3, 4, 6 and 7 we will make very specific assumptions regarding the statistics of arrivals to the queueing system and for the time it takes for the server to process a customer.

Shorthand Notation

There is a standard shorthand notation used to describe queueing systems containing a single queue. As an example, consider the M/M/m/n queueing system. The characters refers to, respectively, the statistics of the arrival process, the statistics of the server, the number of servers, and the number of customers that the queueing system can hold.

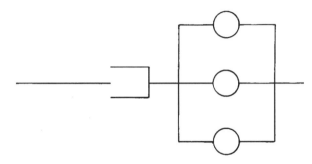

Fig. 1.2: An M/M/3 Queueing System

For instance, an M/M/3 queueing system is illustrated in Figure 1.2. This system might be used to model three tellers at a bank with a single common waiting line. There are three servers in parallel. Once a customer leaves a server, the customer at the head of the waiting line immediately takes his place at the server. There is no limit to the number of customers in such a queueing system as the fourth descriptor is absent from "M/M/3".

The characters in the first two positions indicate:

M: Markovian statistics,

D: Deterministic timing,

G: General (arbitrary) statistics,

Geom: Geometric statistics.

Networks of Queues

Many practical systems can be modeled as networks of queues. An "open" queueing network accepts and loses customers from/to the outside world. Thus the total number of customers in an open network varies with time. A "closed" network does not connect with the outside world and has a constant number of customers circulating throughout it.

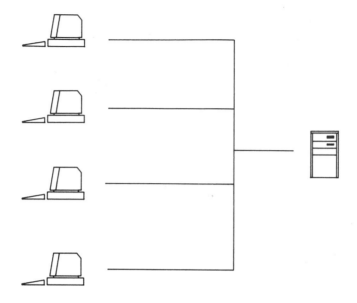

Fig. 1.3: Time-shared Computer and Terminals

For instance, in Figure 1.3 we have four terminals and a time-shared computer. An abstracted, closed, queueing model of this system appears in Figure 1.4. Each terminal is modeled by a single server that holds at most one customer. A "customer" that circulates in this queueing network represents the control of the computer terminal. That is, the presence of a customer at a terminal server indicates that the person at the terminal is thinking or typing and the presence of a customer at the queue modeling the computer indicates that the person has hit "return" and is waiting for the prompt (exit from the computer queue). Naturally, in this specific queueing model a customer leaving a terminal server must eventually return to the same server.

Service Discipline

Most everyday queueing systems are used to serve customers in the order that they arrive. This is a First In First Out (FIFO) service discipline.

Other service disciplines are sometimes encountered in practical systems. In the Last In First Out (LIFO) service discipline the most recent arrival enters the server, either displacing the customer in service or waiting for it to finish. When service is resumed for the displaced customer at the point where service was disrupted, we have the LIFO service discipline with pre-emptive resume (LIFOPR). The LIFO service discipline could be useful, for instance, for modeling an emergency message system where recent messages receive priority treatment.

In the processor sharing (PS) discipline the processor (server) has a fixed rate of service it can provide that is equally distributed among the customers in the system. That is, there is no waiting queue, with each customer immediately

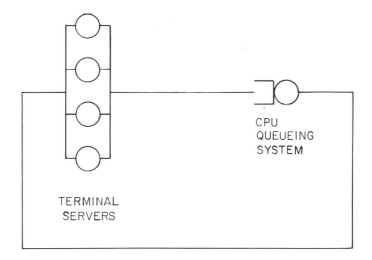

Fig. 1.4: Queueing Model of Time-shared Computer

receiving some service. Of course, if there are many customers in the system, each receives a small fraction of the total processing effort. The PS discipline is useful for modeling multiprogramming in computers where the processor splits its effort among several jobs.

State Description

In the simplest state description, the state of a queueing network is a vector indicating the total number of customers in each queue at a particular time instant. Markovian statistics are often used in modeling queues as this state description is sufficient to describe completely a queueing network at an instant of time. That is, it is not necessary to include in the state description elapsed times in service or the time since the last arrival.

If there are several "classes" of customers, the state description can be more detailed. Class may be used to decide arrival or service statistics by class or routing by class. One may now indicate the number of each class in each queue in the state description. At a more detailed level one may indicate the class of each customer at each waiting line position in the queue.

Unfortunately, as we shall see, even small queueing networks may have a number of states so large as to produce computational problems. A great deal of cleverness has gone into devising techniques to overcome this difficulty.

The Rest of This Book

In chapter 2 we will look at single queueing systems using continuous time models. Chapter 3 discusses networks of continuous time queues. Chapter 4 explains numerical algorithms for continuous time queueing networks. Chapter 5 discusses the rather different and more recent paradigm of the Petri Net. In chapter 6 discrete time queueing systems are examined. Finally, chapter 7 covers network traffic modeling.

We conclude this chapter with four case studies of practical modeling drawn from the technical literature. They are included to show that the study and use of queueing models is a most practical enterprise.

1.4 Case Study I: Performance Model of a Distributed File Service

The following section discusses the performance modeling of a distributed file service. It specifically deals with the Digital Equipment Corp. VAX Distributed File Service (DFS). The file service provides remote file access for VAX/VMS systems. To a client DFS simply appears to be a local file service. The following is excerpted from "Design and Implementation of the VAX Distributed File Service" by William G. Nichols and Joel S. Emer in the Digital Technical Journal, No. 9, June 1989, Copyright, 1989, Digital Equipment Corporation (VAX is a DEC trademark).

Performance Analysis Model

For contrasting design alternatives, it is too time consuming to build many systems with different designs. Even for a given design, large multiple-client testing is very time consuming. Therefore, we developed a queueing network model to assess the performance of the file design alternatives quickly and quantitatively. The model represents a distributed system comprising the file server and multiple single-user workstations as client machines. Since we were primarily interested in the performance of the file server, the queueing network model and the workload we describe here represent multiple clients making requests only to a single DFS server.

Some important characteristics of the DFS communication mechanism affect the model of the system. The DFS protocols are connection oriented, with long-living connections between client and server machines. Also, these protocols generally have a strict request-response nature, so that at any time a client will usually have only one outstanding request to a server. As a result, the number of requests outstanding in the overall system (that is, requests being processed at the server or being processed or generated at the client) corresponds to the number of single-user client workstations in the system. Because of this characteristic, it is convenient to use a closed queueing network model to represent a distributed system in which multiple clients request service from a DFS server. In this model, the number of customers in the queueing network corresponds to the number of workstations.

Client requests to the server may experience queueing at each of the individual resources at the server node. The delays caused by queueing reduce the system performance as seen by the user. Therefore, we represent the server in substantial detail; each service center that represents a resource at the server also has a queue for requests that may await service at that resource. The server resources being modeled are:

→ The network interface (The transmit and receive portions are each modeled separately),

→ The CPU, and

→ The disk subsystem at the server.

Since no queueing occurs at the client, the various segments of processing taking place at the client are represented by delay servers, i.e., service centers with no associated queue. A simple view of the distributed system is shown in Figure 1.5.

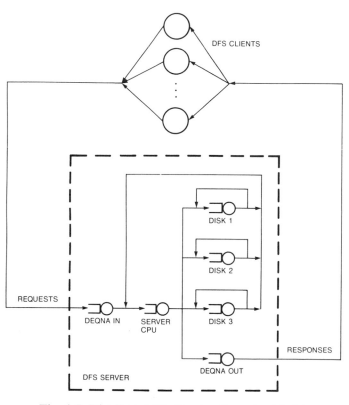

Fig. 1.5: Distributed File Service Queueing Model
© 1989 DEC

A particular system design and workload may be modeled by substituting the appropriate parameters into the model. For determining many of the system design parameters, we use a prototype file service that was developed in corporate research [RAMA 86].

The prototype had already demonstrated the feasibility of a VMS file service and some of its performance potential. This prototype itself represented only one design alternative. To estimate the other design alternatives, we used the prototype to provide parameters for our models.

All the models have the same set of service centers, but the centers are distributed differently in the various models. Therefore, we measured the time taken to execute the parts of the prototype that correspond to the service centers in our model. We could then analyze the models with real parameters to predict the performance of various design alternatives.

We constructed an artificial workload to drive our model. The workload was based on our understanding of typical user behavior and was also derived from measurements of typical time sharing environments [JAIN], [OUST]. A user at a client workstation was considered to invoke program images repeatedly. The programs invoked by the user require access to some number of files. The programs alternate between performing local computation and doing system service operations for access to these files. The files are assumed to be stored at the file server.

The amount of data transferred and the processing done at the server for each request depends on the type of request. The model distinguishes two types of request: control operations and data access operations. Control operations are operations such as open and close file, which have a high computational component. On the other hand, data access operations are simple reads and writes. Data access operations usually have low computational requirements but require larger data transfers.

In between the program invocations, we assumed that the user spends a certain amount of time thinking or doing processing unrelated to our investigation. All processing at the client is represented in the model as delay servers with the appropriate branching probabilities.

To make the model tractable, we assumed that the service time at each service center is exponentially distributed. Since we are interested primarily in a file server's mean performance, this was an acceptable assumption [LAZO]. Even given this assumption, the model had to distinguish control and data access operations.

Unfortunately, these two classes of operations have different service times at the various service centers. This difference means that the queueing network model does not satisfy the requirements for a product-form solution and may not be solved exactly [BASK 75]. We used a simple approximate-solution technique to solve this closed queueing network model. This technique is an extension of the basic multiclass mean value analysis technique [BRUE 80], [REIS 79]. The model characteristics, parameters, and the solution technique are described in greater detail in reference [RAMA 89]. Additional information appears in [EMER].

1.5 Case Study II: Single-Bus Multiprocessor Modeling

The following is a description of a model of a single bus multiprocessor. That is, in the system in question processors and global memory communicate amongst each other through a common bus. It originally appeared in "Modeling and Performance Analysis of Single-Bus Tightly-Coupled Multiprocessors" by B.L. Bodnar and A.C. Liu in the IEEE Transactions on Computers, Vol. 38, No. 3, March 1989, copyright 1989 IEEE.

Description of the Multiprocessor Model

A tightly-coupled multiprocessor (TCMP) is defined as a distributed computer system where all the processors communicate through a single (global) shared memory. A typical physical layout for such a computer structure is shown in Figure 1.6. Figure 1.7 illustrates our queueing model for this single-bus tightly-coupled multiprocessor (SBTCMP) architecture.

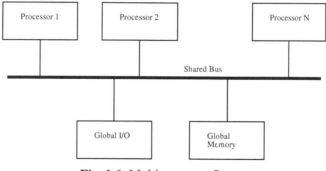

Fig. 1.6: Multiprocessor System
© 1989 IEEE

Each processing element, PE (with identifying index "i"), is modeled as a finite set of tasks (the task pool), a CPU, a bus interface unit (BIU) that allows the PE to access the shared bus, and a set of queues associated with the CPU and BIU. Each CPU and BIU has mean service rate $\mu(i, 1)$ and $\mu(i, 2)$, respectively. Tasks are assumed to have a mean sleep time in the task pool of $\mu(i, 0)^{-1}$ time units. We also assume the CPU and BIU operate independently of each other, and that all the BIU's in the multiprocessor can be lumped together into a single "equivalent BIU".

The branching probabilities are $p(i,1)$, $p(i,2)$ and $p(i,3)$. The branching probability $p(i,3)$ is interpreted as the probability that a task associated with PE i will join the CPU queue at PE i after using the BIU; $1-p(i,3)$ is then the probability that a task associated with PE i will wait for an interrupt acknowledgment at PE i. It is *not* the probability that a task will migrate from one PE to another. The problem of task migration is treated later in this paper.

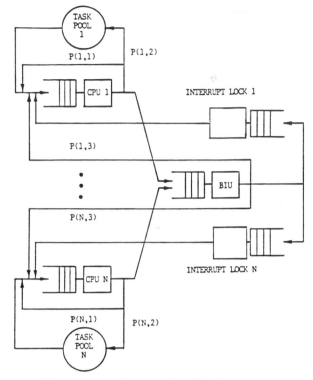

Fig. 1.7: Queueing Model of Multiprocessor
© 1989 IEEE

Interrupts to the PE are assumed to occur at a mean rate $\mu(i, 3)$. In general, the interrupt rate to PE i will be a function of task location, intertask communication probability, speeds of resources on other PE's, global I/O devices, etc. The operation of the multiprocessor is now described. In this discussion, we will assume that the workload is fixed.

A task is assumed initially asleep in the task pool. When it awakens, it will queue for CPU usage in the ready list (the "ready list" is a list of tasks that are eligible to use the CPU as a result of either being awakened or as a result of having finished using some resource). Interrupt-driven jobs *are not* placed on the ready list; instead, they have their own (higher priority) queue. After using the CPU, the task may request more CPU service (in which case it will requeue in the ready list), it may go back to sleep in the task pool, or it may request bus usage.

If the task requests bus usage, it will enter the BIU queue. If the bus is free, the task will begin using the bus; otherwise, it will wait until the bus is released. After using the bus, the task will either request CPU time (it will be placed at the back of the ready list) or it will wait for an interrupt (i.e., an acknowledgment from a task residing on another PE or an acknowledgment from a globally accessible I/O device). *We assume that only one of the preceding events can occur at a given moment.* That is, CPU and interrupt processes cannot occur concurrently.

We also assume that a task queued for bus access will not continue using the CPU.

Tasks waiting for an interrupt or undergoing interrupt servicing will be referred to as "interrupt-driven tasks". Tasks waiting for interrupts are modeled by a queue feeding a "lock". The lock is drawn using a triangle to signify that it is enabled via an external stimulus. The purpose of the lock is to allow only one task to pass by it in response to an interrupt.

Upon receipt of an interrupt, an interrupt-driven task will usurp the CPU from *any other task*. If the task that was forced to release the CPU was an interrupt-driven task, then this pre-empted task will become the first entry on a last-come-first-served queue (i.e., a stack). If the pre-empted task was not an interrupt driven task, then it will become the first entry on the ready list.

We assume that *all* interrupt-driven tasks on the stack must be serviced before any noninterrupt-driven ones obtain service. We will further assume that a pre-empted task can be restarted at the point where it was pre-empted. Hence, the CPU operates under two distinct service modes; namely, a low priority mode (such as round-robin, first-come-first-served, etc.) and a high priority last-come-first-served pre-emptive resume (LCFSPR).

A task residing in a given processor may also migrate to another processor. This could be done, for instance, in order to equalize load on the multiprocessor. Task migration from one PE to another has to be via the single time-shared bus (as this is the only communication medium). Note: The task migration problem is covered in Section IV of this paper.

In summary, this model takes into account the generalized overall behavior of the workload present on a single-bus tightly coupled multiprocessor. Specifically, it not only considers the bus contention caused by multiple processors attempting to access the shared memory, but also attempts to include realistically the local behavior of the various tasks running on these same processors.

1.6 Case Study III: TeraNet, A Lightwave Network

A recent concern in engineering communication networks has been making use of the very large transmission capacities of fiber optics. Before the introduction of fiber optics the bottleneck in most communications systems was the network links. Now, one often finds that the limiting bottleneck to throughput is the speed of nodal electronics. There has thus been an interest in new network architectures that exploit the unique properties of fiber optics.

TeraNet is a network for integrated traffic being constructed at Columbia University by the university's Center for Telecommunications Research and a number of private companies. The goal is to examine network architectures with high access rates (as high as 1 Gbps) and which can support a variety of services including video and graphics. The following discussion is based on that in [GIDR].

TeraNet utilizes two new ideas for network design. One is the use of a fiber optic star coupler. A star coupler is a passive fiber optic device. Each node in the network has an input fiber to send signals into the star coupler and an output fiber from the star coupler to receive signals from other nodes. A signal at carrier frequency i is transmitted into the star coupler from node i (at power P). The star coupler divides the signal equally and redistributes on each of the N output fibers (at power P/N) to all of the nodes.

Therefore, if each node ($i=1,2...N$) is transmitting into the star coupler at carrier frequency i, each node receives on its output fiber all of the signals from every other node. To receive the signal of a particular node, a node will tune to the transmitting node's unique carrier frequency. It should be noted that the number of nodes that can be supported by a star coupler is limited by the $1/N$ power division in the coupler.

Coordinating the transmission and reception of messages in a star coupler based network requires some care [GOOD]. With present technology tunable laser receivers are relatively slow. Therefore, TeraNet uses an elegant multi-hop approach to simplify implementation. Each node transmits into the star coupler at only two fixed carrier frequencies and receives on only two (different) fixed frequencies. These frequencies can be assigned in such a way that, at worse, a message from node A to node B must "hop" between a number of intermediate nodes before reaching B [ACAM]. The beauty of this scheme is that it only requires two fixed frequency receivers and two fixed frequency receivers per station. Naturally the hopping does increase delay. However this is traded off against the high bandwidth of the fiber optic links and the simplicity of implementation.

Figure 1.8 illustrates the conceptual diagram of a node's Network Interface Unit (NIU). The NIU is the interface between local traffic generated and received at the node and the network itself. The NIU also serves to connect neighboring nodes through itself. In the diagram there are two inputs (into optical filters) from two neighboring nodes and two outputs (thru lasers) to two different neighboring nodes. There is also a "User Input" for traffic originating at the node that is to be launched into the network and a "User Output" that is the entry way for network traffic destined for this particular node. The "Switch" system serves to move traffic from each of the three inputs to each of the three outputs. The Switch is illustrated in more detail in Figure 1.9.

In Figure 1.9 one can see the extensive placement of queues or "buffers". For instance, each of the three inputs feeds into a FIFO queue. These queues allow timing on the switch inputs (each of which has its own synchronization) to be isolated from the timing on the switch outputs, which is based on a NIU oscillator. There is also a FIFO queue ("Processor Output FIFO") that allows control packets from the NIU control processor to enter the Switch.

Words in the Switch consist of 40 bits. This accounts for the 40-bit wide paths in Figure 1.9. A word leaving an input FIFO queue is switched via one of three buses (heavy vertical lines) to one of three multiport queues. The Channel Filter Lookup Table is used to decide which is the correct output queue for each word.

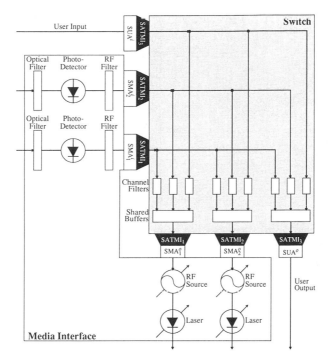

Fig. 1.8: Network Interface Unit (NIU)
© 1991 IEEE

These three output queues (labeled as Four Port RAM's) can each store 2000 40 bit words. Each port can receive and transmit data at up to 1 Gbps (that is, with 40 lines at 25 MHz each). Each queue is organized into twelve areas distinguished by the input that the word arrived over and the class of the corresponding packet of information. This is an example of a multiclass queue. In TeraNet, specifically, each of the four classes of traffic is distinguished by its requirements in terms of network performance (i.e., delay, loss probability etc...). Each Buffer Controller and Output Scheduler implements an algorithm known as ATS [LAZA 90] to decide which packet to transmit on the output link next.

What should be emphasized about the TeraNet switch is the key role that queues play. In particular their scheduling and control is not at present completely understood and is an intriguing research area [HYMA 91,92][LAZA 91].

1.7 Case Study IV: Performance Model of a Shared Medium Packet Switch

The following case study is written by R. Guerin of IBM and describes a high speed packet switch architecture and its analysis.

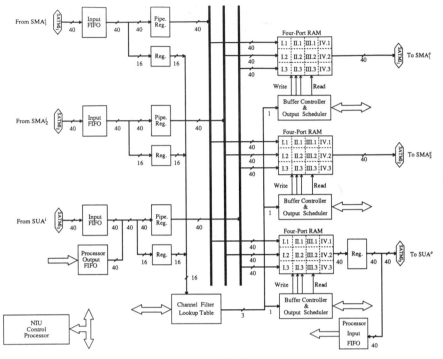

Fig. 1.9: TeraNet NIU Block Diagram
© 1991 IEEE

Introduction

In this section we discuss a model used to analyze the performance of a fast packet switch relying on a shared medium technology. In particular, the switch under study is modeled after the IBM's plaNET switch [CIDO], which is to be deployed and tested in a number of field trials, e.g., [CLAR]. The switch relies on a simple bus-based structure shared by all input adapters and operating at a high enough rate so that the bulk of the queueing takes place at the output. An important design issue is the appropriate sizing of the output buffers, which requires accurate modeling of both the shared bus and the incident traffic. The results quoted in this section are extracted from [CHEN c].

Switch Model

The plaNET switch [CIDO] consists of a series of link adapters interconnected by means of a high-speed active bus (see Figure 1.10), which can transfer data at a rate approximately equal to the aggregate speed from all incoming links. A bus arbiter controls how the different link adapters are granted transmission access to the bus. In particular, the arbiter implements a gated round-robin policy where bus access is cyclically passed from adapter to adapter, with each adapter only allowed to transmit those packets that arrived during the current service cycle. This policy can be made to approximate closely a global FIFO policy, provided the service cycle is chosen small enough so that only few packets can arrive during any one cycle. This arbitration scheme can be shown to result in small bounded input queues [CIDO 88][BIRM 89]. Therefore, the main performance issue for this system is the appropriate sizing of the output buffers. In particular, these buffers must be sized to accommodate the bursts of packets that may be delivered to an adapter by the high-speed bus. The sizing of these buffers is especially important not only to ensure adequate performances but also because the cost of the FIFOs which interface to the high-speed bus contribute significantly to the overall system cost. It is, therefore, critical that the performance model of the system capture the key operating features of the implementation so that the required output buffers be sized accurately.

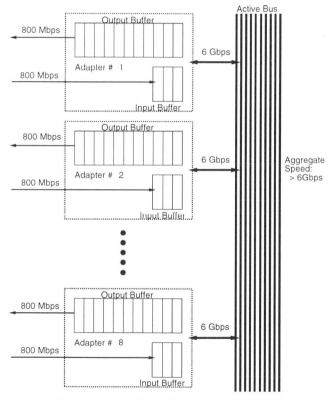

Fig. 1.10: plaNET Switching Configuration

System operation is essentially as follows. Packets arrive at the input adapters where some small buffering space is provided, so that they can be stored while waiting for access to the bus. As mentioned above, this waiting is minimal as the high speed of the bus and the arbitration policy guarantee that an adapter with complete packets in its input buffers can transmit them within a small and bounded amount of time. The bus is an "active" 64 bit wide bus. That is the signal is regenerated at each adapter in a ring type configuration. The clock rate is about 80 MHz, and ECL logic is used.

Once an adapter has been granted bus access, it proceeds with the uninterrupted transmission of its packets (those that are entitled to transmission in this access cycle). Note that packets have variable size and are assumed available for transmission only after they have been fully received in the input adapter buffer. The shared bus is responsible for delivering complete packets to the appropriate destination adapters, where they are stored in the associated output buffer prior to being forwarded on the transmission link. The link can be viewed as a second service stage in tandem with the shared bus but operating at a lower service rate. It is this lower service rate that necessitates buffering, so that incoming packets can wait while the packets ahead of them are being transmitted. This is the main queueing point in the system, and the goal of the model is to determine the amount of buffering needed to achieve a given performance level, i.e., guarantee a certain packet loss.

There are a number of important system characteristics that have the potential to influence heavily the sizing of the output buffers. Foremost among all is the fact that the bus does not forward packets instantaneously to the outputs. Rather, the transmission of data is progressive, and although the shared bus is faster than the attached transmission links (the switch of [CIDO] is 8x8, so the bus is approximately 8 times faster than the links), a reasonable amount of data can still be forwarded before a full packet has been received, i.e., cut-through is implemented. Similarly, the progressive nature of packet arrivals also bounds the amount of data that can be generated to a link buffer in any time interval. The resulting smoother arrival process is known [FEND][CIDO 93] to lower, especially at high loads, the buffering needed to achieve a given overflow or loss probability. Given the cost of the transmission buffers for high-speed links, it is therefore important to capture this smoothing effect accurately.

Another important feature likely to contribute to lowering the buffering requirements in the output adapters is the fact that the bus is shared by all outputs. This sharing implies that no two outputs can be receiving data simultaneously. Hence, while the server is busy forwarding data to an output, the others are given the opportunity to empty their buffers. It is again important that this be accurately reflected in the model of the system.

As described in Figure 1.11, the model used to represent the switch is that of a shared central server that alternates between a number of inputs and whose output is "randomly" switched among a number of possible destinations upon completion of each service time or packet transmission. Because of the complexity of such a general model, a number of simplifying steps are required in order to make its analysis tractable. The main issue is to achieve this simplification while preserving the key features of the system.

The first simplifying step is to aggregate the packet arrival processes from all the inputs to the central server so that a single arrival process can be assumed. Such an aggregation is appropriate because of the very limited input queueing and the approximately FIFO global service policy. This aggregate process is derived

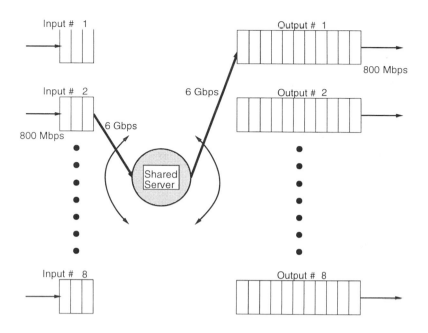

Fig. 1.11: Model of Switch

from the individual input processes through the use of standard approximation techniques based on moments characterization and fitting [WHIT 81][WHIT 82][ALBI] (see [CHEN c] for details). Once the aggregate input process has been obtained, the next step is to characterize the output process from the shared server as it determines the input process to the output adapters. In particular, it is important to account for the sequential delivery of packets to different outputs.

An underlying Markov chain is used to capture the changes in how the output from the shared server is forwarded to different outputs. At any given time, the state of the Markov chain determines the current destination output adapter. In other words, when the chain is in state i, the server is forwarding data to output i. State (output) 0 is defined to be the idle state of the server, i.e., there are no packets waiting in the input adapters. State transitions occur either at the end of an idle period or upon completion of a packet transmission. The accuracy and adequacy of using a Markov chain to describe the output process of the shared server is discussed in [CHEN c].

While the state transitions of the Markov chain reflect how the server directs its output to different destination adapters, the progressive nature of data arrival and transmission must still be accounted for. This is achieved by using a flow model, e.g., [ANIC], where the arrival and transmission of bits to and from an output adapter emulate that of a fluid. Specifically, when an adapter is the destination of the packet(s) currently transmitted on the bus, its buffer is being filled at the rate of data transfer on the bus (e.g., 6 Gbps) while it is emptied at the transmission rate of the link (e.g., 800 Mbps). This model captures the progressive nature of packets arrivals and transmissions, which results in a more accurate representation of the actual amount of data received and transmitted in any time period by an output adapter. As mentioned earlier, this is important when estimating the buffering requirements of the system.

Based on this fluid model for the arrival and service processes in an output adapter, it is then possible to estimate accurately the amount of buffering needed to achieve a given buffer overflow probability. Results on the application of this model and the techniques it relies on are reported in [CHEN c]. Numerical examples are provided and compared to values obtained using simpler models that do not represent as precisely many of the above features. The resulting differences are found often to be significant, which can translate into rather large differences in system cost. This illustrates the need for careful identification and modeling of key system features, when trying to provide design guidelines through performance analysis.

Chapter 2: Single Queueing Systems

2.1 Introduction

In this chapter we will look at the case of the single queue. Even though chapters 3 and 4 will discuss networks of queues, it is probably safe to say that the majority of modeling work that has been done for queues has involved a single queue. One reason is that for any investigation into queueing theory the single queueing system is a natural starting point. Another reason is tractability: much more can easily be ascertained about single queues than about networks of queues.

We will start by examining the background of the most basic of queueing systems, the M/M/1 queue. This "Markovian" queue is distinguished by a Poisson arrival process and exponential service times. It is also distinguished by a one dimensional, and infinite in extent, state transition diagram. The coordinate of the one dimensional state transition diagram is simply the number of customers in the queueing system.

Later, by modifying the rates associated with this diagram and the diagram's extent we will arrive at a number of other useful queueing systems such as systems with multiple servers and systems with finite waiting line capacity.

We will spend some time looking at a non-Markovian queue of great interest, the M/G/1 queueing system. For this generalized queueing system a surprising amount of analytic information can be obtained.

For some queueing systems it is easier to calculate state probabilities in the z domain rather than directly. We will look at this z-transform technique in terms of two examples, the transient analysis of the M/M/1 queueing system and the M/G/1 queueing system.

Finally, there has been recent interest in systems carrying multiple classes of traffic. In the last section of this chapter the existing literature on multiclass queueing models is reviewed.

2.2 The M/M/1 Queueing System

Consider the queueing system of Figure 2.1. It is the queueing system from which all queueing theory proceeds. This M/M/1 queueing system has two crucial assumptions. One is that the arrival process is a Poisson process, and the other is that the single server has a service time which is an exponentially distributed random variable. These assumptions lead to a very tractable model and are reasonable for a wide variety of situations. We will start by examining these assumptions.

Fig. 2.1: M/M/1 Queueing System

2.2.1 The Poisson Process

The *Poisson process* is the most basic of arrival processes. Interestingly, it is a purely random arrival process. One way of looking at it goes like this. Suppose that the time axis is divided into a large number of small time segments of width Δt. We will let the probability of a single customer arriving in a segment be proportional to the length of the segment, Δt, with a proportionality constant, λ, which represents the mean arrival rate:

$$P(exactly\ 1\ arrival\ in\ [t,t+\Delta t]) = \lambda \Delta t, \qquad (2.1)$$

$$P(no\ arrivals\ in\ [t,t+\Delta t]) = 1 - \lambda \Delta t, \qquad (2.2)$$

$$P(more\ than\ 1\ arrival\ in\ [t,t+\Delta t]) = 0. \qquad (2.3)$$

Here we ignore higher order terms in Δt.

Now we can make an analogy between the arrival process and coin flipping. Each segment is like a coin flip with $\lambda \Delta t$ being the probability of an arrival (say, heads) and $1 - \lambda \Delta t$ being the probability of no arrival (say, tails).

As $\Delta t \rightarrow 0$ we form the continuous time Poisson process. From the coin flipping analogy one can see that arrivals are independent of one another as they can be thought of as simply the positive results of a very large number of independent coin flips. Moreover, one can see that no one instant of time is any more or less likely to have an arrival than any other instant.

Two ideas that extend the usefulness of the Poisson process are the concept of a random split of a Poisson process and a joining of Poisson processes. Figure 2.2 illustrates a queueing system with a random split. Arrivals are randomly split between three queues with independent probabilities p_1, p_2, and p_3. A situation involving a joining of independent Poisson processes is illustrated in Figure 2.3. Arrivals from a number of independent Poisson processes are joined or merged to form an aggregate process.

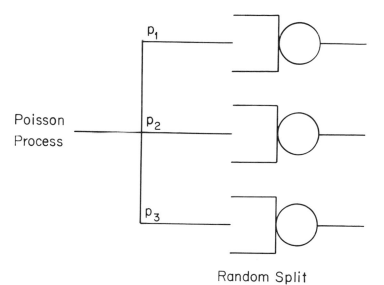

Random Split

Fig. 2.2: Random Split of a Poisson Process

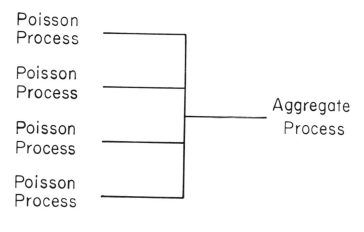

A Joining

Fig. 2.3: Joining of Poisson Processes

It turns out that random splits of a Poisson process are Poisson and that a joining of independent Poisson processes is also Poisson. A little thought with the coin flipping analogy will show that this is true.

One of the original applications of the Poisson process in communications was to model the arrivals of calls to a telephone exchange. The use of each telephone, at least in a first analysis, can be modeled as a Poisson process. These are joined into the aggregate Poisson stream of calls arriving at the exchange. It is very reasonable to model the aggregate behavior of a large number of independent users in this manner. However, this model may not always be valid. For instance, if there are correlations between an individual users calls - say, given that the user has made one call he/she is more likely to make a second - then the Poisson assumption is not valid. Another example is that of a heavily loaded exchange. Calls are blocked (busy signal), which results in repeated attempts by individual users to get through. This would also invalidate the Poisson assumption.

Thus the life of a modeler can become complicated, but we will continue to discuss elegant models with which we can make analytical progress.

2.2.2 Foundation of the Poisson Process

We will derive the underlying differential equation of the Poisson process by using difference equation arguments and letting $\Delta t \to 0$. We will start by letting

$$P_n(t) \equiv P(\# \; Arrivals = n \; at \; time \; t \;), \tag{2.4}$$

and let $p_{ij}(\Delta t)$ be the probability of going from i arrivals to j arrivals in a time interval of Δt seconds. The approach will be similar to that of [KLEI 75].

The number of arrivals is the "state" of the system. It contains all the information necessary to describe the system completely. We can write

$$P_n(t + \Delta t) = P_n(t)p_{n,n}(\Delta t) + P_{n-1}(t)p_{n-1,n}(\Delta t). \tag{2.5}$$

Again, we have neglected higher order terms in Δt. What this equation says is that one can arrive at a situation with n customers at time $t + \Delta t$ from either having n or $n-1$ customers at time t. Notice that we still assume that Δt is small enough so that only one customer at most may arrive during this interval.

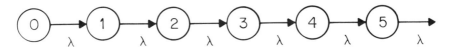

Fig. 2.4: Poisson Process State Transition Diagram

For this system we have the state transition diagram of Figure 2.4. Here the circles represent the states of the system (number of arrivals) and the transition rate λ is associated with each transition. We need the following special equation for the state 0 to complete our difference equation description:

$$P_0(t+\Delta t) = P_0(t)p_{0,0}(\Delta t). \tag{2.6}$$

If we substitute in our previous expressions for the probability of exactly one arrival and the probability of no arrivals in an interval $[t, t+\Delta t]$, we have

$$P_n(t+\Delta t) = P_n(t)(1 - \lambda\Delta t) + P_{n-1}(t)(\lambda\Delta t), \tag{2.7}$$

$$P_0(t+\Delta t) = P_0(t)(1 - \lambda\Delta t). \tag{2.8}$$

By multiplying out these expressions and re-arranging one can arrive at

$$\frac{P_n(t+\Delta t) - P_n(t)}{\Delta t} = -\lambda P_n(t) + \lambda P_{n-1}(t), \tag{2.9}$$

$$\frac{P_0(t+\Delta t) - P_0(t)}{\Delta t} = -\lambda P_0(t). \tag{2.10}$$

If we let $\Delta t \to 0$, the set of difference equations becomes a set of differential equations:

$$\boxed{\begin{aligned} \frac{dP_n(t)}{dt} &= -\lambda P_n(t) + \lambda P_{n-1}(t), \\[2mm] \frac{dP_0(t)}{dt} &= -\lambda P_0(t). \qquad (2.11) \end{aligned}}$$

Here $n \geq 1$. We now naturally wish to find the solution to these differential equations, $P_n(t)$. For the second equation, a knowledge of differential equations enables us to see that

$$P_0(t) = e^{-\lambda t}. \tag{2.12}$$

This makes the next equation

$$\frac{dP_1(t)}{dt} = -\lambda P_1(t) + \lambda e^{-\lambda t}, \tag{2.13}$$

which has a solution of

$$P_1(t) = \lambda t e^{-\lambda t}. \tag{2.14}$$

Hence the next equation is

$$\frac{dP_2(t)}{dt} = -\lambda P_2(t) + \lambda^2 t e^{-\lambda t}, \tag{2.15}$$

which has a solution of

$$P_2(t) = \frac{\lambda^2 t^2}{2} e^{-\lambda t}. \tag{2.16}$$

Continuing, we would find by induction that

$$\boxed{P_n(t) = \frac{(\lambda t)^n}{n!} e^{-\lambda t}. \tag{2.17}}$$

This is the *Poisson distribution*. It tells us the probability of n arrivals in an interval of t seconds for a Poisson process of rate λ. A plot of the Poisson distribution appears in Figure 2.5.

Example: A telephone exchange receives 100 calls a minute on average, according to a Poisson process. What is the probability that no calls are received in an interval of five seconds?

The solution is simply

$$P_0(t) = e^{-\lambda t} = e^{-100 \times \frac{1}{12}} = 0.00024.$$

∇

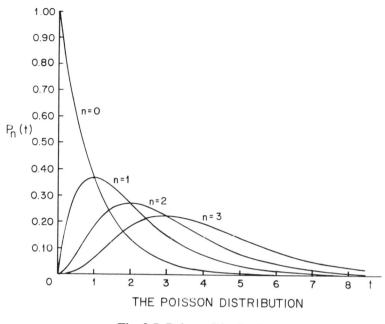

THE POISSON DISTRIBUTION

Fig. 2.5: Poisson Distribution

2.2.3 Poisson Distribution Mean and Variance

We will calculate the mean and variance of the Poisson distribution. Let $\bar{n}$ be the mean number of arrivals in an interval of length t. Then:

$$\bar{n} = \sum_{n=0}^{\infty} n P_n(t), \tag{2.18}$$

$$\bar{n} = \sum_{n=0}^{\infty} n \, \frac{(\lambda t)^n}{n!} \, e^{-\lambda t}, \tag{2.19}$$

$$\bar{n} = e^{-\lambda t} \sum_{n=1}^{\infty} \frac{(\lambda t)^n}{(n-1)!}. \tag{2.20}$$

With a change of variables this is

$$\bar{n} = e^{-\lambda t} \sum_{n=0}^{\infty} \frac{(\lambda t)^{n+1}}{n!}, \tag{2.21}$$

$$\bar{n} = e^{-\lambda t} \lambda t \sum_{n=0}^{\infty} \frac{(\lambda t)^n}{n!} = e^{-\lambda t} \lambda t \, e^{\lambda t} = \lambda t. \tag{2.22}$$

The equation $\bar{n} = \lambda t$ makes intuitive sense. The mean number $\bar{n}$ is proportional to the length of the interval t with rate constant λ.

For the variance, following the treatment in [THOM] and [KLEI 75] we will first calculate

$$E(n(n-1)) = \sum_{n=0}^{\infty} n(n-1)\frac{(\lambda t)^n}{n!} e^{-\lambda t}, \tag{2.23}$$

$$E(n(n-1)) = \sum_{n=2}^{\infty} \frac{(\lambda t)^n}{(n-2)!} e^{-\lambda t}. \tag{2.24}$$

Making a change of variables,

$$E(n(n-1)) = e^{-\lambda t}(\lambda t)^2 \sum_{n=0}^{\infty} \frac{(\lambda t)^n}{n!}, \tag{2.25}$$

$$E(n(n-1)) = e^{-\lambda t}(\lambda t)^2 e^{\lambda t} = (\lambda t)^2. \tag{2.26}$$

Now

$$E(n(n-1)) = E(n^2) - E(n) = (\lambda t)^2, \tag{2.27}$$

or

$$E(n^2) = (\lambda t)^2 + \lambda t. \tag{2.28}$$

We also make use of the fact that the variance of n is

$$\sigma_n^2 = E(n - \bar{n})^2 = E(n^2) + E(\bar{n})^2 - 2\bar{n} E(n), \tag{2.29}$$

$$\sigma_n^2 = E(n^2) - \bar{n}^2. \tag{2.30}$$

Substituting one has

$$\sigma_n^2 = (\lambda t)^2 + \lambda t - (\lambda t)^2 = \lambda t. \tag{2.31}$$

So the mean and the variance of the Poisson distribution are both equal to λt.

2.2.4 The Inter-Arrival Times

The times between successive events in an arrival process are called the inter-arrival times. For the Poisson process these times are independent exponentially distributed random variables. To see that this is true, we can write the probability that the time between arrivals is $\leq t$ as

$$P\,(\textit{time between arrivals} \leq t) = 1 - P\,(\textit{time between arrivals} > t),$$

$$P\,(\textit{time between arrivals} \leq t) = 1 - P_0(t),$$

$$P\,(\textit{time between arrivals} \leq t) = 1 - e^{-\lambda t}.$$

If we differentiate this we have

$$\textit{Inter--arrival Density}\ (t) = \lambda e^{-\lambda t}.$$

An exponential random variable is the only continuous random variable that has the *memoryless property*. Loosely speaking, this means that the previous history of the random variable is of no help in predicting the future. To be more specific, a complete state description of an M/M/1 queueing system at an instant of time consists only of the number of customers in the system at that instant. We do not have to include in the state description the time since the last arrival because the M/M/1 queueing system has a Poisson arrival process with exponential inter-arrival times. Furthermore, we also do not have to include the time since the last service completion in the state description since the service time in an M/M/1 queueing system is exponentially distributed.

In terms of the Poisson arrival process the memoryless property means that the distribution of time until the next event is the same no matter what point in time we are at. Let us give an intuitive analogy. Suppose that trains arrive at a station according to a Poisson process with a mean time between trains of 20 minutes. Suppose now that you arrive at the station and are told by someone standing at the platform that the last train arrived nineteen minutes ago. What is the expected time until the next train arrives?

One would naturally like to answer one minute, and this would be true if trains arrived deterministically exactly twenty minutes apart. However, we are dealing with a Poisson arrival process. The answer is twenty minutes!

To see that this is true, we have to go back to our coin flipping explanation of the Poisson process. An event is the result of a positive outcome of a "coin flip" in an infinitesimally small interval. As time proceeds there are a very large number of such flips. When we put the Poisson process in this context it is easy to see that the distribution until the next positive outcome does not depend at all on when the last positive outcome occurred. That is, you cannot predict when you will flip heads by knowing the last time when heads occurred.

We can make this idea more precise. If an arrival occurs at time t=0, we know from before that the distribution of time until the next arrival is $\lambda e^{-\lambda t}$, $t \geq 0$, and the cumulative distribution function is $1-e^{-\lambda t}$, $t \geq 0$. Suppose now that no arrivals occur up until time t_0. Given this information what is the distribution of time until the next arrival? We can write it in the following way [KLEI 75] with the dummy variable $t*$:

$$P(t* \leq t + t_0 \mid t* > t_0) = \frac{P(t_0 < t* \leq t + t_0)}{P(t* > t_0)}, \qquad (2.32)$$

$$P(t* \leq t + t_0 \mid t* > t_0) = \frac{\int_{t_0}^{t+t_0} \lambda e^{-\lambda t*} dt*}{\int_{t_0}^{\infty} \lambda e^{-\lambda t*} dt*}, \qquad (2.33)$$

$$P(t* \leq t + t_0 \mid t* > t_0) = \frac{-e^{-\lambda t*} \Big|_{t_0}^{t+t_0}}{-e^{-\lambda t*} \Big|_{t_0}^{\infty}}, \qquad (2.34)$$

$$P(t* \leq t + t_0 \mid t* > t_0) = \frac{-e^{-\lambda(t+t_0)} + e^{-\lambda t_0}}{-0 + e^{-\lambda t_0}}, \qquad (2.35)$$

$$P(t* \leq t + t_0 \mid t* > t_0) = 1 - e^{-\lambda t}. \qquad (2.36)$$

This is the same cumulative distribution function we had when we asked what the distribution was at time $t = 0$!

We close this section by noting the *discrete* distribution with the memoryless property is the geometric distribution [FELL] (see chapter 6).

2.2.5 The Markov Property

The memoryless property is more accurately referred to as the *Markov property*. It says, intuitively, that one can predict the future state of the system based on the present state as well as if one had the past history of the state available. In terms of a continuous, non-negative, random time T, this Markov property can be expressed as [COOP]

$$P(T>t_0+t \mid T>t_0) = P(T>t) \tag{2.37}$$

for every $t_0>0$ and every $t>0$. In terms of a discrete random variable $X(t_n)$ one can say [KLEI 75] that

$$P(X(t_{n+1})=x_{n+1} \mid X(t_n)=x_n, X(t_{n-1})=x_{n-1}, \dots, X(t_1)=x_1)$$

$$= P(X(t_{n+1})=x_{n+1} \mid X(t_n)=x_n) \tag{2.38}$$

where the t_i are monotonically increasing and the x_n take on some values from a discrete state space.

Markovian systems are memoryless systems. This makes them relatively simple to analyze since when one describes the state of the system one does not need to take into account the time since the last arrival or the time that service has been underway. When the arrival process is not Poisson or the service time is not exponential then we must take these quantities into account. This makes understanding the behavior of even a single queue under these conditions a non-trivial undertaking.

A concept that we will come across later is that of a *Markov chain*. This is a Markov process with a discrete state space. The appearance of the state transition diagram of the M/M/1 queueing system which we are about to see (Figure 2.6), accounts for the use of the word "chain".

2.2.6 Exponential Service Times

In an M/M/1 queueing system the service times are independent exponentially distributed random variables. That is, they occur according to the probability density $\mu e^{-\mu t}$ where μ is the mean service rate and $1/\mu$ is the mean time for service. As has been discussed, this type of server is memoryless and thus produces a tractable analysis.

How realistic is an exponential server? It has been found to be reasonable for modeling the duration of telephone conversations. But it has also been used in situations where the physical basis for its use is weaker. For instance, the time to transmit fixed length packets of information in computer networks is often modeled by an exponential random variable. Sometimes this sort of assumption is made for reasons of tractability, even if some degree of realism is sacrificed.

The exponential server will be with us for a long time because, in conjunction with the Poisson process, it leads to Markovian systems and the associated elegant network wide results of the next two chapters.

2.2.7 Foundation of the M/M/1 Queueing System

Let us take Figure 2.4 and add transitions that represent servicing to create the state transition diagram of Figure 2.6. This state transition diagram represents what is called a *birth-death process*. That is, if we think of the state as being a population size, it may increase by one member at a time ("birth") or it may decrease by one member at a time ("death"). There are other models that may change by several members at a time (called "batch arrival" or "batch departure" processes). In the context of the performance evaluation of information systems the state of the system can actually be thought of as the number of packets in a communication processor, the number of new calls in a telephone exchange computer, or the number of jobs in a computer. The approach we will take is similar to that in [KLEI 75].

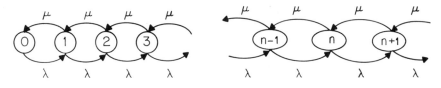

Fig. 2.6: M/M/1 System State Transition Diagram

When we were discussing the Poisson process we had the events

$$P\,(exactly\ 1\ arrival\ in\ [t,t+\Delta t]) = \lambda \Delta t,$$

$$P\,(no\ arrivals\ in\ [t,t+\Delta t]) = 1 - \lambda \Delta t.$$

Add to this the following two events:

$$P\,(1\ service\ completion\ in\ [t,t+\Delta t]) = \mu \Delta t,$$

$$P\,(no\ service\ completions\ in\ [t,t+\Delta t]) = 1 - \mu \Delta t.$$

Also again we will define

$$P_n(t) \equiv P(\#\ Arrivals = n\ at\ time\ t\,)$$

and let $p_{ij}(\Delta t)$ be the probability of going from i arrivals to j arrivals in a time interval of Δt seconds.

Whereas previously the state of the system was the number of Poisson arrivals, it is now the number in the queueing system, including the one in service. We can write:

$$P_n(t+\Delta t) = \tag{2.39}$$

$$P_n(t)p_{n,n}(\Delta t) + P_{n-1}(t)p_{n-1,n}(\Delta t) + P_{n+1}(t)p_{n+1,n}(\Delta t).$$

Higher-order terms in Δt are neglected.

This equation says that one can arrive at a situation with n customers at time $t+\Delta t$ from either having $n-1$, n or $n+1$ customers at time t. Because this is a birth-death process, these are the only possibilities. For the state 0 we will need the following special equation:

$$P_0(t+\Delta t) = P_0(t)p_{0,0}(\Delta t) + P_1(t)p_{1,0}(\Delta t). \tag{2.40}$$

We will now substitute the previous expressions for the probabilities of the various events in an interval $[t,t+\Delta t]$ into this equation to yield

$$P_n(t+\Delta t) = \tag{2.41}$$

$$P_n(t)(1-\lambda\Delta t)(1-\mu\Delta t) + P_{n-1}(t)(\lambda\Delta t) + P_{n+1}(t)(\mu\Delta t),$$

$$P_0(t+\Delta t) = P_0(t)(1-\lambda\Delta t) + P_1(t)(\mu\Delta t).$$

By multiplying out these expressions, re-arranging and letting $\Delta t \to 0$ one has

$$\boxed{\begin{aligned} \frac{dP_n(t)}{dt} &= -(\lambda+\mu)P_n(t) + \lambda P_{n-1}(t) + \mu P_{n+1}(t), \\ \frac{dP_0(t)}{dt} &= -\lambda P_0(t) + \mu P_1(t). \end{aligned}} \tag{2.42}$$

Here $n \geq 1$.

These differential equations describe the evolution in time of the state probabilities of the M/M/1 queueing system. But what do these equations really mean? To answer this question we will need the following discussion of flows and balancing.

2.2.8 Flows and Balancing

One of the most intuitively pleasing concepts in queueing theory is the way in which one can make an analogy between the flow of current in an electric circuit and the flows in the state transition diagram of a queueing system. But what is "flowing" in the state transition diagram?

The answer is that *probability flux* flows from one state to another along the transitions. The probability flux along a transition, numerically, is the product of the probability of the state at which the transition originates and the transition rate associated with the transition. The probability flux along a transition, physically, is the mean number of times per second that the event corresponding to the transition occurs. For instance, if a transition has a rate of $10\ sec^{-1}$ and the probability of the state at which the transition originates is 0.1 then the transition is actually traversed by the system state once a second, on average.

The analogy with electric circuits is straightforward. The state probabilities correspond to nodal voltages, and the transition rates correspond to conductance (inverse resistance) values. There are some differences though. Specifically:

There is no analog of a voltage source or battery in the queueing world. Rather we have the normalization equation that holds that at any instant of time the sum of the state probabilities sums to one. This normalization equation "energizes" the state transition diagram in the sense that it guarantees that there will be non-zero flows.

The direction of flow is preset in the state transition diagram by the transition direction. As has been mentioned what "flows" is the product of the probability of the state at which the transition originates and the transition rate. This is different from the electric circuit case where a branch current's magnitude and direction depend on the difference of voltages of the two nodes to which the branch is connected.

The most useful concept that carries over from the electric circuit world to the queueing theory world is the idea of looking at what flows across boundaries in the circuit or state transition diagram. For instance, suppose we draw a boundary around state n in the state transition diagram of the M/M/1 queueing system, as in Figure 2.7. What we are going to do is equate the mean number of times a second the state is entered with the mean number of times one leaves the state. What flows out of the state at time t is

$$-(\lambda + \mu)P_n(t)$$

and what flows into the state from states $n-1$ and $n+1$ is

$$\lambda P_{n-1}(t) + \mu P_{n+1}(t).$$

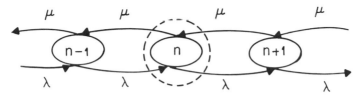

Fig. 2.7: Global Balance Boundary Around a State

The difference between these two quantities is just the change in the probability of state n or:

$$\frac{dP_n(t)}{dt} = -(\lambda+\mu)P_n(t) + \lambda P_{n-1}(t) + \mu P_{n+1}(t). \tag{2.43}$$

But this is just the same differential equation that we found using difference equation arguments in the last section! The equation for $dP_0(t)/dt$ can be derived in a similar manner.

So for the M/M/1 queueing system we have a set of differential equations to describe the evolution of the state probabilities as a function of time. As with electric circuits, this is useful if one wants a *transient analysis* of the queueing system. This is the study of the queueing system over a limited time period. For instance, we may be interested in the queueing system's behavior for the first ten minutes after service commences for an initially empty queue. Then the initial conditions of the set of differential equations becomes

$$P_0(0) = 1.0, \tag{2.44}$$

$$P_n(0) = 0.0, \quad n \ge 1,$$

and the differential equations can be solved. We will look at this situation in section 2.11.

An interesting thing happens when we let $dP_n(t)/dt = 0$. This is the situation that occurs when the queueing system has been running for a long time, transient behavior has settled out and the queue is in equilibrium. *Equilibrium or steady-state analysis* deals with this situation.

Since the probabilities do not change with time we can write:

$$0 = -(\lambda+\mu)p_n + \lambda p_{n-1} + \mu p_{n+1},$$

$$0 = -\lambda p_0 + \mu p_1. \qquad (2.45)$$

Here $n \geq 1$. Note that we have changed probability to lower case.

These equations are now a set of linear equations rather than a set of differential equations. Each equation can be arrived at by drawing a boundary completely around a specific state and equating what flows into the state with what flows out of the state. An equation arrived at in such a manner is known as a *global balance equation*. The reader familiar with electric circuits will be reminded of Kirchoff's current conservation law which holds that the *net* flow of current into and out of a circuit node equals zero as charge can not build up at a node. Similarly, in equilibrium the mean rate at which a state is entered should equal the mean rate at which the state is left.

We can calculate the equilibrium state probabilities of any queueing system with a finite number of states in the following manner. For each of the N states we write one global balance equation. This gives us a set of N linear equations with N unknowns. Any one of the equations is redundant and should be replaced with the equation

$$p_0 + p_1 + p_2 + \ldots\ldots + p_N = 1,$$

which normalizes the solution. The resultant set of linear equations can then be solved using standard numerical techniques. The reason that we have emphasized the calculation of the state probabilities is that many performance measures of interest are functions of the state probabilities.

The difficulty with this approach is that realistic queueing models often have extremely large state spaces making numerical solution computationally infeasible.

Example: In [KAUF] L. Kaufman, B. Gopinath and E. F. Wunderlich present a Markovian model of packet switches. A packet switch is a dedicated processor which transmits, relays and receives packets of digital information. It requires four dimensions to describe each packet switch in this model:

$h = $ *# of local packets in input buffer*,

$i = $ *# of foreign packets in input buffer*,

$j = $ *# of packets in output buffer*,

$$k = \# \ of \ outstanding \ acknowledgments.$$

Here local packets are headed for local destinations while foreign packets are headed for foreign packet switches. In the two packet switch case that is considered in this paper the number of states is tabulated as in Table 2.1 (copyright 1981 IEEE).

Table 2.1: Number of Model States			
Buffer Size	Window Size	6-D Model	8-D Model
4	2	961	4,225
5	2	2,116	12,321
6	2	4,096	30,625
6	3	5,476	38,025
8	4	21,025	211,600
9	4	34,225	416,025

Here the window size refers to the allowable number of outstanding acknowledgments. The growth in the number of states for the two packet switch, eight-dimensional model is impressive given the relatively small buffer and window size. A reduced six-dimensional model is possible if one does not distinguish between the number of local and foreign packets in each switch but only looks at the total number of packets in the input buffer. While this reduces the state space size, modest increases in the buffer and window sizes would negate this savings.
∇

Luckily, there is a large class of useful networks which satisfy a more specific type of balancing that leads to an analytic solution. Suppose we redraw the M/M/1 queueing system state transition diagram as in Figure 2.8.

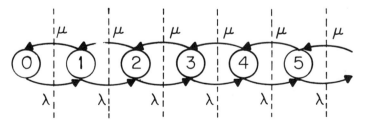

Fig. 2.8: Local Balance Boundaries Between States

Here we have drawn vertical boundaries between each adjacent pair of states. Each boundary actually separates the state transition diagram into two halves. We can equate the flow from left to right across such a boundary with the flow from right to left. We thus obtain a set of *local balance equations*:

$$\lambda p_0 = \mu p_1, \tag{2.46}$$

$$\lambda p_1 = \mu p_2,$$

$$\lambda p_2 = \mu p_3,$$

$$\lambda p_{n-1} = \mu p_n.$$

From these equations we can easily write the recursions:

$$p_1 = \frac{\lambda}{\mu} p_0, \tag{2.47}$$

$$p_2 = \frac{\lambda}{\mu} p_1,$$

$$p_3 = \frac{\lambda}{\mu} p_2,$$

$$p_n = \frac{\lambda}{\mu} p_{n-1}.$$

Substituting one into each other leads to the equations

$$p_n = \left[\frac{\lambda}{\mu}\right]^n p_0, \tag{2.48a}$$

$$p_0 = \frac{1}{\sum_{n=0}^{\infty} \left[\frac{\lambda}{\mu}\right]^n}. \tag{2.48b}$$

Here we used the normalization (conservation of probability) equation to solve for p_0. This result is the one-dimensional version of what is known as the product form solution for the state probabilities. The generalization appears in chapter 3.

2.2.9 The M/M/1 Queueing System in Detail

We can further simplify the expression for p_0 if we make use of the identity

$$\sum_{n=0}^{\infty} \rho^n = \frac{1}{1-\rho}, \qquad 0 \le \rho < 1, \tag{2.49}$$

so that

$$p_n = \rho^n p_0, \tag{2.50}$$

$$p_0 = (1-\rho), \tag{2.51}$$

or

$$\boxed{p_n = \rho^n(1-\rho) \qquad (2.52)}$$

where we are letting $\rho = \lambda/\mu$. The variable ρ is known as *utilization*. This makes sense since for an M/M/1 queueing system the utilization of the queueing system is the probability that the queueing system is nonempty (Figure 2.9):

$$\boxed{U = 1 - p_0 = \rho \qquad (2.53)}$$

The quantity ρ can also be viewed as the normalized offered load. That is, λ can vary from near zero (light load) to near, but not greater than, μ (heavy load). Thus ρ can vary between 0 and 1.

The state space of the M/M/1 queueing system is infinite in extent. Physically, we are saying that the queueing system has a potentially infinite-sized waiting line. However, since p_n is proportional to ρ^n, the probability of the nth state

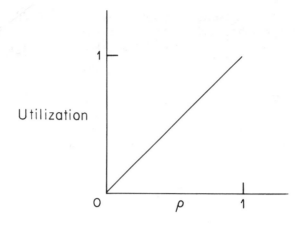

Fig. 2.9: M/M/1 System Utilization

is smaller than the probability of the $n-1$st state by a factor of ρ. Thus the probability of a larger waiting line is smaller than the probability of a smaller waiting line. This can be seen in Figure 2.10 where $\rho = 1/2$.

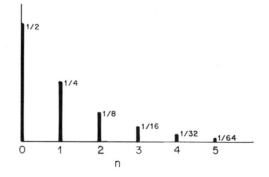

Fig. 2.10: p_n for M/M/1 system when $\rho=1/2$

Example: Suppose that ρ equals either 0.1, 0.5, or 0.9. Table 2.2 illustrates the probabilities of different waiting line sizes

Table 2.2: Probs. of Queue System Occupancy			
	$\rho=.1$	$\rho=.5$	$\rho=.9$
$P\,(0\leq\ \leq 3)$	.9999	.9375	.3439
$P\,(4\leq\ \leq 7)$	1×10^{-4}	.0586	.2256
$P\,(8\leq\ \leq 11)$	1×10^{-8}	.00366	.1480
$P\,(\geq 12)$	1×10^{-12}	.000249	.2825

From the table we can see that for a lightly loaded queueing system ($\rho=0.1$) there are usually less than four customers in the system. When $\rho=0.5$ the probability that there are less than four customers drops to 94%. At $\rho=0.9$ it is only 34% and there is a 28% chance of twelve or more customers.
∇

What we have implied in all of this and what we can state directly now is that the arrival rate can never be greater than the service rate for an M/M/1 queueing system. Put another way, ρ can never be greater than one since the queue size would be indefinitely increasing. That is, the queueing system would no longer be in equilibrium.

It is easy to calculate the average number in the queueing system [GROS], [KLEI 75] as follows. The average number is

$$\bar{n} = E\,(n) = \sum_{n=0}^{\infty} np_n, \qquad (2.54)$$

$$\bar{n} = \sum_{n=0}^{\infty} n(1-\rho)\rho^n, \qquad (2.55)$$

$$\bar{n} = (1-\rho) \sum_{n=0}^{\infty} n\rho^n. \qquad (2.56)$$

But:

$$\sum_{n=0}^{\infty} n\rho^n = \rho \sum_{n=1}^{\infty} n\rho^{n-1}, \tag{2.57}$$

$$\sum_{n=0}^{\infty} n\rho^n = \rho \frac{d}{d\rho} \sum_{n=0}^{\infty} \rho^n, \tag{2.58}$$

$$\sum_{n=0}^{\infty} n\rho^n = \rho \frac{d}{d\rho} \frac{1}{1-\rho}, \tag{2.59}$$

$$\sum_{n=0}^{\infty} n\rho^n = \frac{\rho}{(1-\rho)^2}. \tag{2.60}$$

So by substituting

$$\boxed{\bar{n} = \frac{\rho}{1-\rho}. \tag{2.61}}$$

This function is plotted in Figure 2.11. What is most noticeable is the large non-linear increase in the waiting line size as $\rho \rightarrow 1$. Interestingly, this function is often used as a "penalty function" in computer network optimization algorithms to penalize large waiting line sizes [GERL].

We can also calculate the variance in this waiting line size. This is defined as

$$\sigma_n^2 = \sum_{n=0}^{\infty} (n-\bar{n})^2 p_n, \tag{2.62}$$

$$\sigma_n^2 = \sum_{n=0}^{\infty} n^2 p_n + \bar{n}^2 \sum_{n=0}^{\infty} p_n - 2\bar{n} \sum_{n=0}^{\infty} n p_n, \tag{2.63}$$

$$\sigma_n^2 = \sum_{n=0}^{\infty} n^2 p_n + \bar{n}^2 - 2\bar{n}^2, \tag{2.64}$$

$$\sigma_n^2 = \sum_{n=0}^{\infty} n^2 p_n - \bar{n}^2. \tag{2.65}$$

But:

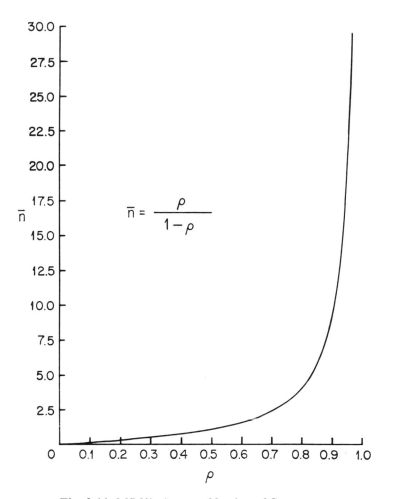

Fig. 2.11: M/M/1: Average Number of Customers

$$\sum_{n=0}^{\infty} n^2 p_n = \sum_{n=0}^{\infty} n^2 (1-\rho)\rho^n, \tag{2.66}$$

$$\sum_{n=0}^{\infty} n^2 p_n = (1-\rho) \sum_{n=0}^{\infty} n^2 \rho^n. \tag{2.67}$$

And:

$$\sum_{n=0}^{\infty} n^2 \rho^n = \frac{\rho(1+\rho)}{(1-\rho)^3}. \tag{2.68}$$

So substituting yields

$$\sigma_n{}^2 = \frac{\rho(1+\rho)}{(1-\rho)^2} - \frac{\rho^2}{(1-\rho)^2}. \qquad (2.69)$$

And:

$$\sigma_n{}^2 = \frac{\rho}{(1-\rho)^2}. \qquad (2.70)$$

This function is plotted in Figure 2.12. What we can see is that the variability of the number in the queueing system increases drastically as $\rho \to 1$.

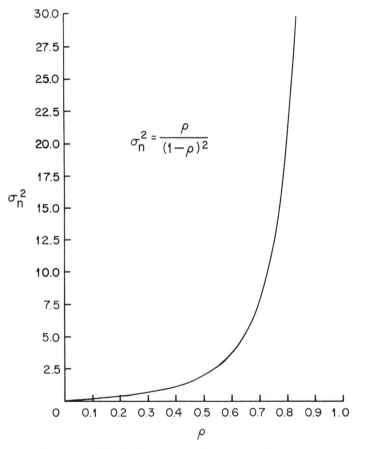

Fig. 2.12: M/M/1: Variance of Number of Customers

2.3 Little's Law

We now take some time out from our discussion of specific queueing systems to look at a simple result that holds under a surprisingly large variety of situations. This is Little's Law. It relates the mean number of customers in a queueing system, $\bar{n}$; the mean arrival rate to the queueing system, λ; and the mean waiting time to get through the queueing system, τ:

$$\boxed{\bar{n} = \lambda\bar{\tau}. \quad (2.71)}$$

Intuitively this equation makes sense. For example, if an assembly line accepts twenty chassis an hour and takes three hours to produce a finished product, then at any time there are sixty units in various states of assembly in the assembly line.

Proofs of this deceptively simple result can become quite involved. We will give three proofs of Little's Law.

Proof 1: We will first consider a proof, due to [GROS], for a queueing system with Poisson arrivals that is FIFO with a single server and a general service distribution. This is the M/G/1 queueing system. We start by noting that the probability that there are n customers in the system when a customer departs is equal to the probability that n customers arrive during the time spent in the queueing system by this customer. It turns out that the probability of n customers in the queueing system at a departure instant is equal to the probability of there being n customers in the queueing system at any time. We will return to this point in section 2.12.3. So

$$p_n = \int_0^\infty Prob(n \; arrivals \; during \; [0,T] \,|\, T{=}t)d\bar{\tau}(t). \quad (2.72)$$

Here $\bar{\tau}(t)$ is the cumulative distribution function of waiting time. Using the Poisson distribution

$$p_n = \int_0^\infty \frac{(\lambda t)^n}{n!}e^{-\lambda t}d\bar{\tau}(t). \quad (2.73)$$

Here $n \geq 0$. Now

$$\bar{n} = E(n) = \sum_{n=0}^\infty np_n. \quad (2.74)$$

Substituting,

$$\bar{n} = \sum_{n=1}^{\infty} \frac{n}{n!} \int_0^\infty (\lambda t)^n e^{-\lambda t} d\bar{\tau}(t). \tag{2.75}$$

The integration and summation can be interchanged:

$$\bar{n} = \int_0^\infty d\bar{\tau}(t) \sum_{n=1}^{\infty} e^{-\lambda t} \frac{(\lambda t)^n}{(n-1)!}, \tag{2.76}$$

$$\bar{n} = \int_0^\infty \lambda t e^{-\lambda t} d\bar{\tau}(t) \sum_{n=1}^{\infty} \frac{(\lambda t)^{n-1}}{(n-1)!}, \tag{2.77}$$

$$\bar{n} = \int_0^\infty \lambda t e^{-\lambda t} d\bar{\tau}(t) \sum_{n=0}^{\infty} \frac{(\lambda t)^n}{n!}, \tag{2.78}$$

$$\bar{n} = \int_0^\infty \lambda t e^{-\lambda t} d\bar{\tau}(t) e^{\lambda t}, \tag{2.79}$$

$$\bar{n} = \lambda \int_0^\infty t d\bar{\tau}(t), \tag{2.80}$$

$$\boxed{\bar{n} = \lambda \bar{\tau}. \quad (2.81)}$$

Proof 2: This is a heuristic proof due to P. J. Burke and related by Cooper [COOP]. We start by assuming that the mean values $\bar{n}$ and $\bar{\tau}$ exist and we have an equilibrium situation. We will look at a long interval $[0,t]$. The mean number of customers who arrive is λt. Let each customer bring his/her waiting time with them. The mean amount of waiting time brought in during the interval is $\lambda t \bar{\tau}$.

Now each customer in the queueing system "uses" up waiting time linearly with time. So if $\bar{n}$ is the mean number of customers present during the interval then $\bar{n}t$ is the mean amount of time used up during $[0,t]$. If we let $t \to \infty$, we can expect that the amount of waiting time brought into the queueing system must equal the amount of waiting time used up. Thus

$$\lim_{t \to \infty} \frac{\lambda_t t \bar{\tau}}{\bar{n} t} = 1 \qquad (2.82)$$

and so

$$\boxed{\bar{n} = \lambda \bar{\tau}. \qquad (2.83)}$$

In this heuristic proof no special assumptions were made.

Proof 3: This popular type of proof is based on a diagram like that of Figure 2.13. Versions of it appear in [KOBA], [SCHW 87], [GROS] and [KLEI 75]. We will follow the approach in [KLEI 75]. As in proof 2, we will consider an interval $[0,t]$.

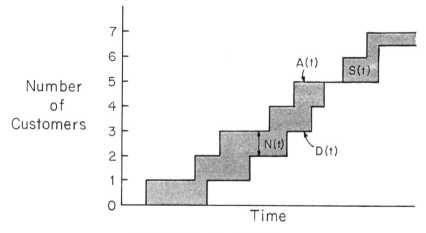

Fig. 2.13: Proving Little's Law

In Figure 2.13 we have $A(t)$ as the number of arrivals in $[0,t]$ and $D(t)$ as the number of departures. Then the number in the system is:

$$N(t) = A(t) - D(t). \qquad (2.84)$$

We will now present three simple relations amongst the quantities in Figure 2.13 that can be brought together to form Little's Law. The first is

$$\lambda_t = \frac{A(t)}{t} \qquad (2.85)$$

where λ_t is the mean arrival rate during $[0,t]$. The second is

$$\bar{\tau}_t = \frac{S(t)}{A(t)} \tag{2.86}$$

where $\bar{\tau}_t$ is the mean time spent in the queueing system by each customer and $S(t)$ is the cumulative area between the curves $A(t)$ and $D(t)$ at time t. This area, at t, is the total time spent in the queueing system by all customers up to time t. Finally, the third equation is

$$\bar{n}_t = \frac{S(t)}{t}, \tag{2.87}$$

which gives the mean number of customers in the queueing system at time t. Now if we substitute these three equations into one another we have

$$\bar{n}_t = \frac{S(t)}{t} = \frac{\bar{\tau}_t A(t)}{t} = \lambda_t \bar{\tau}_t. \tag{2.88}$$

If we then go to the limit $t \to \infty$

$$\bar{n} = \lim_{t \to \infty} \bar{n}_t, \tag{2.89}$$

$$\lambda = \lim_{t \to \infty} \lambda_t,$$

$$\bar{\tau} = \lim_{t \to \infty} \bar{\tau}_t.$$

Then

$$\boxed{\bar{n} = \lambda\bar{\tau}. \tag{2.90}}$$

In this derivation no specific assumptions have been made about the arrival process, the service distribution, the number of servers, or the queueing discipline within the queueing system. Thus Little's Law holds even for a G/G/m queueing system.

Example: Suppose that one wishes to calculate the mean waiting time in an M/M/1 queueing system. We know that

$$\bar{n} = \frac{\rho}{(1-\rho)}, \tag{2.91}$$

so using Little's Law we have

$$\boxed{\bar{\tau} = \frac{1/\mu}{(1-\rho)}. \qquad (2.92)}$$

∇

Little's Law also holds if we consider only the queue and not the server. Then

$$\bar{n}_Q = \lambda \bar{\tau}_Q \tag{2.93}$$

where $\bar{n}_Q$ and $\bar{\tau}_Q$ are the analogous quantities for the waiting queue.

Little's Law existed informally long before J. D. C. Little formalized it in [LITT]. Following Little's notation, it is often written as $L = \lambda W$. Other works on this subject are mentioned at the end of this chapter in To Look Further.

2.4 Reversibility and Burke's Theorem

2.4.1 Introduction

The input to the M/M/1 queueing system is a Poisson process. But what can we say of its output? According to Burke's theorem, which was published in 1956, that output is also Poisson. This result is not immediately intuitive, despite its appealing symmetry. After all, consider the inter-departure times of the M/M/1 queueing system. As long as the queueing system is non-empty, and because we have an exponential server, these are exponentially distributed with mean duration $1/\mu$. But the problem is that sometimes the queueing system *is* empty. Then the time interval between successive departures from the queueing system is the sum of the time it takes for a customer to arrive at the empty queueing system (exponential with mean duration $1/\lambda$) and the time for that customer to be serviced (exponential with mean $1/\mu$). Thus the inter-departure times of the output process alternates between these two types of intervals. Despite this structure, it turns out that the output process is indeed Poisson with rate λ.

To prove Burke's theorem simply, we will introduce a new concept called *reversibility*. Reversibility for a stochastic process means that when the direction of time is reversed, that is, if time flows backwards, the statistics of the process are the same as in the time normal case. As J. Walrand puts it [WALR 88]: "The technique of time reversal has proved to be a valuable tool for gaining insight into the behavior of networks. Among the successful applications of that method are simpler proofs and new examples of product-form results, as well as new results on sojourn times in networks".

We will first introduce and prove a theorem whose purpose is to allow us to establish that the number of customers in an M/M/1 queueing system as a function of time, $n(t)$, is a reversible process. This will be used to prove Burke's theorem.

In this discussion of reversibility and Burke's theorem we will be largely following the treatment in [KELL]. This is an excellent source for more information on these concepts.

2.4.2 Reversibility

Let us make the following definition:

Definition: A stochastic process, X(t), is reversible if the samples $(X(t_1),X(t_2),X(t_3) \cdots X(t_m))$ has the same distribution as $(X(\tau-t_1),X(\tau-t_2)X(\tau-t_3) \cdots X(\tau-t_m))$ for every real τ (we are thinking of continuous processes) and for every $t_1,t_2, \cdots t_m$.

This definition formally states that the time reversed samples of the process X(t) are statistically the same as the time normal samples for a reversible process.

We will now relate the reversibility of a process to the satisfaction of a set of detailed balance equations. Formally we define the transition rate from state i to state j as

$$q_{ij} = \lim_{\tau \to 0} \frac{P(X(t+\tau)=j \mid X(t)=i)}{\tau} \qquad (2.94)$$

for $i \neq j$ and $q_{jj}=0$.

Equilibrium will be naturally achieved if

$$\sum_{j \varepsilon S} p_i q_{ij} = \sum_{j \varepsilon S} p_j q_{ji} \qquad (2.95)$$

or

$$p_i \sum_{j \varepsilon S} q_{ij} = \sum_{j \varepsilon S} p_j q_{ji} \qquad (2.96)$$

where p_j is the equilibrium probability of the jth state and S is the set of states. What is being described above is a form of global balance. What is involved in the next theorem is a form of local balancing:

Theorem: A stationary Markov chain is reversible if and only if there is a collection of positive numbers $p_i, i \varepsilon S$, which sum to one and satisfy the detailed balance equations

$$p_i q_{ij} = p_j q_{ji} \qquad (2.97)$$

for i,j ε S. These p_i are naturally the equilibrium state probabilities.

Proof: If the process is reversible then

$$P(X(t){=}i, X(t{+}\tau){=}j) = P(X(t){=}j, X(t{+}\tau){=}i). \qquad (2.98)$$

But this can be rewritten as

$$P(X(t){=}i)P(X(t{+}\tau){=}j \mid X(t){=}i) = \qquad (2.99)$$
$$P(X(t){=}j)P(X(t{+}\tau){=}i \mid X(t){=}j)$$

or

$$p_i P(X(t{+}\tau){=}j \mid X(t){=}i) = p_j P(X(t{+}\tau){=}i \mid X(t){=}j). \qquad (2.100)$$

Dividing both sides by τ results in

$$p_i q_{ij} = p_j q_{ji}, \qquad (2.101)$$

which is the desired result.

Proving the converse is a little more involved. We start by assuming that the p_i exist which satisfy the detailed balance equations. These have to be the equilibrium probabilities since if we sum

$$p_i q_{ij} = p_j q_{ji} \qquad (2.102)$$

over j we have

$$\sum_{j \epsilon S} p_i q_{ij} = \sum_{j \epsilon S} p_j q_{ji} \qquad (2.103)$$

which is just a form of global balance equation.

We will look now at the interval $t \; \epsilon \; [-T,T]$. Suppose that the process is in state i_1, at time -T and stays in this state for a period k_1 and then jumps to state i_2 which it stays in for a period k_2 and so on, until the state is i_m for a period k_m until time T.

We will now develop the tools we will need so that we can determine the probability density of this situation. The probability density of the random variable k_1 is

$$q_{i_1} e^{-q_{i_1} k_1} \qquad (2.104)$$

where

$$q_i = \sum_{j \epsilon S} q_{ij}. \qquad (2.105)$$

Also, the probability that i_2 is the system state following i_1 is

$$\frac{q_{i_1 i_2}}{q_{i_1}}. \qquad (2.106)$$

Similar expressions can be developed for the probability density of k_2 and the probability that i_3 follows i_2 and so forth. Finally, the probability that the state is i_m for an interval of at least k_m is

$$\int_{k_m}^{\infty} q_{i_m} e^{-q_{i_m} t} \, dt = -e^{-q_{i_m} t} \Big|_{k_m}^{\infty} \qquad (2.107)$$

or

$$e^{-q_{i_m} k_m}. \qquad (2.108)$$

The probability density of this situation can now be written

$$p_{i_1} e^{-q_{i_1} k_1} q_{i_1 i_2} e^{-q_{i_2} k_2} \cdots q_{i_{m-1} i_m} e^{-q_{i_m} k_m}. \tag{2.109}$$

This is a probability density in the variables $k_1, k_2, \cdots k_m$. One can calculate a probability of some region in these variables if one integrates over the region. Naturally, $\sum k_i = 2T$.

The last equation needed can be obtained by expanding the detailed balance equation

$$p_i q_{ij} = p_j q_{ji} \tag{2.110}$$

into

$$p_{i_1} q_{i_1 i_2} q_{i_2 i_3} \cdots q_{i_{m-1} i_m} = p_{i_m} q_{i_m i_{m-1}} q_{i_{m-1} i_{m-2}} \cdots q_{i_2 i_1}.$$

This last expression tells us that the previous probability density for the situation on $[-T,T]$ is the same as the probability density that the process starts at $-T$ in state i_m, remains there for a period k_m, and then jumps to state i_{m-1}. This continues until the process is in state i_1 for a period k_1 until T.

We now know that:

1. $X(t)$ and $X(-t)$ are statistically the same.

2. This means that $(X(t_1), X(t_2), X(t_3) \cdots X(t_m))$ and $(X(-t_1), X(-t_2), X(-t_3) \cdots X(-t_m))$ have the same distributions.

3. Since $X(t)$ is stationary $(X(-t_1), X(-t_2), X(-t_3) \cdots X(-t_m))$ has the same distribution as $(X(\tau-t_1), X(\tau-t_2), X(\tau-t_3) \cdots X(\tau-t_m))$.

Thus the process is reversible and the proof is complete.

∇

The detailed balance condition says that for a reversible process states must be connected by transitions in either direction, if they are connected at all. How restrictive is this characterization? While a great many queueing systems would not fit this characterization, some of the most important - including the M/M/1 queueing system - do.

2.4.3 Burke's Theorem

To see that the number of customers in the M/M/1 queueing system, n(t), is a reversible process, we need only to recall the existence of the local balance equation, which will satisfy the condition of the theorem:

$$\lambda p_{n-1} = \mu p_n. \tag{2.111}$$

This was previously justified by equating the probability flux flowing across a vertical boundary through the state transition diagram. In fact this same argument can be extended to show that any stationary Markov process whose state transition diagram is a tree is reversible [KELL].

This brings us to Burke's theorem [BURK]. It says that:

Theorem: The departure process from an M/M/1 queueing system, in equilibrium, is a Poisson process.

Proof: A realization of $n(t)$, the number of customers in the M/M/1 queueing system, is shown in Figure 2.14. The points at which the process $n(t)$ jumps upward corresponds to the (Poisson) arrival process of rate λ. Since $n(t)$ is reversible, the points at which the process $n(-t)$ jumps upward must also represent a Poisson process of rate λ. But these points also correspond to the departure events of the queueing system for $n(t)$. Thus the output process is Poisson with rate λ.
∇

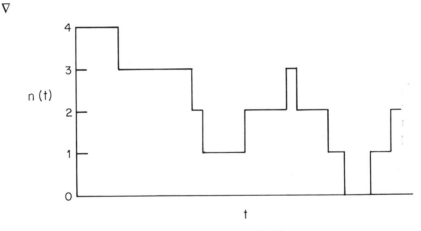

Fig. 2.14: Proving Burke's Theorem

Note how simple the concept of reversibility makes the proof. This sort of argument will work with any queueing system with a Poisson arrival process and for which $n(t)$ is a birth-death process.

2.5 The State-dependent M/M/1 Queueing System

2.5.1 The General Solution

In all that has been said so far in this chapter the arrival rate and the service rate have been constants. Specifically, they have been independent of the number of customers in the queueing system. Many useful queueing systems can be developed if the arrival rate and/or service rate is a function of, or dependent on, the number of customers in the queueing system. This is illustrated in Figure 2.15. The state transition diagram appears in Figure 2.16.

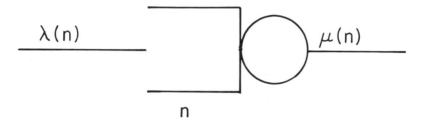

Fig. 2.15: State-dependent Queueing System

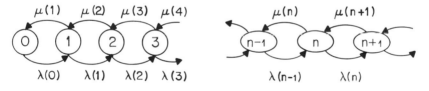

Fig. 2.16: State-dependent State Transition Diagram

By drawing vertical boundaries between each pair of adjacent states and equating the flow of probability flux from left to right with the flow of probability flux from right to left yields a set of local balance equations:

$$\lambda(0)p_0 = \mu(1)p_1, \tag{2.112}$$

$$\lambda(1)p_1 = \mu(2)p_2,$$

$$\lambda(2)p_2 = \mu(3)p_3,$$

$$\lambda(n-1)p_{n-1} = \mu(n)p_n. \tag{2.113}$$

From these we can write the recursions:

$$p_1 = \frac{\lambda(0)}{\mu(1)}p_0, \tag{2.114}$$

$$p_2 = \frac{\lambda(1)}{\mu(2)}p_1,$$

$$p_3 = \frac{\lambda(2)}{\mu(3)}p_2,$$

$$p_n = \frac{\lambda(n-1)}{\mu(n)}p_{n-1}. \tag{2.115}$$

Substituting these recursions into each other leads to the equation

$$p_n = \left[\prod_{i=1}^{n}\frac{\lambda(i-1)}{\mu(i)}\right]p_0. \tag{2.116}$$

To calculate p_0 we make use of the fact that

$$\sum_{n=0}^{\infty}p_n = 1 \tag{2.117}$$

so that we can write out

$$p_0 + \frac{\lambda(0)}{\mu(1)}p_0 + \frac{\lambda(0)\lambda(1)}{\mu(1)\mu(2)}p_0 + \frac{\lambda(0)\lambda(1)\lambda(2)}{\mu(1)\mu(2)\mu(3)}p_0 + \cdots = 1. \tag{2.118}$$

Solving for p_0 we have

$$p_0 = \cfrac{1}{1 + \sum\limits_{n=1}^{\infty} \prod\limits_{i=1}^{n} \cfrac{\lambda(i-1)}{\mu(i)}}. \qquad (2.119)$$

2.5.2 Performance Measures

The previous two boxed equations gives us a way of calculating the equilibrium state probabilities. But why are these so important? Mainly because many practical performance measures are functions of these state probabilities. For instance, the mean *throughput* of the queueing system, or the mean rate at which customers pass through it, is

$$\overline{Y} = \sum_{n=1}^{\infty} \mu(n)p_n. \qquad (2.120)$$

Note that the index starts at 1 since when $n=0$ the queueing system is empty and there is no contribution to throughput. The mean throughput is a weighted average of the service rates where the state probabilities serve as the weights. A second performance measure is the mean number of customers in the queueing system or

$$\overline{n} = \sum_{n=1}^{\infty} np_n. \qquad (2.121)$$

Again, this is a weighted average, of the number of customers in the queueing system with the state probabilities serving as weights.

Since for an infinite buffer M/M/1 queueing system the mean throughput is equal to the arrival rate, λ (recall Burke's theorem), we can use Little's Law to write an expression for the mean *time delay*:

$$\tau = \frac{\bar{n}}{\bar{Y}} = \frac{\displaystyle\sum_{n=1}^{\infty} n p_n}{\displaystyle\sum_{n=1}^{\infty} \mu(n) p_n}. \qquad (2.122)$$

As a check, the numerator is in units of customers and the denominator has units of customers per second, so τ has units of seconds.

Finally a very simple performance measure is utilization, or the probability that the queueing system is nonempty and the server is busy:

$$U = 1 - p_0. \qquad (2.123)$$

Recall from section 2.2.9 that for the infinite buffer M/M/1 queueing system with state-independent service and arrival rates that $U = \lambda/\mu$.

In closing it should be noted that these performance measures can be used directly on queueing systems consisting of a single queue. They can also be used for a queue in a network of queues where the state probability is the marginal state probability of that queue.

In the next several sections we will look at specific instances of state-dependent queueing systems. These include some queueing systems where, by setting certain transition rates to zero, we wind up with a finite number of states. An earlier, somewhat more extensive, collection of such queueing systems is in [KLEI 75].

2.6 The M/M/1/N Queueing System: The Finite Buffer Case

In this scenario we have a queueing system that has a limit, N, on the number of customers that can be in the system. If there are N customers in the queueing system (including the one in the server), an arriving customer is said to be "turned away" or "lost" or "blocked". This means that the arriving customer really is lost and does not return at a later time. If such blocked customers were to return later, the arrival process would no longer be Poisson. The state-dependent arrival and service rates are now:

$$\lambda(n) = \lambda, \quad n = 0, 1, 2, \cdots N-1, \qquad (2.124)$$

$$\mu(n) = \mu, \quad n = 1,2,3, \cdots N.$$

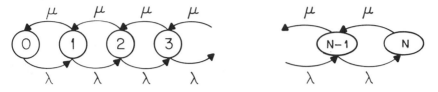

Fig. 2.17: Finite Buffer State Transition Diagram

The state transition diagram, which has a finite number of states, is shown in Figure 2.17. Substituting the transition rates into the last section's expressions for the state probabilities yields

$$p_n = \left[\frac{\lambda}{\mu}\right]^n p_0, \quad 0 \le n \le N, \tag{2.125}$$

$$p_0 = \frac{1}{\sum\limits_{n=0}^{N}\left[\frac{\lambda}{\mu}\right]^n}. \tag{2.126}$$

This expression for p_0 can also be written as:

$$p_0 = \frac{1 - \dfrac{\lambda}{\mu}}{1 - \left[\dfrac{\lambda}{\mu}\right]^{N+1}}. \tag{2.127}$$

The numerator of the expression for p_0 is the same as the expression for p_0 for the M/M/1 queueing system with infinite buffer. The denominator accounts for the fact that the state space is now truncated to a finite number of states. Note that p_n is monotonically increasing or decreasing in n, depending on the relative values of λ and μ.

Therefore,

$$p_n = \frac{1 - \dfrac{\lambda}{\mu}}{1 - \left[\dfrac{\lambda}{\mu}\right]^{N+1}} \left[\frac{\lambda}{\mu}\right]^n \qquad 0 \leq n \leq N. \qquad (2.128)$$

Example: The blocking probability, p_N, is the probability that the queueing system is full. The mean number of customers turned away per second is then λp_N. Table 2.3 shows the blocking probability for various values of offered load, λ/μ, when N=5

Table 2.3: Blocking Probability	
λ/μ	Block. Prob.
0.10	9×10^{-6}
0.50	0.016
0.75	0.072
1.00	0.166
2.00	0.508
5.00	0.800

As λ/μ increases, so does the blocking probability. Here the value at $\lambda/\mu = 1.00$ has to be calculated using L'Hopital's rule. For the M/M/1 queueing system with infinite buffer, λ can never be greater than μ so λ/μ is never greater than 1 since this would lead to an ever increasing waiting line. With a finite buffer λ can be greater than μ since excess customers are simply turned away.

As a continuation of this exercise let us tabulate in Table 2.4 the blocking probability when $\lambda/\mu=0.9$ and the buffer size is varied:

Table 2.4: Blocking Probability	
N	Block. Prob.
1	0.474
2	0.299
3	0.212
5	0.126
10	0.051
20	0.014
100	2.7×10^{-6}

As N increases, the blocking probability decreases.

∇

It can be seen that the analysis of a single queueing system with blocking is quite tractable. However, it turns out that the analysis of networks of blocking queueing systems is quite difficult. The basic problem is that the presence of blocking introduces dependencies between the queueing systems that are not easily taken into account. That is, when one queueing system is full it blocks departures from the queueing system(s) which feed into it. The networks of Markovian queueing systems for which nice analytic (chapter 3) and computational algorithms (chapter 4) exist are ones in which there is ample buffer space. Networks of queueing systems with blocking represent an active research area [PERR 84,89,89b,90][ONVU 90,93].

2.7 The M/M/∞ Queueing System: Infinite Number of Servers

In this queueing system every arriving customer is assigned to its own server of rate μ. Therefore, if there are n customers, the aggregate output rate is $n\mu$. The queueing system is illustrated in Figure 2.18 and the state transition diagram appears in Figure 2.19.

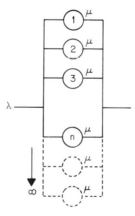

Fig. 2.18: M/M/∞ Queueing System

The state dependent arrival and service rates are

$$\lambda_n = \lambda, \quad n=0,1,2,3,\cdots, \tag{2.129}$$

$$\mu_n = n\mu, \quad n=1,2,3,4,\cdots.$$

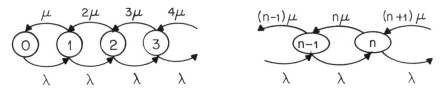

Fig. 2.19: M/M/∞ State Transition Diagram

Clearly, the equilibrium state probabilities are then

$$p_n = \left[\prod_{i=1}^{n}\frac{\lambda}{i\mu}\right] p_0 \qquad (2.130)$$

or

$$p_n = \frac{1}{n!}\left[\frac{\lambda}{\mu}\right]^n p_0. \qquad (2.131)$$

From this it follows that

$$p_0 = \frac{1}{1 + \sum_{i=1}^{\infty}\frac{1}{i!}\left[\frac{\lambda}{\mu}\right]^i}. \qquad (2.132)$$

This can be simplified further. We recall the familiar series expansion

$$\sum_{i=0}^{\infty}\frac{1}{i!} x^i = e^x. \qquad (2.133)$$

Therefore

$$p_0 = e^{-\lambda\mu}. \qquad (2.134)$$

Thus the equilibrium state probabilities are given by

$$p_n = \frac{1}{n!} \left[\frac{\lambda}{\mu}\right]^n e^{-\lambda\mu}. \qquad (2.135)$$

This is just a Poisson distribution with mean $\bar{n} = \lambda/\mu$. For this model $0 < \lambda/\mu < \infty$. It is the unlimited supply of servers that allows λ/μ to exceed 1.

2.8 The M/M/m Queueing System: m Parallel Servers with a Queue

This queueing system is illustrated in Figure 2.20. There are m servers in parallel and a queue for additional customers. The state transition diagram appears in Figure 2.21.

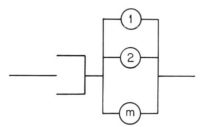

Fig. 2.20: M/M/m Queueing System

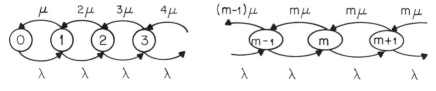

Fig. 2.21: M/M/m State Transition Diagram

The state dependent transition rates are

$$\lambda_n = \lambda, \quad n = 0, 1, 2, 3 \cdots; \qquad (2.136)$$

$$\mu_n = \begin{cases} n\mu & 0 \le n \le m, \\ m\mu & n \ge m. \end{cases}$$

We can solve for the equilibrium state probabilities for $n \leq m$ and $n \geq m$ separately. For $n \leq m$, that is, the number of customers is less than or equal to the number of servers, we have

$$p_n = \left[\prod_{i=1}^{n} \frac{\lambda}{i\mu} \right] p_0 \qquad (2.137)$$

or

$$p_n = \frac{1}{n!} \left[\frac{\lambda}{\mu} \right]^n p_0, \quad n \leq m. \qquad (2.138)$$

This is the same result as for the M/M/∞ system as there is one server for each customer. For $n \geq m$, or the number of customers is greater than or equal to the number of servers, we have

$$p_n = \left[\frac{1}{m!} \left[\frac{\lambda}{\mu} \right]^m \right] \left[\prod_{i=m+1}^{n} \frac{\lambda}{m\mu} \right] p_0 \qquad (2.139)$$

or

$$p_n = \frac{1}{m! \, m^{n-m}} \left[\frac{\lambda}{\mu} \right]^n p_0, \quad n \geq m. \qquad (2.140)$$

Summing up over all the states one has

$$p_0 = \left[1 + \sum_{n=1}^{m-1} \frac{1}{n!} \left[\frac{\lambda}{\mu} \right]^n + \sum_{n=m}^{\infty} \frac{1}{m! \, m^{n-m}} \left[\frac{\lambda}{\mu} \right]^n \right]^{-1} \qquad (2.141)$$

or

$$p_0 = \left[1 + \sum_{n=1}^{m-1} \frac{1}{n!} \left[\frac{\lambda}{\mu} \right]^n + \frac{1}{m!} \left[\frac{\lambda}{\mu} \right]^m \sum_{n=m}^{\infty} \frac{1}{m^{n-m}} \left[\frac{\lambda}{\mu} \right]^{n-m} \right]^{-1} \qquad (2.142)$$

or with a change of variables

$$p_0 = \left[1 + \sum_{n=1}^{m-1} \frac{1}{n!} \left[\frac{\lambda}{\mu} \right]^n + \frac{1}{m!} \left[\frac{\lambda}{\mu} \right]^m \sum_{n=0}^{\infty} \frac{1}{m^n} \left[\frac{\lambda}{\mu} \right]^n \right]^{-1} \qquad (2.143)$$

or:

$$p_0 = \left[1 + \sum_{n=1}^{m-1} \frac{1}{n!} \left[\frac{\lambda}{\mu} \right]^n + \frac{1}{m!} \left[\frac{\lambda}{\mu} \right]^m \left[\frac{1}{1-\rho} \right] \right]^{-1}. \qquad (2.144)$$

In the above $\rho = \lambda/m\mu$.

In telephone engineering one quantity of interest is the probability that all the servers in this system are busy. In this context customers are actually calls. This quantity is found from what is known as Erlang's C formula (or in Europe, Erlang's formula of the second kind). We have

$$P[queueing] = \sum_{n=m}^{\infty} p_n \qquad (2.145)$$

or

$$P[queueing] = \left[\sum_{n=m}^{\infty} \frac{1}{m!m^{n-m}} \left[\frac{\lambda}{\mu} \right]^n \right] p_0. \qquad (2.146)$$

Using the same technique as before this is

$$P[queueing] = \frac{1}{m!} \left[\frac{\lambda}{\mu} \right]^m \left[\frac{1}{1-\rho} \right] p_0. \qquad (2.147)$$

Substituting for p_0 this becomes

$$P[queueing] = \frac{\dfrac{1}{m!} \left[\dfrac{\lambda}{\mu} \right]^m \left[\dfrac{1}{1-\rho} \right]}{\left[1 + \displaystyle\sum_{n=1}^{m-1} \frac{1}{n!} \left[\dfrac{\lambda}{\mu} \right]^n + \frac{1}{m!} \left[\dfrac{\lambda}{\mu} \right]^m \left[\dfrac{1}{1-\rho} \right] \right]}. \qquad (2.148)$$

2.9 The M/M/m/m Queue: A Loss System

Suppose that we take the M/M/m queueing system of the previous section and do not allow customers/calls to wait. That is, if a customer arrives and finds all the servers busy, it is lost.

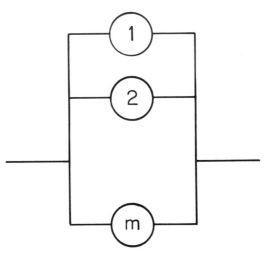

Fig. 2.22: M/M/m/m Queueing System

As Figure 2.22 shows, the queueing system simply consists of m parallel servers. The state transition diagram is like the one for the M/M/m queueing system except that it is truncated at the mth state (Figure 2.23).

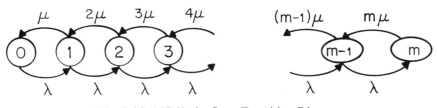

Fig. 2.23: M/M/m/m State Transition Diagram

Therefore, the state-dependent transition rates are

$$\lambda_n = \lambda, \quad n=0,1,2,\cdots m-1, \tag{2.149}$$

$$\mu_n = n\mu, \quad n=1,2,3,\cdots m.$$

We have

$$p_n = \left[\prod_{i=1}^{n} \frac{\lambda}{i\mu}\right] p_0 \qquad (2.150)$$

or

$$p_n = \frac{1}{n!}\left[\frac{\lambda}{\mu}\right]^n p_0, \quad n \leq m. \qquad (2.151)$$

Summing over all the $m+1$ states leads to

$$p_0 = \frac{1}{1 + \displaystyle\sum_{n=1}^{m} \frac{1}{n!}\left[\frac{\lambda}{\mu}\right]^n}. \qquad (2.152)$$

A quantity of great interest in telephone engineering is p_m, the probability that all m servers are busy. For instance, the mean number of calls turned away per second is then λp_m. The expression for p_m is known as Erlang's B formula (or Erlang's formula of the first kind, in Europe) and is clearly

$$p_m = \frac{\dfrac{1}{m!}\left[\dfrac{\lambda}{\mu}\right]^m}{1 + \displaystyle\sum_{n=1}^{m} \frac{1}{n!}\left[\frac{\lambda}{\mu}\right]^n}. \qquad (2.153)$$

2.10 Central Server CPU Model

This model involves a group of m terminals which access a central CPU (Central Processing Unit). Control, in the queueing form of a customer, rotates between each terminal and the CPU. The terminals are modeled as m exponential servers, each with service rate λ. The CPU is modeled as a queueing system with a single server of rate μ. This is illustrated in Figure 2.24.

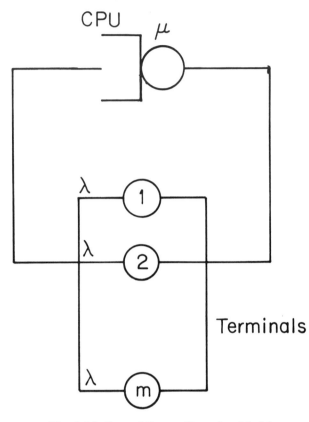

Fig. 2.24: Central Server Queueing Model

A single customer will move from its unique terminal server to the CPU queue and back to the same terminal server. The time spent by the customer at the terminal server represents the time a person spends thinking and typing at the terminal between receiving the prompt and hitting the carriage return. The time spent by the customer at the CPU is, to the user, the time between hitting the carriage return and the return of the prompt.

Even though there are really two queueing systems involved, the CPU and the terminals, the state transition diagram is still one dimensional because we have a closed queueing network. That is, the state variable can be the number of customers in the CPU, n, and then the number of active terminals is a simple function of this, $m-n$.

A word is necessary concerning the state description. In using the number of customers in the CPU as the state variable, we are not really specifying which customer leaves the queueing system at a departure instant, so long as some customer leaves. Thus the service discipline may be First In First Out (FIFO) or, for instance, Processor Sharing (PS). In Processor Sharing each of the n customers in the CPU receives μ/n of the service effort. This models how a multiprogrammed CPU works on a little bit of each "job" in turn. The state transition diagram, which appears in Figure 2.25, does not really expose the fine structure of the service discipline.

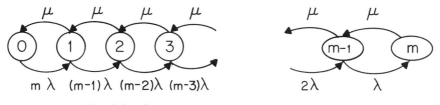

Fig. 2.25: Central Server State Transition Diagram

To calculate the state probabilities of this system we have

$$\lambda_n = \lambda(m-n), \quad 0 \leq n \leq m. \tag{2.154}$$

The state probabilities are then

$$p_n = \left[\prod_{i=1}^{n} \frac{\lambda(m-i+1)}{\mu} \right] p_0. \tag{2.155}$$

A little thought will show that this can be rewritten as

$$p_n = \left[\left[\frac{\lambda}{\mu} \right]^n \frac{m!}{(m-n)!} \right] p_0, \quad (2.156)$$

$$p_0 = \frac{1}{\left[\sum_{n=0}^{m} \left[\frac{\lambda}{\mu} \right]^n \frac{m!}{(m-n)!} \right]}.$$

Here 0! is defined as 1.

2.11 Transient Solution of the M/M/1/∞ Queueing System

2.11.1 The Technique

The differential equations for the M/M/1/∞ queueing system are

$$\frac{dP_n(t)}{dt} = -(\lambda+\mu)P_n(t) + \lambda P_{n-1}(t) + \mu P_{n+1}(t), \tag{2.157}$$

$$\frac{dP_0(t)}{dt} = -\lambda P_0(t) + \mu P_1(t). \tag{2.158}$$

These equations are easy to solve in the equilibrium case as the derivatives are then equal to zero and a set of equations which are linear in the state probabilities results. In this section we will look at the more complicated transient case. Specifically, we will look at the case when there are initially i customers in the queueing system. The question is then what is the behavior of the queueing system as time proceeds.

The solution technique we will use is one that can be applied to a variety of queueing systems. It makes use of the moment-generating function

$$P(z,t) = \sum_{n=0}^{\infty} P_n(t)z^n. \tag{2.159}$$

Basically, the state probabilities form a sequence of numbers that can be transformed, like any numerical sequence, into a "frequency" like domain. It turns out to be relatively easy to obtain the frequency domain solution. This solution is then inverted to provide the solution to the state probabilities.

The details of the solution technique are:

A. Take the nth equation and multiply both sides by z^n. Then sum the equations up to n= ∞.

B. Take the resulting single equation and identify the moment-generating function throughout the equation. It may be necessary to add and subtract a finite number of terms to obtain explicitly the moment-generating function in places.

C. Use any "boundary" equation(s) to eliminate unknowns (i.e., state probabilities).

D. Solve for the moment-generating function. If a differential equation results, repeat the analogs of steps A-D using Laplace transforms.

E.   Invert the moment-generating function, or if this is not possible, calculate moments

We will now make this methodology a little more real in terms of the M/M/1/∞ transient analysis problem. We mention in passing that the approach for this problem first appeared in [BAIL]. Our treatment will be similar to that appearing in [GROS] and [KLEI 75].

2.11.2 The Solution

We first take the differential equation of the previous section, multiply both sides by z^n, and sum from $n=1$ to ∞:

$$\sum_{n=1}^{\infty} \frac{dP_n(t)}{dt} z^n = -\lambda \sum_{n=1}^{\infty} P_n(t)z^n - \mu \sum_{n=1}^{\infty} P_n(t)z^n \qquad (2.160)$$

$$+ \lambda \sum_{n=1}^{\infty} P_{n-1}(t)z^n + \mu \sum_{n=1}^{\infty} P_{n+1}(t)z^n.$$

Recall that to get proper moment generating functions the indexes must start at $n=0$ so we have

$$\sum_{n=0}^{\infty} \frac{dP_n(t)}{dt} z^n - \frac{dP_0(t)}{dt} = \qquad (2.161)$$

$$-\lambda \sum_{n=0}^{\infty} P_n(t)z^n + \lambda P_0(t) - \mu \sum_{n=0}^{\infty} P_n(t)z^n + \mu P_0(t)$$

$$+ \lambda \sum_{n=1}^{\infty} P_{n-1}(t)z^n + \mu \sum_{n=0}^{\infty} P_{n+1}(t)z^n - \mu P_1(t).$$

But three of the terms here are just the boundary differential equation at n=0 and so drop out, leaving:

$$\sum_{n=0}^{\infty} \frac{dP_n(t)}{dt} z^n = \qquad (2.162)$$

$$-\lambda \sum_{n=0}^{\infty} P_n(t)z^n - \mu \sum_{n=0}^{\infty} P_n(t)z^n + \mu P_0(t)$$

$$+ \lambda \sum_{n=1}^{\infty} P_{n-1}(t)z^n + \mu \sum_{n=0}^{\infty} P_{n+1}(t)z^n.$$

We can rewrite this further

$$\sum_{n=0}^{\infty} \frac{dP_n(t)}{dt} z^n = \tag{2.163}$$

$$-\lambda \sum_{n=0}^{\infty} P_n(t)z^n - \mu \sum_{n=0}^{\infty} P_n(t)z^n + \mu P_0(t)$$

$$+\lambda z \sum_{n=1}^{\infty} P_{n-1}(t)z^{n-1} + \mu z^{-1} \sum_{n=0}^{\infty} P_{n+1}(t)z^{n+1}.$$

Now identifying the moment-generating functions (which sometimes requires adding and subtracting terms) one has

$$\frac{\partial P(z,t)}{\partial t} = \tag{2.164}$$

$$-\lambda P(z,t) - \mu(P(z,t) - P_0(t))$$
$$+\lambda z P(z,t) + \mu z^{-1}(P(z,t) - P_0(t)).$$

This can be simplified

$$\frac{\partial P(z,t)}{\partial t} = \tag{2.165}$$

$$(1-z)(\mu z^{-1} - \lambda)P(z,t) + (1-z)(-\mu z^{-1})P_0(t),$$

$$\boxed{\frac{\partial P(z,t)}{\partial t} = (1-z)z^{-1}\left[(\mu - \lambda z)P(z,t) - \mu P_0(t)\right].} \tag{2.166}$$

The initial condition to this equation involves i customers at time 0:

$$P_n(0+) = \begin{cases} 1, & n=i, \\ 0, & n \neq i ; \end{cases} \tag{2.167}$$

or:

$$P(z, 0+) = \sum_{n=0}^{\infty} P_n(0+)z^n = z^i \tag{2.168}$$

To solve this partial differential equation, one can use Laplace transforms which are defined as

$$L\{P_i(t)\} = P_i^*(s) = \int_0^\infty P_i(t)e^{-st}dt, \tag{2.169}$$

$$L\{P(z,t)\} = P^*(z,s) = \int_0^\infty P(z,t)e^{-st}dt.$$

For the left-hand side of the boxed equation one can integrate by parts to find its transform:

$$\int u\,dv = uv - \int v\,du. \tag{2.170}$$

Specifically,

$$\int_0^\infty e^{-st}\frac{\partial P(z,t)}{\partial t}dt = \left[e^{-st}P(z,t)\right]_0^\infty + \int_0^\infty se^{-st}P(z,t)dt \tag{2.171}$$

$$= -P(z,0) + sP^*(z,s) \tag{2.172}$$

$$= -z^i + sP^*(z,s) \tag{2.173}$$

Now take the Laplace transform of the boxed differential equation and multiply out the constants so that

$$-z^{i+1} + szP^*(z,s) = \tag{2.174}$$
$$(\mu - \lambda z - \mu z + \lambda z^2)P^*(z,s) - \mu(1-z)P_0^*(s).$$

We can now solve for $P^*(z,s)$:

$$P^*(z,s) = \frac{z^{i+1} - \mu(1-z)P_0^*(s)}{-\lambda z^2 + (s+\lambda+\mu)z - \mu}. \qquad (2.175)$$

At this point one must determine $P_0^*(s)$ and invert the transform. The interested reader can see [GROS] for the details. The well-known result for the time evolution of the state probabilities is

$$P_n(t) =$$

$$e^{-(\lambda+\mu)t}\left\{\rho^{(n-i)/2}I_{n-i}(2t\sqrt{\lambda\mu}) + \rho^{(n-i-1)/2}I_{n+i+1}(2t\sqrt{\lambda\mu})\right.$$

$$\left. (1-\rho)\rho^n \sum_{l=n+i+2}^{\infty} \rho^{-l/2}I_l(2t\sqrt{\lambda\mu})\right\} \qquad (2.176)$$

where

$$\rho = \lambda/\mu,$$

$$n(0) = i \; customers,$$

and $I_l(\;)$ is the modified Bessel function of the first kind of order l:

$$I_l(x) = \sum_{k=0}^{\infty} \frac{\left[\dfrac{x}{2}\right]^{l+2k}}{(l+k)!k!}. \qquad (2.177)$$

2.11.3 Speeding Up the Computation

In calculating the expressions for $P_n(t)$ in the previous box the main difficulty is in calculating the infinite summation of Bessel functions. We will now present an efficient procedure for the calculation of $P_n(t)$ which appeared in 1980 [JONE] and is due to Jones, Cavin, and Johnston. The approach is to show that all but a finite number of terms in the infinite summation can be expressed in terms of the Q functions. This function is normally associated with the field of hypothesis testing.

If

$$Q(a,b) = \int_b^{\infty} \exp\left[-\frac{a^2+x^2}{2}\right] I_0(ax)x \, dx \qquad (2.178)$$

then

$$1 - Q(a,b)$$

is the circular coverage function. The function $Q(a,b)$ is known as the Q-function [SCHW 66] in the areas of radar and communication. It is the probability that $\{(X^2+Y^2)^{1/2} > b\}$ where X and Y are independent normal (Gaussian) random variables with variances of one and means which satisfy $(\bar{X}^2+\bar{Y}^2)^{1/2} = a \geq 0$. Note that these are the same a and b that appear in the argument of $Q(a,b)$.

The Q-function was first developed by Marcum [MARC]. There are existing algorithms for its calculation [DIDO].

The key to efficient computation is to make use of the expansion

$$Q(a,b) = \exp\left[-\frac{(a^2+b^2)}{2}\right] \sum_{m=0}^{\infty} \left[\frac{a}{b}\right]^m I_m(ab). \qquad (2.179)$$

This is done by first letting $a=\sqrt{2\mu t}$ and $b=\sqrt{2\lambda t}$ so that

$$Q(\sqrt{2\mu t},\sqrt{2\lambda t}) =$$
$$\exp(-(\lambda+\mu)t) \sum_{m=0}^{\infty} \left[\frac{\mu}{\lambda}\right]^{m/2} I_m(2t\sqrt{\lambda\mu}). \qquad (2.180)$$

The relevant infinite summation can then be set equal to a finite number of terms:

$$\exp(-(\lambda+\mu)t) \sum_{l=n+i+2}^{\infty} \rho^{-l/2} I_l(2t\sqrt{\lambda\mu}) = \qquad (2.181)$$
$$Q(\sqrt{2\mu t},\sqrt{2\lambda t}) - \exp(-(\lambda+\mu)t) \sum_{l=0}^{n+i+1} \rho^{-l/2} I_l(2t\sqrt{\lambda\mu}).$$

Here $\rho=\lambda/\mu$. Substituting this into the boxed expression for $P_n(t)$ of the previous section leads to

$$P_n(t) =$$

$$(1-\rho)\rho^n \, Q(\sqrt{2\mu t}, \sqrt{2\lambda t}) +$$

$$\exp(-(\lambda+\mu)t)\left\{\rho^{(n-i)/2}I_{n-i}(2t\sqrt{\lambda\mu}) + \rho^{(n-i-1)/2}I_{n+i+1}(2t\sqrt{\lambda\mu}) - \right.$$

$$\left. (1-\rho)\rho^n \sum_{l=0}^{n+i+1} \rho^{-l/2}I_l(2t\sqrt{\lambda\mu})\right\}. \qquad (2.182)$$

Computational results reported in [JONE] indicate that this expression is up to seven times faster to evaluate than the original expression with the infinite summation. Some additional references dealing with computation in the transient case appear in To Look Further at the end of this chapter.

2.12 The M/G/1 Queueing System

2.12.1 Introduction

We have been able to get some very specific results throughout this chapter by assuming Poisson arrivals and an exponential server(s). This has been due to the memoryless nature of these models. Surprisingly, some very simple and powerful results can be derived for the M/G/1 queueing system where one assumes that arrivals are again Poisson but that service times are described by an arbitrary ("general") probability distribution. These are the famous Pollaczek-Khinchin formulas [POLL], [KHIN]. The first one at which we will look is for the mean number of customers in the queueing system. The second one at which we will look is a transform equation for the state probabilities.

2.12.2 Mean Number in the Queueing System

Introduction

Our basic approach will be to write a difference equation for the number in the queueing system immediately after the ith departure. By taking the expectation of the square of this expression we will be able to develop the Pollaczek-Khinchin mean value formula. A key point is that the state probabilities at these departure instants are in fact equal to the state probabilities at any time instant. This is something that will be assumed true for now but is proved explicitly in section 2.12.3. This approach that we take was first described by David Kendall in 1951 [KEND]. However, the actual mean value expression goes back to Pollaczek in 1930 [POLL] and Khinchin in 1932 [KHIN] where it was found by means of more complex derivations. As many other authors have done ([GROS], [HAMM], [KLEI 75]) we will expand on Kendall's approach.

The Recursion

Let us start. We will look at the M/G/1 queueing system at the instants during which departures occur. The number in the system *immediately after* the $(i+1)$st departure instant is n_{i+1}. This number includes both those customers in the queue and the customer in service. This number is also equal to the number in the system immediately after the ith departure instant minus 1 (because of the departure of a customer at the $(i+1)$st departure instant) plus the number that arrive into the system between the ith and $(i+1)$st departure instants. This number of arrivals we will call a_{i+1}. For this equation to work there must be at least one customer in the system at the ith departure instant. Thus

$$n_{i+1} = n_i - 1 + a_{i+1}, \qquad n_i > 0. \tag{2.183}$$

Of course, we still have to take care of the case $n_i=0$. This is actually simpler than the case $n_i>0$. The number in the queueing system at the $(i+1)$st departure instant is just the number of arrivals during the service period of the first customer to arrive to the empty queueing system. This first customer is the one that departs at the $(i+1)$st departure instant and is not counted in determining a_{i+1}. So

$$n_{i+1} = a_{i+1}, \qquad n_i = 0. \tag{2.184}$$

Bringing this together we have

$$n_{i+1} = \begin{cases} n_i - 1 + a_{i+1}, & n_i > 0, \\ a_{i+1}, & n_i = 0. \end{cases} \tag{2.185}$$

We would like to express this as a single equation. To do this, we will introduce the unit step function:

$$u(n_i) = \begin{cases} 1, & n_i > 0, \\ 0, & n_i = 0. \end{cases} \tag{2.186}$$

Now we can rewrite the equations for n_{i+1} as

$$n_{i+1} = n_i - u(n_i) + a_{i+1}. \qquad (2.187)$$

If we square both sides of this recursion and take expectations we have

$$E[(n_{i+1})^2] = E[(n_i - u(n_i) + a_{i+1})^2], \qquad (2.188)$$

$$E[(n_{i+1})^2] = \qquad (2.189)$$

$$E[n_i^2] + E[(u(n_i))^2] + E[a_{i+1}{}^2]$$

$$- 2E[n_i u(n_i)] + 2E[n_i a_{i+1}] - 2E[u(n_i)a_{i+1}].$$

In order to solve for the Pollaczek-Khinchin formula we need to solve for each of these terms, one by one. We will now do this

Sundry Results

$E[(n_{i+1})^2]$, $E[n_i{}^2]$:

These two terms should be equal in equilibrium so that they will cancel in (2.189).

$E[(u(n_i))^2]$:

Because of the nature of the step function $u(n_i)^2 = u(n_i)$, we can replace $E[(u(n_i))^2]$ with $E[u(n_i)]$. But what is $E[u(n_i)]$? We will have to digress in order to solve for it. Now

$$E[u(n_i)] = \sum_{n=0}^{\infty} u(n_i)P[n_i = n], \qquad (2.190)$$

$$E[u(n_i)] = \sum_{n=1}^{\infty} P[n_i{=}n], \qquad (2.191)$$

$$E[u(n_i)] = P[busy] \qquad (2.192)$$

Luckily, one can easily determine $P[busy]$ for any queueing system. An argument appearing in [KLEI 75] will be used. Consider an interval I. The server will be busy for a time period equal to

$$I - Ip_0.$$

The number of customers served is

$$(I - Ip_0)\mu$$

where μ is the mean service rate of the queueing system. In equilibrium the quantity above should be equal to the number of arrivals:

$$\lambda I = (I - Ip_0)\mu. \qquad (2.193)$$

Rearranging this one has

$$\boxed{\rho = 1 - p_0. \qquad (2.194)}$$

Note that no restrictive assumptions have been made about either the arrival process or the service times. Thus this result is valid for a G/G/1 queueing system, that is, one with an arbitrary arrival process and service time distribution.

We can now bring all this together to say that

$$E[(u(n_i))^2] = E[u(n_i)] = P[busy] = 1 - p_0 = \rho. \qquad (2.195)$$

$2E[n_i u(n_i)]$:

Clearly $n_i u(n_i) = n_i$, so we can say

$$2E[n_i u(n_i)] = 2E[n_i]. \qquad (2.196)$$

$2E[u(n_i)a_{i+1}]$:

The number of customers that arrive into the system between the ith and $(i+1)$st departure instant, a_{i+1}, is independent of the number in the queueing system immediately after the ith departure instant, n_i. Thus

$$E[u(n_i)a_{i+1}] = E[u(n_i)]E[a_{i+1}].   \qquad (2.197)$$

We already know that $E[u(n_i)]=\rho$. To determine $E[a_{i+1}]$ one can look at the original recursion for $E[n_{i+1}]$

$$n_{i+1} = n_i - u(n_i) + a_{i+1}   \qquad (2.198)$$

and take the expectation of both sides:

$$E[n_{i+1}] = E[n_i] - E[u(n_i)] + E[a_{i+1}].   \qquad (2.199)$$

In equilibrium $E[n_{i+1}]=E[n_i]$, so one is left with

$$E[a_{i+1}] = E[u(n_i)] = \rho.   \qquad (2.200)$$

One can now say that:

$$2E[u(n_i)a_{i+1}] = 2E[u(n_i)]E[a_{i+1}] = 2\rho^2.   \qquad (2.201)$$

$2E[n_i a_{i+1}]$:

Using the same argument as before, the two quantities are statistically independent, so

$$2E[n_i a_{i+1}] = 2E[n_i]E[a_{i+1}] = 2E[n_i]\rho.   \qquad (2.202)$$

Putting It All Together

Now we can substitute these sundry results into the equation:

$$E[(n_{i+1})^2] = \qquad (2.203)$$

$$E[n_i^2] + E[(u(n_i))^2] + E[a_{i+1}^2]$$

$$- 2E[n_i u(n_i)] + 2E[n_i a_{i+1}] - 2E[u(n_i)a_{i+1}].$$

This results in

$$0 = 0 + \rho + E[a_{i+1}^2] - 2E[n_i] + 2E[n_i]\rho - 2\rho^2. \qquad (2.204)$$

Solving for $E[n_i]$ we have

$$E[n_i] = \frac{\rho + E[a_{i+1}^2] - 2\rho^2}{2(1-\rho)}. \qquad (2.205)$$

We will assume for now, and prove in section 2.12.3, that the expected number of customers in the queueing system at the departure instants is the same as the expected number of customers at *any* instant of time:

$$E[n_i] = E[n]. \qquad (2.206)$$

We can also say, because the input is Poisson, that the expected number of arrivals between two successive departure instants is a time-invariant quantity:

$$E[a_{i+1}^2] = E[a^2]. \qquad (2.207)$$

We thus have

$$\boxed{E[n] = \frac{\rho + E[a^2] - 2\rho^2}{2(1-\rho)}. \qquad (2.208)}$$

We almost have a simple expression for the mean number in the M/G/1 queueing system in equilibrium or $E[n]$. What is lacking is a simple expression for $E[a^2]$

What is $E[a^2]$?

The quantity "a" is the number of arrivals during a typical servicing period in equilibrium. Proceeding as in [GROS] we can write

$$E[a^2] = VAR[a] + (E[a])^2, \tag{2.209}$$

$$E[a^2] = VAR[a] + \rho^2. \tag{2.210}$$

We need to solve for VAR [a]. Let s be the service time. Then an interesting relationship [GROS] for two random variables X and Y that we can make use of is [PARZ]

$$VAR[Y] = E[VAR(Y \mid X)] + VAR[E(Y \mid X)]. \tag{2.211}$$

In terms of our application this becomes

$$VAR[a] = E[VAR(a \mid s)] + VAR[E(a \mid s)]. \tag{2.212}$$

One can solve for the two components of this sum:

$$E[VAR(a \mid s)] = E[\lambda s] = \lambda E[s] = \rho, \tag{2.213}$$

$$VAR[E(a \mid s)] = VAR[\lambda s] = \lambda^2 \sigma_s^2. \tag{2.214}$$

Here σ_s^2 is the variance of the service time. So we have

$$VAR[a] = \rho + \lambda^2 \sigma_s^2 \tag{2.215}$$

and

$$\boxed{E[a^2] = \rho + \lambda^2 \sigma_s^2 + \rho^2. \qquad (2.216)}$$

The Pollaczek-Khinchin Mean Value Formula

From before we have

$$E[n] = \frac{\rho + E[a^2] - 2\rho^2}{2(1-\rho)}.$$ (2.217)

Substituting (2.216) for $E[a^2]$ leads to

$$E[n] = \frac{\rho + \rho + \lambda^2\sigma_s^2 + \rho^2 - 2\rho^2}{2(1-\rho)}$$ (2.218)

or:

$$E[n] = \frac{2\rho - \rho^2 + \lambda^2\sigma_s^2}{2(1-\rho)}.$$ (2.219)

This is one form of the Pollaczek-Khinchin mean value formula. We can get another form by expressing $E[n]$ as

$$E[n] = \frac{2\rho - 2\rho^2 + \rho^2 + \lambda^2\sigma_s^2}{2(1-\rho)}$$ (2.220)

or

$$E[n] = \rho + \frac{\rho^2 + \lambda^2\sigma_s^2}{2(1-\rho)}.$$ (2.221)

What is amazing about the Pollaczek-Khinchin mean value formula is that it expresses $E[n]$ as a function of the mean arrival rate, mean service rate, and variance of the service time distribution. One might think that since the service time distribution is general, the formula should include higher moments of the service time distribution. But this is not the case. The formula depends only on the mean and variance of the service time distribution.

We note in passing that Little's Law can be used in conjunction with this mean value formula to develop expressions for the mean waiting time, $\bar{\tau}$.

Example: In the M/D/1 queueing system the arrivals are Poisson and the service time is a fixed, deterministic quantity. This is actually a good model for the processing of fixed length packets in a communications network.

Let us make a comparison. For the M/M/1 queueing system the Pollaczek-Khinchin mean value formula reduces to the result we had previously:

$$E[n] = \frac{\rho}{1 - \rho}.$$

For the M/D/1 queueing system $\sigma_s^2 = 0$, so

$$E[n] = \rho + \frac{\rho^2}{2(1 - \rho)}.$$

Table 2.5 compares E[n] for both types of queueing systems for various values of ρ

Table 2.5: E[n]		
ρ	M/M/1	M/D/1
0.1	0.111	0.106
0.2	0.250	0.225
0.3	0.429	0.364
0.4	0.667	0.533
0.5	1.00	0.750
0.6	1.50	1.05
0.7	2.33	1.52
0.8	4.00	2.40
0.9	9.00	4.95
0.99	99.0	50.0

Note that the expected number in Table 2.5 is smaller for the M/D/1 queueing system. In fact, we can show, with some algebraic manipulation, that for the M/D/1 queueing system

$$E[n] = \frac{\rho}{1 - \rho} - \frac{\rho^2}{2(1 - \rho)}.$$

The $E[n]$ is equal to that for the M/M/1 queueing system, less a positive quantity. Thus $E[n]$ is always smaller for an M/D/1 queueing system compared to an M/M/1 queueing system.

It is also straightforward to show, as Table 2.5 suggests, that as $\rho \to 1$ the expected number of customers in an M/M/1 queueing system is twice that in an M/D/1 queueing system.

Although the M/D/1 queueing system is a good model for packet processing, the problem is that we only have simple analytical results for a single such system. Because networks with Poisson arrivals and deterministic service times are analytically intractable, the memoryless M/M/1 queueing system is often used as an approximation in its place.

∇

2.12.3 Why We Use Departure Instants

An essential part of the previous derivation of the Pollaczek-Khinchin mean value formula is our assumption that the state probabilities at departure instants are the same as the state probabilities at any instant. Following the method used in [GROS 85], we will show that this is true.

We will take a realization of the queueing process over a long interval. The number in the queueing system at time t will be $n(t)$. The interval will stretch from time 0 to time T.

The number of arrivals in the interval $(0,T)$ that occur when the number of customers in the system is n will be called $a_n(T)$. The number of departures bringing the system from state $n+1$ to state n in the interval $(0,T)$ will be called $d_n(T)$.

If one thinks about the movement of the system state in the usual one-dimensional state transition diagram for a birth-death system, one can see that

$$|a_n(T) - d_n(T)| \leq 1. \qquad (2.222)$$

It should also be true from first principles that

$$d(T) = a(T) + n(0) - n(T) \qquad (2.223)$$

where $d(T)$ and $a(T)$ are the total number of departures and arrivals, respectively, over $(0,T)$. One can now define the state probabilities for the departure instants as

$$\lim_{T \to \infty} \frac{d_n(T)}{d(T)}, \qquad n = 1,2,3,.... \qquad (2.224)$$

This empirical definition of the state probability becomes exact in the limit. Now

$$\frac{d_n(T)}{d(T)} = \frac{a_n(T) + d_n(T) - a_n(T)}{a(T) + n(0) - n(T)}. \qquad (2.225)$$

The denominator can be recognized as the equation we just discussed, (2.223). The numerator was created by adding and subtracting $a_n(T)$ to $d_n(T)$. As $T \to \infty$, $d_n(T) - a_n(T)$ will never be greater than one, as has been mentioned. Also, because we have a stationary process if $n(0)$ is finite, so will be $n(T)$. Thus as $T \to \infty$, $a(T)$ and $a_n(T)$ will $\to \infty$, and

$$\lim_{T \to \infty} \frac{d_n(T)}{d(T)} = \lim_{T \to \infty} \frac{a_n(T)}{a(T)}. \tag{2.226}$$

What this proves directly is that the state probabilities seen at the departure instants are the same as the state probabilities seen at the arrival instants. But going back to our Poisson process coin-flipping analogy, the arrival instants are random points in time and independent of the queueing system state. Thus it should seem reasonable to say that the state probabilities at arrival instants are the same as the state probabilities seen at any time. This completes the proof that looking at state probabilities at departure instants for the M/G/1 queueing system is the same as looking at state probabilities for this queueing system at *any* instant.

∇

In Kendall's original 1951 paper [KEND] he called the instants of departures "regeneration points". As he put it: "An epoch is a point of regeneration if a knowledge of the state of the process at that particular epoch has the characteristic Markovian consequence that a statement of the past history of the process loses all its predictive value." He went on further to say that "A Markov process, therefore, is precisely a process for which 'every' epoch is a point of regeneration."

In the printed discussion that followed this paper D. V. Lindley formally expressed a regeneration point as a point, t_0, where for all $t > t_0$:

$$DIST\{X(t) \mid X(t_0)\} \equiv DIST\{X(t) \mid X(\tau) \text{ for } all \ \tau \le t_0\}.$$

There is more than one set of regeneration points. Kendall used the set of departure instants in his paper. In the subsequent discussion Lindley suggested the instants when the queueing system becomes empty. He also mentioned the use of the time instants when a customer had been in service x seconds. For simplicity of solution though, the choice of departure instants is probably best.

2.12.4 Probability Distribution of the Number in the Queueing System

What does the Markov chain imbedded at the times of departure look like? Let us define a matrix of transition probabilities as

$$\mathbf{P} = [P_{rs}] \equiv P[n_{i+1}=s \mid n_i=r]. \tag{2.227}$$

Here n_i is the number in an M/G/1 queueing system immediately after the ith departure. To get from $n_i=r$ to $n_{i+1}=s$ there must be s-r+1 arrivals between the ith and the $(i+1)$st departure instants. The "+1" in s-r+1 is necessary because the number in the queueing system is reduced by one due to the departure at the $(i+1)$st departure instant. Let us set

$$[P_{rs}] = k_{s-r+1} = k_{\# \, arrivals} \tag{2.228}$$

where "# arrivals" is the number of customers that arrive between the ith and the i+1st departure instants. We are using k's because this is the original notation appearing in Kendall's 1951 paper. What we now have is a state transition matrix that looks like Table 2.6:

Table 2.6: $\mathbf{P} = [p_{rs}] =$						
	0	1	2	3	4	·
0	k_0	k_1	k_2	k_3	k_4	·
1	k_0	k_1	k_2	k_3	k_4	·
2	0	k_0	k_1	k_2	k_3	·
3	0	0	k_0	k_1	k_2	·
4	0	0	0	k_0	k_1	·
·		·	·	·	·	·

The sum of the k's in each row must be 1 as this covers the probability of going from state r to *some* (and possibly the same) state. The matrix is also of infinite dimension, but this will not cause any problems in what follows.

How was the placement of matrix entries chosen? In the top row one goes from 0 customers in the queueing system at the ith departure instant to 0,1,2,3,... customers at the $(i+1)$st departure instant. Thus we need k_0, k_1, k_2, k_3...... arrivals. In the second row, for instance, to go from 1 customer at the ith departure instant to 3 customers at the $(i+1)$st departure instant there must be 3 arrivals during this time. That is, we start with 1 customer at the end of the ith departure instant, three more customers arrive for a total of 4 customers, and one customer leaves at the end of the $(i+1)$st departure instant for a final total of 3 customers.

We can express k_j, the probability of j arrivals between the ith and the $(i+1)$st departure instants, explicitly as

$$k_j = \int_0^\infty Prob[j \; arrivals \mid time \; s]b(s)ds \qquad (2.229)$$

where $b(s)$ is the service time density. The variable s is the time between the ith and the $(i+1)$st departure instants. This time is a constant in equilibrium. Now since the arrival stream is Poisson, we can substitute the Poisson distribution into the integral for

$$k_j = \int_0^\infty \frac{(\lambda s)^j e^{-\lambda s}}{j!} b(s)ds. \qquad (2.230)$$

The basic idea is that for a known service time density, $b(s)$, one can plug it into the above equation and perform the integration to solve for k_j. We began this section by asking what the Markov chain imbedded at the times of departures would look like. The answer is illustrated in Figure 2.26. Here are shown all the transitions leaving the nth state. The k_j associated with each transition is the probability that one leaves the state via that transition as opposed to another transition at the ith departure instant. It is *not* a transition rate, as in the previous state transition diagrams.

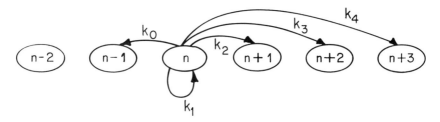

Fig. 2.26: Imbedded Markov Chain for M/G/1

Let us now ask a second question. Now that we have this Markov chain, can we calculate the probability of there being j customers in the queueing system at the departure instants? The answer is yes. We will use an approach described by [GROS]. An alternate approach appears in [KLEI].

We know from the previous section that the state probabilities at these instants are in fact the same as the state probabilities at any instant. The first step is to write the equilibrium equations

$$\pi = \pi \mathbf{P} \qquad (2.231)$$

where

$$\pi = [\pi_0 \ \pi_1 \ \pi_2 \ \pi_3 \ \cdots \ ] \tag{2.232}$$

are the state probabilities. We know the structure of **P**. Equation (2.231) gives rise to equations of the form:

$$\pi_0 = \pi_0 k_0 + \pi_1 k_0, \tag{2.233}$$

$$\pi_1 = \pi_0 k_1 + \pi_1 k_1 + \pi_2 k_0,$$

$$\pi_2 = \pi_0 k_2 + \pi_1 k_2 + \pi_2 k_1 + \pi_3 k_0,$$

$$\pi_3 = \pi_0 k_3 + \pi_1 k_3 + \pi_2 k_2 + \pi_3 k_1 + \pi_4 k_0.$$

These equations can be compactly summarized, where r and s are integers, as:

$$\pi_r = \pi_0 k_r + \sum_{s=1}^{r+1} \pi_s k_{r-s+1}, \qquad r=0,1,2,3,.... \tag{2.234}$$

The second step is to make use of transforms. First we introduce the change of variables $t=s-1$ or $s=t+1$. This gives us:

$$\pi_r = \pi_0 k_r + \sum_{t=0}^{r} \pi_{t+1} k_{r-t}. \tag{2.235}$$

We will also introduce the transform of the state probabilities

$$\Pi(z) = \sum_{i=0}^{\infty} \pi_i z^i \tag{2.236}$$

and the transform of the k_j's

$$K(z) = \sum_{i=0}^{\infty} k_i z^i. \tag{2.237}$$

The k's, after all, are a sequence of real (positive) numbers like any other sequence so there is no reason why we cannot take their transform. Finally, the summation in the above difference equation for π_r can be seen to be "almost" a convolution. That is, it is similar to

$$x_i = \sum_{j=0}^{\infty} f_j g_{i-j} \tag{2.238}$$

for which we know that

$$X(z) = F(z)G(z) \tag{2.239}$$

where $X(z)$, $F(z)$, and $G(z)$ are the one sided (i=0 to ∞) transforms of x_i, f_i, and g_i, respectively.

Let us now take the one-sided transform of the previous difference equations for π_r. This is

$$\Pi(z) = \pi_0 K(z) + Z\{\pi_{t+1}\}K(z). \tag{2.240}$$

We can take the transform of the convolution, in spite of the finite index, because π_i and k_i are only defined for $i \geq 0$. In the preceding equation

$$Z\{\pi_{t+1}\} = \sum_{t=0}^{\infty} \pi_{t+1} z^t = z^{-1} \sum_{t=0}^{\infty} \pi_{t+1} z^{t+1}, \tag{2.241}$$

$$Z\{\pi_{t+1}\} = z^{-1} \sum_{t=-1}^{\infty} \pi_{t+1} z^{t+1} - z^{-1}\pi_0, \tag{2.242}$$

$$Z\{\pi_{t+1}\} = z^{-1}\Pi(z) - z^{-1}\pi_0. \tag{2.243}$$

So substituting this into the transform equation yields

$$\Pi(z) = \pi_0 K(z) + [z^{-1}\Pi(z) - z^{-1}\pi_0]K(z) \tag{2.244}$$

or

$$\Pi(z)(1 - z^{-1}K(z)) = \pi_0(1 - z^{-1})K(z). \tag{2.245}$$

Solving for $\Pi(z)$ gives us

$$\Pi(z) = \frac{\pi_0(1 - z^{-1})K(z)}{1 - z^{-1}K(z)}.$$ (2.246)

Multiplying by -z/-z results in

$$\Pi(z) = \frac{\pi_0(1 - z)K(z)}{K(z) - z}.$$ (2.247)

To solve for π_0 one can make use of the fact that $\Pi(1)=1$, $K(1)=1$ and $K'(1)=\lambda/\mu$ and use of L'Hopital's rule to obtain $\pi_0 = 1 - \lambda/\mu = 1 - \rho$, so

$$\Pi(z) = \frac{(1-\rho)\,(1-z)\,K(z)}{K(z) - z}.$$ (2.248)

This is the result that we are seeking. It expresses the transform of the probabilities of the number in the M/G/1 queueing system as a function of the transform of the k_j's. Recall that we obtain the k_j's from an integral equation which includes $b(s)$, the service time distribution.

2.13 Priority Systems for Multiclass Traffic

Often, customers can be divided into different classes. For instance, in a computer network there may be a natural distinction between data, voice, and video packets. In a computer system different jobs may be distinguished by their needs for service. In such cases as these it is common to implement service disciplines at the queues that treat customers according to their "priority".

Establishing a priority structure is a natural way to provide different levels of service to different classes of traffic. Each class of traffic then receives service based on its priority and the presence or absence of traffic of higher priority. It is well known that networks with customer priorities do not admit a tractable product form solution (see chapter 3). This can be explained in two ways. One is that a priority system introduces dependencies between customers in a node that is at odds with the necessary independence of customers for a product form solution. The other is that the underlying Markov chains do not have the building block structure which is characteristic of product form networks (see chapter 3). Nonetheless, there has been a growing interest in priority systems for multiclass traffic [KLEI 75], [KIM], [CHIP]. The following discussion provides a summary of some important priority results.

One fact that has been recognized is that in priority systems a form of delay conservation is present. While this can be expressed mathematically, what it basically means is that the preferred treatment received by higher priority classes is at the expense of lower priority classes. Delay averaged over all classes is a constant [SCHW 87]. L. Kleinrock was the first to formalize such conservation [KLEI 75], [KLEI 65].

Priority systems are distinguished by whether they are preemptive or non-preemptive. In the former a customer in service will be displaced by an arriving customer of higher priority. There are several ways in which the displaced customer may resume service. In the nonpreemptive case a customer in service is never displaced. Analysis is available for single-node models using both disciplines, though no network wide results are available.

One priority system in widespread use is known as the Head of the Line (HOL) system [KLEI 75]. The first discussion of it appeared in 1954 by E. Cobham [COBH]. In HOL systems an arriving customer of priority class i joins the end of a segregated queue for class i. The server always serves the highest class available in the queues. One can visualize the system as a single queue with each class occupying adjacent queue positions and ordered with the highest priority customers nearest the server, followed by the next priority class, etc... HOL is attractive as under certain linear cost functions it can be shown to be an optimal policy [KLEI 75].

Variations on HOL are possible. One such, proposed in 1988 by Y. Lim and J. Kobza [LIM] is HOL-PJ: Head of the Line with Priority Jumps. In HOL-PJ there is a maximum queueing delay for each class of packets in the queue. A packet not served by the time that this limit is reached is advanced to the tail of the next highest priority queue. An advantage of HOL-PJ is that by setting different delay limits, delay performance can be controlled. A limitation of existing analysis for both HOL and HOL-PJ is that it is only available for a single node.

Some other priority policies of note are that of Minimum Laxity Threshold (MLT) and Queue Length Threshold (QLT) policies. These were discussed in 1989 by R. Chipalkatti, J. Kurose, and D. Towsley [CHIP] for use in a statistical multiplexer. MLT makes use of the idea of "laxity" used in real time systems. Laxity here is the difference between a packet's real time deadline and the current time. In MLT priority is given to real time traffic when the minimum laxity among queued real time packets is less than or equal to a threshold, otherwise priority is given to nonreal time traffic. Under QLT whenever the number of queued nonreal time packets exceeds a threshold priority is given to non-real time packets, otherwise it is given to real time packets.

By choosing the threshold parameter both MLT and QLT allow the performance of real time and nonreal time traffic to be traded off against each other. In [CHIP] it was found that as both MLT and QLT perform similarly, QLT is to be preferred as it is easier to implement. Again, analytic results are only available for a single node.

For the specific application of voice packet transport Chen, Walrand, and Messerschmitt in 1989 [CHEN b] studied an Oldest Customer First (OCF) policy for reducing the variability in delivery times of voice packets. Under this policy packets in a node are given a priority based on the time spent so far in the network. This priority increases with age. They showed that for a single M/D/1 queue with a work conserving, nonpreemptive discipline OCF minimizes the maximum age of departures, maximizes the minimum age of departures, and minimizes the sample variance of the age of departures.

In 1991 A. Viterbi published [VITE] an expression for mean equilibrium queueing delay for a synchronous packet network node as well as the moment generating function of the delay distribution for the case of N priority classes.

Finally, a number of papers dealing with a four class telecommunications model called Asynchronous Time Sharing (ATS) have been written by A. Lazar and colleagues [LAZA 91], [HYMA 91,92]. In [HYMA 91] an efficient real time scheduling algorithm, MARS, is introduced. In MARS a dynamic cycle of H cells is used. The scheduler, intuitively, picks cells which cannot be further delayed for transmission. In [HYMA 92] joint admission control and scheduling is considered. Again, the problem is formulated in terms of a single node.

To Look Further

An English translation of Markov's original 1907 paper on Markov models appears in an appendix of [HOWA].

Stidham's [STID] 1974 extended proof of Little's Law is discussed in [COOP]. Further work on Little's Law appears in [BRUM], [HEYM], [MIYA], [JEWE], and [EILO]. A good survey of Little's Law work appears in [WHIT].

The proof of Burke's Theorem using the concept of reversibility is due to E. Reich [REIC], 1957.

[GROS] has a wealth of information on transient solutions of queueing systems. [KLEI 75] also describes the transient solution for the M/M/1/∞ queueing system. Ackroyd [ACKR] describes an alternative method to make the calculation of $P_n(t)$ in the M/M/1/∞ transient case more efficient. It involves a discrete Fourier transform. A discussion of its merits relative to the approach in section 2.11.3 appears in [CHIE]. A method for the computation of the transient M/M/1 queueing system cumulative distribution function, probability distribution function and mean, which uses generalized Q-functions, is described by Cantrell in [CANT]. A method for calculating performance measures based on numerical integration appears in [ABAT]. Finally, numerical approaches to the transient analysis problem are discussed in [REIB].

More examples of the use of the moment-generating function appear in chapter 12 of [SCHW 87] and in chapter 4 of [KLEI 75].

Batch arrival and batch departure processes are discussed in [KLEI 75].

Problems

2.1 Use the coin flipping analogy of section 2.2.1 to show intuitively that random splits of a Poisson process and a joining of independent Poisson processes are Poisson.

2.2 Arrive at a set of differential equations describing the time evolution of the Poisson process from

$$P_n(t+\Delta t) = P_n(t)(1-\lambda\Delta t) + P_{n-1}(t)(\lambda\Delta t),$$

$$P_0(t+\Delta t) = P_0(t)(1 - \lambda\Delta t)$$

in a step-by-step manner.

2.3 a) Show that the solution of

$$\frac{dP_1(t)}{dt} = -\lambda P_1(t) + \lambda e^{-\lambda t}$$

for the Poisson process is

$$P_1(t) = \lambda t e^{-\lambda t}.$$

b) Show that the solution of

$$\frac{dP_2(t)}{dt} = -\lambda P_2(t) + \lambda^2 t e^{-\lambda t}$$

for the Poisson process is

$$P_2(t) = \frac{\lambda^2 t^2}{2} e^{-\lambda t}.$$

2.4 Say that a thousand calls per hour arrive to a small telephone exchange. What is the probability that one or more calls arrive in a one second interval?

2.5   Arrive at a set of differential equations describing the M/M/1 queueing system from

$$P_n(t+\Delta t) =$$

$$P_n(t)(1-\lambda\Delta t)(1-\mu\Delta t)+P_{n-1}(t)(\lambda\Delta t)+P_{n+1}(t)(\mu\Delta t),$$

$$P_0(t+\Delta t) = P_0(t)(1-\lambda\Delta t) + P_1(t)(\mu\Delta t)$$

in a step-by-step manner.

2.6   Suppose a single queue does *not* have a Poisson process input and does *not* have an exponential server. What variables are necessary to describe the state of the system at any given instant of time? Again, the interarrival times and the service distribution are nonexponential.

2.7   Derive the differential equations describing the evolution of probability of the system state for an M/M/1 queueing system when the service rate, μ_n, is a function of the number in the queue, n, and the arrival rate, λ_n, is also a function of the number in the queue [KLEI 75].

2.8   Find the equation for an M/M/1 queueing system for $dP_0(t)/dt$ by examining the flows across a boundary about state 0.

2.9   a) What is the difference between local balance and global balance?

b) Suppose that one observes a Poisson process with a packet arriving, on average, every 1 msec. Suppose further that no packet has arrived for 5 consecutive milliseconds. How does this fact affect the probability of seeing an arrival during the next millisecond?

c) Suppose that one observes a Poisson process with a packet arriving, on average, every 1 msec. Suppose that it is $t=0$. Calculate the probability that the first arrival occurs within 3 msec.

2.10 Verify Table 2.2 of section 2.2.9.

2.11 a) Very briefly explain why it is not a good idea to operate a queueing system under a heavy load (arrival rate close to service rate).

b) Very briefly explain why an infinite buffer queue is said to be "unstable" when the mean arrival rate is greater than the mean service rate.

c) On advising day a professor schedules one student appointment every ten minutes. This is based on the assumption that he can advise the average student in ten minutes. Will the professor have a long line outside of his/her office?

2.12 Prove that for a stationary Markov process the probability flux across a cut across a state transition diagram balances [KELL].

2.13 Using the result of problem 2.12, prove that if the state transition diagram of a stationary process is a tree, then the process is reversible [KELL].

2.14 Show that for an M/M/1 queueing system in equilibrium the number of customers in the queue at t_0, $n(t_0)$, is independent of the departure process prior to t_0 [REIC], [KELL].

2.15 For the M/M/1/N queueing system use L'Hopital's rule to calculate the blocking probability when $N=5$ and $\lambda/\mu = 1$.

2.16 a) It is observed for a queueing system that the arrival rate is 10 customers/sec and the mean time for a customer to move through the system is 2 seconds. What is the mean number of customers in the queueing system?

b) An M/M/1 queueing system has an arrival rate of 10 customers per second and a service rate of 20 customers per second. Calculate the utilization of the server.

2.17 For the state-independent M/M/1 queueing system find the ratio p_{100}/p_0 when $\lambda=1.0$ and $\mu=2.0$. Comment on potential numerical problems.

2.18 Consider a state-independent M/M/1 queueing system with $\rho=0.1$, 0.5, 0.9 Calculate the utilization, average number in the queueing system, the variance of the number in the queueing system, and two standard deviations from the average of the number in the queueing system.

2.19 Consider an M/M/1 infinite buffer queueing system with arrival rate λ and service rate μ. Let $\rho=0.3$. There is approximately a 97% probability that there are m or less customers in the queueing system. Find m.

2.20 Consider a finite buffer single server queueing system with arrival and service rates such that the equilibrium probabilities of state are:

n	$p(n)$
0	0.415
1	0.277
2	0.185
3	0.123
≥ 4	0.0

a) Using the values in the table, calculate numerical values for mean throughput, mean number of customers in the queueing system and mean delay. Let $\mu=10.0$. Show all work.

b) Is the following table realizable for a finite buffer state-independent M/M/1 system? Why or why not?

n	$p(n)$
0	0.4
1	0.3
2	0.2
3	0.1
≥ 4	0.0

2.21 Compute the $P[queueing]$ using the Erlang C formula for the values in the following table when $m=3$:

λ/μ	$\lambda/m\mu$	$P[queueing]$
0.5	0.166	
1.0	0.333	
1.5	0.500	
2.0	0.666	
2.5	0.833	
2.9	0.966	

2.22 Compute p_m, the probability that all servers are busy, for an M/M/m/m queueing loss system for various values of λ/μ. Use Erlang's B formula to fill in the following table when $m=3$:

λ/μ	p_m
0.5	
1.0	
2.0	
3.0	
6.0	

2.23 A small business has three outside tie lines for its telephones. Measurements show that the average outside call lasts 1 minute and 2 calls are generated per minute over the whole business. The current probability of not getting an outside line is 0.21, which is considered unacceptable. How many additional outside tie lines should be added to get the blocking probability below 5%. Note: Blocked calls are lost.

2.24 Consider a single Markovian queueing system with discouraged arrivals. That is, the arrival rate decreases as the number of the customers in the queue increases:

$$\mu(n) = \mu,$$

$$\lambda(n) = \frac{\lambda}{n+1}.$$

a) Draw and label the state transition diagram.

b) Express $p(n)$ as a function of λ, μ, and $p(0)$.

c) Find a closed form solution for $p(0)$.

2.25 a) Consider a single phone. A call is made every hour, on average. It lasts 6 minutes, on average. Construct a Markov chain, and solve for the probabilities that the phone is busy and idle.

b) A queueing system can hold at most 5 packets. For this system $\lambda=3.0$ and $\mu=4.0$. Calculate the utilization of the server.

c) One hundred phones in an office feed into two leased lines to Asia. Each phone makes an Asian call every 60 minutes, on average. Each call lasts two minutes, on average. Calculate the fraction of time that both lines are busy.

d) Consider an M/M/∞ queueing system with $\lambda/\mu=0.7$. Calculate the probability that more than 2 servers are busy.

2.26 Consider a queueing network consisting of two service facilities connected in a cyclic fashion (input to output, input to output). One facility is a M/M/2 queueing system. Each server operates at rate μ. The other facility consists of three servers in parallel with no queue (M/M/3/3). The service rate of each of these servers is λ. Let n be the number of customers in the M/M/2 queueing system. The total number of customers in this closed system is 3.

Assume Markovian statistics. In terms of applications the M/M/2 system could be a dual processor unit and the M/M/3/3 system could be a number of terminals. Customers would be job submissions.

a) Draw a schematic of the queueing network.

b) Draw and label the state transition diagram for n=0,1,2,3.

c) Write out expressions for p_n in terms of p_0, λ, μ, and n up to n=3.

d) Let λ=1.0 and μ=1.0. Calculate the average time it takes a customer to circulate around the system (through both queueing systems).

2.27 Consider a cyclic queueing network consisting of two queueing systems. The first consists of three parallel servers, each with service rate μ, modeling modems in a modem pool. The second queueing system consists of five parallel servers modeling five terminals. Each terminal submits a request for a modem at rate λ.

Each server holds at most 1 "customer" and there are five customers in the system (one associated with each terminal server). A "customer" at a terminal server indicates the terminal is active but not using a modem. A "customer" at a modem server indicates that the modem is being used by the terminal associated with the customer.

a) Draw and label the state transition diagram of the system. Let the state variables be the number of active modems (n). Assuming Markovian statistics, write expressions for p_n, n=1,2,3.

b) Suppose that at time t=0 all the modems are idle (no "customers"). What is the *minimum* expected time until *all* modem servers become active (have "customers")?

2.28 A Markovian queueing model of a packet buffer and transmitter receives packets every two seconds which are 1,000 bits long. The channel capacity on the outgoing link is 1,000 bps. Assume that the complete system holds at most 3 packets.

a) Calculate the numerical value of the utilization (fraction of time that the server is busy).

b) If there are two packets in the queue, what is the probability of a packet leaving the queue as opposed to the event of the arrival of a third packet?

2.29 For each of the two situations described below, draw the appropriate queueing schematic, identify the model, and draw and label the state transition diagram.

a) A minicomputer is time shared with four terminals.

b) An airline reservation office has four operators. Calls are queued (listen to music) if all the operators are busy.

2.30 a) There are twelve telephones in an office. They are all tied into a master phone on the receptionist's desk. The master phone has three outgoing lines. If a call from one of the twelve phones to the outside world cannot get one of these three lines, it is blocked. What queueing model/formula can be used for this situation?

b) A new, more advanced phone system is installed in the office. Now if an outgoing call is blocked, the master phone will ring back to the phone that attempted the call when one of the three lines becomes available. Calls are processed in a FIFO manner. What queueing model/formula can be used for this situation?

2.31 a) A packet buffer *and* a single communication server can hold at most five packets. The arrival rate to the system is 20 packets/msec. The communication server can serve 40 packets per msec. Calculate the blocking probability of the system under Markovian statistics. Show all work.

b) A large company has hundreds of phones but only three international phone lines. The international call arrival rate is 10 calls/hour. Each call lasts, on average, 12 minutes. Calculate the probability that all three lines are busy using Markovian statistics.

2.32 Consider a cyclic network of two queueing systems modeling a CPU and three time-shared terminals. One queueing system models the CPU (FIFO with service rate of 20). The other queueing system consists of three parallel servers, each with service rate 5.0. There are thus three "customers" in the network. A customer in this model is really an indication of whether a job has been submitted to the CPU or not. Calculate a numerical value for average time delay for a job to get through the CPU queueing system.

2.33 An office has 100 phones, each of which places an outside call with average rate λ. The mean holding time of a call is $1/\mu$. Using a model and equations explain how one can decide on the minimum number of outgoing lines needed to provide a blocking probability no greater than 0.01. Assume blocked calls are lost. A numerical answer is not necessary.

2.34 Consider two servers in parallel without a queue. The "primary" server has service rate μ_p, and the "secondary" server has service rate μ_s. Arrivals are Poisson with rate λ, and the servers are exponential with the indicated service rates. If the system is empty, an arriving customer is always sent to the primary server. If the system is full (one customer in each server) then arriving customers are turned away.

a) Draw and clearly label the state transition diagram.

b) How could one solve for the equilibrium state probabilities?

c) Write an expression for the mean throughput of the system in terms of the service rates and the equilibrium probabilities.

2.35 Prove [SCHW 87]

$$\sum_{n=1}^{\infty} p_{n-1} z^n = zP(z), \tag{A}$$

$$\sum_{n=0}^{\infty} p_{n+1} z^n = z^{-1}[P(z) - p_0], \tag{B}$$

$$P(1) = \sum_{n=0}^{\infty} p_n = 1.0, \tag{C}$$

$$\frac{dP(z)}{dz}\bigg|_{z=1} = \sum_{n=0}^{\infty} n p_n = E(n), \tag{D}$$

$$\frac{d^2 P(z)}{dz^2}\bigg|_{z=1} = E(n^2) - E(n). \tag{E}$$

2.36 Find $P(z)$, the moment-generating function for the equilibrium state probabilities of a standard M/M/1 queue with arrival rate λ and service rate μ. Solve for p_0. Note: It is not necessary to invert $P(z)$.

2.37 Using the moment-generating function of the M/M/1 queueing system and the relationships from the appendix and above:
a) Find $E(n)$.
b) Find σ_n^2.

2.38 In finding the probability distribution of the number of customers in the M/G/1 queueing system one arrives at

$$\Pi(z) = \frac{\pi_0 (1-z) K(z)}{K(z) - z}.$$

Solve for π_0 using L'Hopital's rule and $\pi(1)=1$, $K(1)=1$ and $K'(1)=\lambda/\mu$.

2.39 a) An M/D/1 queue has an arrival rate of 10 customers per second and a service rate of 20 customers per second. Calculate the mean number of customers in the queue.

b) Prove for an M/D/1 queueing system with ρ close to 1 that the mean number of customers in the queue is half of that for an M/M/1 queueing system.

2.40 Consider a single queue implementing a priority discipline. There are two classes of packets, high priority and low priority. Arrivals are Poisson. The service time is an exponential random variable.

→ Low priority packets are only served if no high priority packets are present. That is, any high priority packets in the queue are served, one at a time, before the server accepts low priority packets.

→ If the queue has only low priority packets, an arriving high priority packet will instantly displace the low priority packet in the server and take its place.

Let the arrival rate be 5 packets per millisecond for each class and the service rate be 20 packets (of either class) per second. Calculate the average number of high priority packets in the queueing system (waiting line plus server). Justify your answer with a sentence or two of explanation.

Hint: Think of how the queueing system treats high priority customers.

2.41 A communication processor can hold at most one packet. If it is a high priority packet, it is sent over a high-speed line whose service rate is $\mu_H = 30$ packets/sec. If it is a low priority packet, it is sent over a low-speed line whose service rate is $\mu_L = 10$ packets/sec. Low priority packets and high priority packets *each* arrive to an idle processor at rate $\lambda = 5$ packet/sec.

a) How many states does the communication processor have? Identify them.

b) Draw and label the state transition diagram. Hint: It does not look like that of an M/M/1 system.

c) Calculate numerical values for the steady state probabilities of each state, assuming Markovian statistics.

d) Show how to calculate the total mean throughput of the communication processor (calculate its numerical value).

2.42 Consider a system of two queues. Arriving customers (such as calls to a telephone exchange) normally enter finite buffer queue Q1. If it is full, they "overflow" to infinite buffer overflow queue Q2. Both queues have a single server. The arrival rate to the system is 5.0, Q1's service rate is 7.0 and Q2's service rate is 10.0. Q1 holds at most four customers (including the one in service). Using Markovian statistics, calculate the *numerical* value of the mean arrival rate of overflow calls to Q2. Overflow processes have been well studied (see [WOLF]). Note: The overflow process in this problem is not a Poisson process.

2.43 **Computer Project:** Write a computer program to generate data for plots of the Erlang B and Erlang C formula.

Erlang B: Plot the probability that all servers are busy versus ρ where $\rho = \lambda/\mu$. Do this for $m=1,2,3$. The vertical scale can be a linear scale. Let ρ vary from 0 to $2m$ on the horizontal scale.

Erlang C: Plot P[queueing] versus ρ where $\rho = \lambda/m\mu$. Plot curves for $m=1,2,3$. The vertical scale for P[queueing] should be a log scale. Let ρ vary from 0 to 1 on the horizontal scale.

Chapter 3: Networks of Queues

3.1 Introduction

In this chapter we will consider networks of queues. Simple analytical results are usually only possible for Markovian queueing networks. We will start by establishing the product form solution for the equilibrium state probabilities for such networks in section 3.2. The existence of the product form solution basically means that the joint state probability can be expressed as a simple product of functions associated with a network's individual queues. In the case of open queueing networks of state-independent queues these functions are simply the marginal state probabilities so that it seems that the queues act *as if* they were independent. This interesting observation was first noted by J. R. Jackson [JACK 57] in the original product form paper for open networks in 1957 and later generalized in [JACK 64]. In 1967, W. J. Gordon and G. F. Newell [GORD] demonstrated the existence of the product form solution for closed networks. In 1975 F. Baskett, K. M. Chandy, R. R. Muntz and F. G. Palacios [BASK 75] generalized the families of queueing networks known to have the product form solution.

We will also look at the intriguing phenomena of local balance. There is always a "local balancing" of flows on pairs of state transition diagram transitions when the product form solution exists.

In section 3.3 a relatively new interpretation of the product form solution will be presented. We will see that the pattern of flow of probability flux on the state transition diagram has a very definite, decomposable structure. In finding the product form solution for a state we are really solving localized building blocks within the diagram in a recursion from a reference state to the state of interest.

Many networks of practical interest do not have a product form solution. However, the equilibrium probabilities of quite a few such non-product form models can be determined recursively, or at least in a decomposed fashion. Two such classes of models are presented in section 3.4.

In section 3.5 novel classes of queueing networks with "negative customers" are described. Applications for such negative customer queueing networks include certain transaction models and certain neural network models. Even though the underlying equations of the models to be described can be nonlinear they do have a product form solution for the equilibrium state probabilities.

3.2 The Product Form Solution

3.2.1 Introduction

We will look at both the case of open and closed networks. In each case we will set up a generalized global balance equation for the typical state. Then it will be shown that this balance equation is satisfied by the "product form" solution. The approach taken here appears in [KOBA] and [SCHW 87].

3.2.2 Open Networks

3.2.2.1 The Global Balance Equation

We will assume that we have a network of M queues with a single external source for customers and a single external destination for customers. The routing through this network is random. The network is illustrated in Figure 3.1. In this figure the probability of a customer leaving queue i for queue j is r_{ij}. The probability of a customer leaving the source for queue i is r_{si}. The probability of a customer leaving queue i for the destination is r_{id}.

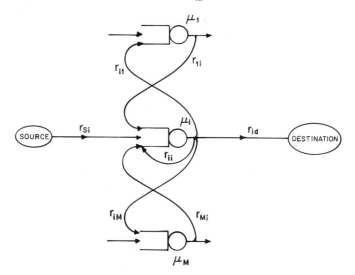

Fig. 3.1: Open Queueing Network

The source produces customers in a Poisson manner with constant rate λ. The service times at the ith queue are independent exponential random variables with rate μ_i.

Now for a little bit about the state description. Let $\underline{1}_i$ be the vector

$$\underline{1}_i = (0,0,0, \cdots 0,1,0, \cdots 0) \tag{3.1}$$

with a 1 in the ith position. The state description is the vector

$$\underline{n} = (n_1, n_2, n_3, \cdots n_i, \cdots n_M). \tag{3.2}$$

Let the vector $\underline{n+1}_i$ be $\underline{n}$ with an additional customer in the ith position. Let $\underline{n-1}_i$ be $\underline{n}$ with one customer removed from the ith position. Let $\underline{n+1}_j\underline{-1}_i$ be $\underline{n}$ with one customer added to queue j and one customer removed from queue i.

If one equates the net flow out of state $\underline{n}$ with the net flow into state $\underline{n}$ one has [SCHW 87] the global balance equation at the state $\underline{n}$:

$$\left[\lambda + \sum_{i=1}^{M} \mu_i\right] p(\underline{n}) = \sum_{i=1}^{M} \lambda r_{si} p(\underline{n-1}_i) \tag{3.3}$$

$$+ \sum_{i=1}^{M} \mu_i r_{id} p(\underline{n+1}_i) + \sum_{i=1}^{M}\sum_{j=1}^{M} \mu_j r_{ji} p(\underline{n+1}_j\underline{-1}_i)$$

The left-hand side of this equation corresponds to leaving state $\underline{n}$ through arrivals or departures for the queues. The three terms on the right-hand side cover all the ways in which one may enter the state: through an arrival, a departure, or a transfer between two queues.

In order to suggest a solution to this equation, we will make a digression:

3.2.2.2 The Traffic Equations

Let the average throughput through queue i be θ_i. Then in equilibrium it must be true for each queue that

$$\theta_i = r_{si}\lambda + \sum_{j=1}^{M} r_{ji}\theta_j, \qquad i=1,2,3,...M. \tag{3.4}$$

This set of simultaneous linear equations are known as the "traffic equations". Their unique solution is the average throughput for each queue. The equation above says that the average throughput through queue i is equal to the average arrival flow from the source plus the flows from all the other queues including queue i.

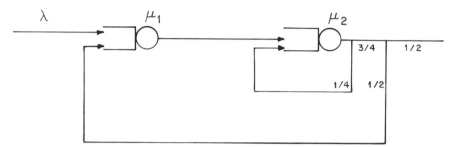

Fig. 3.2: Example Open Queueing Network

Example: Consider the open queueing network of Figure 3.2. The routing probabilities are shown on the diagram. It is left as an exercise to the reader to show that:

$$\theta_1 = 2\lambda, \qquad \theta_2 = \frac{8}{3}\lambda.$$

Each of the queue's throughputs are greater than λ as the feedback paths allow customers to move through each queue several times before leaving the network.

∇

3.2.2.3 The Product Form Solution

To solve the previous global balance equation we will re-arrange the traffic equations as

$$\lambda r_{si} = \theta_i - \sum_{j=1}^{M} r_{ji}\theta_j, \qquad i=1,2,3,...M. \tag{3.5}$$

and use it to eliminate λr_{si} in the global balance equation [CHEN]. Substituting, one has

$$\left[\lambda + \sum_{i=1}^{M}\mu_i\right]p(\underline{n}) = \sum_{i=1}^{M}\theta_i p(\underline{n}-\underline{1}_i) - \sum_{i=1}^{M}\sum_{j=1}^{M}r_{ji}\theta_j p(\underline{n}-\underline{1}_i)$$

$$+ \sum_{i=1}^{M} \mu_i r_{id} p(\underline{n}+\underline{1}_i) + \sum_{i=1}^{M} \sum_{j=1}^{M} \mu_j r_{ji} p(\underline{n}+\underline{1}_j-\underline{1}_i) \tag{3.6}$$

We will now propose using the identity $\theta_i p(\underline{n}-\underline{1}_i) = \mu_i p(\underline{n})$ in this equation. This is really a local balance equation for the ith queue. It is a multidimensional generalization of the local balance equations used in the previous chapter for the M/M/1 queueing system. It is equivalent with $\theta_j p(\underline{n}-\underline{1}_i) = \mu_j p(\underline{n}-\underline{1}_i+\underline{1}_j)$. Substituting these equations:

$$\left[\lambda + \sum_{i=1}^{M} \mu_i \right] p(\underline{n}) = \sum_{i=1}^{M} \mu_i p(\underline{n}) - \sum_{i=1}^{M} \sum_{j=1}^{M} r_{ji} \mu_j p(\underline{n}-\underline{1}_i+\underline{1}_j)$$

$$+ \sum_{i=1}^{M} \mu_i r_{id} p(\underline{n}+\underline{1}_i) + \sum_{i=1}^{M} \sum_{j=1}^{M} \mu_j r_{ji} p(\underline{n}+\underline{1}_j-\underline{1}_i). \tag{3.7}$$

Cancelling terms results in

$$\lambda p(\underline{n}) = \sum_{i=1}^{M} \mu_i r_{id} p(\underline{n}+\underline{1}_i). \tag{3.8}$$

Using a third, equivalent, local balance equation $\mu_i p(\underline{n}+\underline{1}_i) = \theta_i p(\underline{n})$ results in

$$\lambda p(\underline{n}) = \sum_{i=1}^{M} r_{id} \theta_i p(\underline{n}) \tag{3.9}$$

or

$$\lambda = \sum_{i=1}^{M} r_{id} \theta_i. \tag{3.10}$$

This equation, which relates the net flow into the network with the net flow out of the network is obviously true. Thus what has been demonstrated is that the local balance equation $\theta_i p(\underline{n}-\underline{1}_i) = \mu_i p(\underline{n})$ satisfies the previous global balance equation. To get a closed form solution we can rearrange the local balance equation as

$$p(\underline{n}) = \frac{\theta_i}{\mu_i} p(\underline{n}-\underline{1}_i), \tag{3.11}$$

$$p(\underline{n}) = \frac{\theta_i}{\mu_i} p(n_1, n_2, n_3, \cdots n_i{-}1, \cdots n_M). \qquad (3.12)$$

Continuing the recursion to zero out the ith term results in

$$p(\underline{n}) = \left[\frac{\theta_i}{\mu_i}\right]^{n_i} p(n_1, n_2, n_3, \cdots 0, \cdots n_M). \qquad (3.13)$$

If one does this for each of the queues one has

$$p(\underline{n}) = p(0) \prod_{i=1}^{M} \left[\frac{\theta_i}{\mu_i}\right]^{n_i}. \qquad (3.14)$$

To solve for $p(0)$ we can use the normalization of probability [SCHW 87]:

$$\sum_{\underline{n}} p(\underline{n}) = p(\underline{0})S = 1, \qquad (3.15)$$

$$S = \sum_{\underline{n}} \left[\prod_{i=1}^{M} \left[\frac{\theta_i}{\mu_i}\right]^{n_i}\right]. \qquad (3.16)$$

Interchanging the summation and product

$$S = \prod_{i=1}^{M} \sum_{n_i=0}^{\infty} \left[\frac{\theta_i}{\mu_i}\right]^{n_i}, \qquad (3.17)$$

$$S = \prod_{i=1}^{M} \left[1 - \frac{\theta_i}{\mu_i}\right]^{-1}. \qquad (3.18)$$

So the *marginal* probability that a queue is empty is

$$p_i(0) = \left[1 - \frac{\theta_i}{\mu_i}\right]. \qquad (3.19)$$

Note the similarity to equation (2.53). And

$$p(\underline{n}) = \prod_{i=1}^{M} p_i(n_i), \quad (3.20)$$

$$p_i(n_i) = \left[1 - \frac{\theta_i}{\mu_i}\right]\left[\frac{\theta_i}{\mu_i}\right]^{n_i}.$$

This is the "product form solution" for the equilibrium state probabilities first observed by Jackson. Naturally one must first solve the traffic equations for the θ_i's in order to compute $p(\underline{n})$. One special case worth mentioning is a number of queues in series (tandem) fed by a Poisson source of rate λ. Then $\theta_i = \lambda$ for i=1,2,3...M.

It should also be mentioned that the product form solution is still valid if there are multiple sources and destinations.

Example: Let us calculate the product form solution for the network of Figure 3.2. From the previous example it is known that $\theta_1 = 2\lambda$ and $\theta_2 = \frac{8}{3}\lambda$. Thus

$$p(\underline{n}) = p(0)\left[\frac{2\lambda}{\mu_1}\right]^{n_1}\left[\frac{(8/3)\lambda}{\mu_2}\right]^{n_2}$$

or

$$p(\underline{n}) = \left[1 - \frac{2\lambda}{\mu_1}\right]\left[\frac{2\lambda}{\mu_1}\right]^{n_1}\left[1 - \frac{(8/3)\lambda}{\mu_2}\right]\left[\frac{(8/3)\lambda}{\mu_2}\right]^{n_2}.$$

∇

What has always attracted attention about the product form solution is that its similarity to the decomposition of a joint probability in terms of its marginal probabilities for independent random variables suggests that, in a product form network, queues act "independently" of one another. Actually what it really means is that if you took a "snapshot" of the network at a time instant the number of customers found in each queue are independent of one another. However if you took snapshots of the network at two, close, time instants you would find strong correlations between the numbers of customers in the queues at the two time instants [DISN].

3.2.3 Local Balance

We just succeeded in solving the global balance equation for state n of an open queueing network by using the local balance equation $\theta_i p(n-1_i) = \bar{\mu}_i p(n)$. Local balance is a phenomena peculiar to queueing networks. Suppose, for simplicity of argument, that we have a series (tandem) network so that $\theta_i = \lambda$. Then $\lambda p(n-1_i) = \mu_i p(n)$. What this says is that the amount of probability flux flowing on pairs of transitions incident to a state are equal. This is a stronger form of balancing than global balance where we simply equate the flow into a state totaled over *all* the incoming transitions with the flow out of the state totaled over *all* the outgoing transitions. With local balance the more specific action is being taken of equating flows on individual pairs of transitions.

Global balance has an analog in resistive circuit theory in the form of Kirchoff's current conservation law. Local balance has no such analog. Intuitively, this is because transition values, unlike resistor values, are labeled in a patterned manner that gives rise to local balance.

If one has a state with several transitions, how does one know which ones are "paired" for local balance? The simple rule to follow is [SAUE 81]:

> The flow of probability flux into a state due to an arrival to a queue equals the flow out of the same state due to a departure from the same queue.

While we have used the case of series queues, what has been said holds generally for product form networks. In section 3.3 local balance will be explained in an integrated fashion.

3.2.4 Closed Queueing Networks

Consider a closed queueing network of M queues with a finite population of N customers. Let μ_i be the service rate of the ith exponential queue. Routing is, again, random with r_{ij} being the probability of a customer departing queue i entering queue j. We have the same state

$$n = (n_1, n_2, n_3, \cdots n_i, \cdots n_M) \tag{3.21}$$

as for the open network and the same use of the unit vector:

$$1_i = (0,0,0, \cdots 1, \cdots 0). \tag{3.22}$$

If one equates the net flow out of state n with the net flow into state n, one has [KOBA] the global balance equation at state n:

$$\sum_{i=1}^{M} \mu_i p(\underline{n}) = \sum_{i=1}^{M} \sum_{j=1}^{M} \mu_j r_{ji} p(\underline{n+1}_{\ j}-\underline{1}_{\ i}). \quad (3.23)$$

The left-hand side of this equation corresponds to customers leaving state $\underline{n}$ through queue departures while the right-hand side corresponds to entering state $\underline{n}$ from state $\underline{n+1}_{\ j}-\underline{1}_{\ i}$ through a transfer of customers from queue j to queue i.

The traffic equations take the form

$$\theta_i = \sum_{j=1}^{M} r_{ji}\theta_j, \qquad i=1,2,3,...M. \quad (3.24)$$

Interestingly, these traffic solutions do not have a unique solution. Any solution $(\theta_1,\theta_2,\theta_3,\cdots\theta_M)$ multiplied by a real number k, $(k\theta_1,k\theta_2,k\theta_3,\cdots k\theta_M)$ is also a solution.

To solve the global balance equation rewrite the traffic equations as

$$1 = \sum_{j=1}^{M} r_{ji}\frac{\theta_j}{\theta_i}, \qquad i=1,2,3,...M. \quad (3.25)$$

Now multiply μ_i in the global balance equation by this identity for one:

$$\sum_{i=1}^{M}\sum_{j=1}^{M} \mu_i r_{ji}\frac{\theta_j}{\theta_i}p(\underline{n}) = \sum_{i=1}^{M}\sum_{j=1}^{M}\mu_j r_{ji}p(\underline{n+1}_{\ j}-\underline{1}_{\ i}). \quad (3.26)$$

This can be rewritten as

$$\sum_{i=1}^{M}\sum_{j=1}^{M} r_{ji}\left\{\frac{\theta_j}{\theta_i}\mu_i p(\underline{n}) - \mu_j p(\underline{n+1}_{\ j}-\underline{1}_{\ i})\right\} = 0. \quad (3.27)$$

Clearly this equation will be satisfied if the following local balance equation holds:

$$\frac{\theta_j}{\theta_i}\mu_i p(\underline{n}) = \mu_j p(\underline{n+1}_{\ j}-\underline{1}_{\ i}), \qquad i,j=1,2,3,...M. \quad (3.28)$$

Rearranging one has

$$p(\underline{n}) = \left[\frac{\theta_i}{\mu_i}\right]\left[\frac{\theta_j}{\mu_j}\right]^{-1} p(\underline{n}+\underline{1}_j-\underline{1}_i), \qquad i,j=1,2,3,...M. \tag{3.29}$$

This is equivalent to

$$p(\underline{n}) = \left[\frac{\theta_i}{\mu_i}\right] p(\underline{n}-\underline{1}_i). \tag{3.30}$$

Expanding

$$p(\underline{n}) = \left[\frac{\theta_i}{\mu_i}\right] p(n_1,n_2,n_3, \cdots n_i-1, \cdots n_M). \tag{3.31}$$

Continuing the recursion to zero out the ith term one obtains

$$p(\underline{n}) = \left[\frac{\theta_i}{\mu_i}\right]^{n_i} p(n_1,n_2,n_3, \cdots 0, \cdots n_M). \tag{3.32}$$

Doing this for each queue yields

$$p(\underline{n}) = \prod_{i=1}^{M} \left[\frac{\theta_i}{\mu_i}\right]^{n_i} p(\underline{0}). \tag{3.33}$$

Clearly all the queues cannot be empty in a closed network. Instead state $\underline{0}$ here can be thought of as a reference state. This is usually a state where all of the customers are in one queue, leaving the remaining queues empty.

We will use the normalization of probability to find an explicit expression for $p(\underline{0})$:

$$\sum_{\underline{n}} p(\underline{n}) = p(\underline{0})G(N) = 1, \tag{3.34}$$

$$p(\underline{0}) = \frac{1}{G(N)} = \frac{1}{\sum_{\underline{n}}\left[\prod_{i=1}^{M}\left[\frac{\theta_i}{\mu_i}\right]^{n_i}\right]}. \tag{3.35}$$

Here $G(N)$ is a "normalization constant" that is computed over the finite state space of the closed queueing network. Because the summation for $G(N)$ does not go over an infinite number of states, as it does open queueing networks, it does *not* factor into a product $p_1(0)p_2(0)p_3(0) \cdots p_M(0)$.

We thus have the product form solution for closed queueing networks:

$$p(\underline{n}) = \frac{1}{G(N)} \prod_{i=1}^{M} \left[\frac{\theta_i}{\mu_i} \right]^{n_i}. \qquad (3.36)$$

Though not explicitly stated, $G(N)$ is also a function of M. As discussed in chapter four, the efficient computation of $G(N)$ is important in the convolution algorithm for calculating closed queueing network performance measures.

Finally, it should be mentioned that even though the θ_i's do not have a unique solution, as long as the same values are used for the θ_i's consistently in (3.36) for both $G(N)$ and the term in brackets, $p(\underline{n})$ will be correct.

3.2.5 The BCMP Generalization

In 1975 F. Baskett, K. M. Chandy, R. R. Muntz, F. G. Palacios published a generalization of the product form solution in terms of networks consisting of servers with four possible service disciplines. The four service disciplines that lead to the product form solution are [BASK 75]:

Service discipline 1: The service discipline is First In First Out (FIFO). All customers, regardless of class, have the same exponential service time distribution. The service rate can be a function of the *total number of customers in the individual queue.*

Service Discipline 2: The service discipline is processor sharing. That is, each of n customers receives $1/n$ of the service effort. Each class of customer may have a distinct service time distribution (with a rational Laplace transform).

Service Discipline 3: In this service discipline there is one server for each customer. Each class of customer may have a distinct service time distribution (with a rational Laplace transform).

Service Discipline 4: The service discipline is preemptive-resume Last In First Out (LIFO). Each class of customer may have a distinct service time distribution (with a rational Laplace transform).

The state description of these product form networks can be quite detailed (jump ahead to Figure 3.16 for an example), state-dependent arrivals are allowed, there are different classes of customers, and some classes may be closed and some may be open. The product form solution for such a complex model appears in [BASK 75].

3.3 Algebraic Topological Interpretation of the Product Form Solution

3.3.1 Introduction

Why does the product form solution exist, and how can one explain its simple form? During the mid 1980s an interesting and aesthetically pleasing interpretation of the existence of the product form solution became possible. The basic idea is the realization that the state transition diagrams of product form networks can be decomposed into an aggregation of elementary "building blocks". Because these building blocks do not interact with each other in a substantial way, they can be solved in isolation for the relative state probabilities. The product form solution, then, for a specific equilibrium state probability in terms of a reference probability is really a recursion that patches together these building block solutions along a "path" from the state of interest back to the reference state.

This algebraic topological interpretation of the product form solution is important for several reasons. It gives a constructive means for arriving at the product form solution. It helps to explain why certain models have a product form solution and why others do not. It becomes possible to modify or select models so that they have a product form solution [HAMI 86a,86b,89] or to construct product form models which provide performance bounds for the more difficult to analyze non-product form networks [VAND].

In what follows these ideas will be expanded upon. The sources for this material are from the work of A. Lazar [LAZA 84a, 84b, 84c], I. Wang [WANG 86] and the author.

3.3.2 A First Look at Building Blocks

Consider the cyclic three queue network in Figure 3.3. The service times are exponential random variables. The state transition diagram is illustrated in Figure 3.4. The state description for this model is two dimensional since this is a closed network. That is, if n_1 and n_2 are the numbers of customers in the upper queues, then $N-n_1-n_2$ is automatically the number in the lower queue where N is the total number of customers in the network.

Now let us look at the local balance equations of this product form network. Recalling the rule that allows us to pair transitions for local balance, in Figure 3.5 the paired transitions which exhibit local balance are shown. Observing the lower left corner of the state transition diagram one can see that the local balance pairings mean that the flow from $(0,0)$ to $(1,0)$ equals that from $(1,0)$ to $(0,1)$. But the

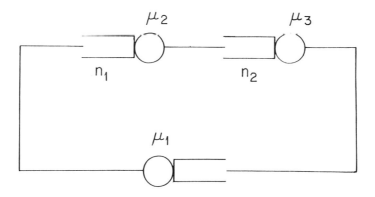

Fig. 3.3: Cyclic Queueing Network

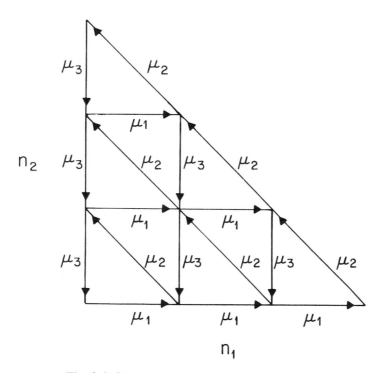

Fig. 3.4: State Transition Diagram of Figure 3.3

flow from (1,0) to (0,1) must equal that from (0,1) to (0,0). Finally the flow from (0,1) to (0,0) must equal that from (0,0) to (1,0).

Thus an alternative view of local balance is that there is a circular flow of probability flux along the edges of the triangular subelement: (0,0) to (1,0) to (0,1) and back to (0,0). With this circular flow, each edge has the same magnitude

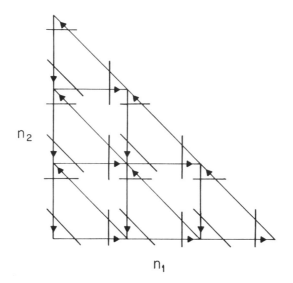

Fig. 3.5: Local Balance Pairings

of probability flux.

Continuing with this logic, each triangular subelement (or building block) in the state transition diagram has a circular flow about its edge. This is illustrated in Figure 3.6.

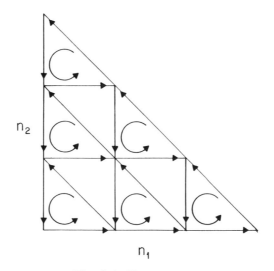

Fig. 3.6: Circular Flows

Because of the decomposition of the flow of probability flux that is thus possible, one can effectively remove one building block at a time from the remaining building blocks without disturbing its flow pattern, except for a renormalization.

The renormalization is necessary so that the probabilities of the remaining states sum to one. A completely decomposed state transition diagram is shown in Figure 3.7.

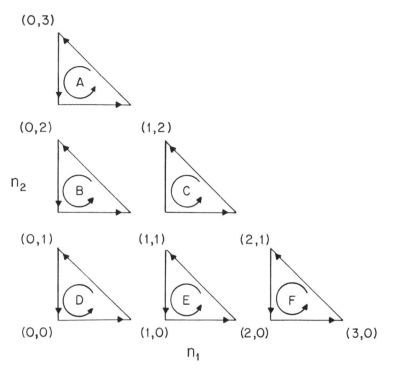

Fig. 3.7: Decomposed State Transition Diagram

If one wants to solve for p_{21} for instance, one solves building block C in this last figure to obtain the local balance equation

$$\mu_1 p_{11} = \mu_2 p_{21} \tag{3.37}$$

and the solution

$$p_{21} = \frac{\mu_1}{\mu_2} p_{11}. \tag{3.38}$$

Solving building block E one obtains the local balance equation

$$\mu_3 p_{11} = \mu_1 p_{10} \tag{3.39}$$

and the solution

$$p_{11} = \frac{\mu_1}{\mu_3} p_{10}.$$ (3.40)

Finally, the building block D yields the local balance equation

$$\mu_1 p_{00} = \mu_2 p_{10}$$ (3.41)

and the solution

$$p_{10} = \frac{\mu_1}{\mu_2} p_{00}.$$ (3.42)

Substituting these solutions back into one another yields

$$p_{21} = \frac{\mu_1}{\mu_2} \frac{\mu_1}{\mu_3} \frac{\mu_1}{\mu_2} p_{00} = \left[\frac{\mu_1}{\mu_2}\right]^2 \left[\frac{\mu_1}{\mu_3}\right] p_{00}$$ (3.43)

or the product form solution. What we have done is solved the building blocks along a path from the desired state to the reference state. Actually this is a conservative system. Any path between the two states would have produced the same results.

If we solve for all the state probabilities we can deduce that

$$p_{n_1 n_2} = \left[\frac{\mu_1}{\mu_2}\right]^{n_1} \left[\frac{\mu_1}{\mu_3}\right]^{n_2} p_{00}.$$ (3.44)

One more point needs to be made. The state transition diagram of Figure 3.4 is the same as that for an open network of two queues in series, with an upper bound on the *total* number of customers in both queues. In the diagram the upper bound, N, takes the form of the boundary $n_1 + n_2 \leq N$. As $N \rightarrow \infty$ and we go to an open network of two infinite buffer queues in series the state transition diagram becomes infinite in extent but everything that has been said about the building block structure still holds.

Example: Consider the (open) M/M/1 queueing system. The state transition diagram is constituted of building blocks consisting of two transitions each. For the *infinite* buffer M/M/1 queueing system there is an infinite number of such building blocks. In the case of the M/M/1/N *finite* buffer system, Figure 3.8, a finite number of such building blocks is pasted together. Note that the same state transition diagram models a (closed) cyclic queueing system of two queues.

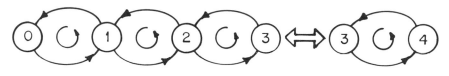

Fig. 3.8: M/M/1 System Building Blocks

Naturally there is only one path from any state to the usual reference state, p_0.

∇

Example: Consider a closed cyclic network of four queues with exponential servers and with only two customers in the network. The state transition diagram is three dimensional and appears in Figure 3.9. The basic building block consists of four edges along a tetrahedron-shaped volume of space. One could show that there are circular and equal-valued flows of probability flux along the edges of these tetrahedrons and that any state probability can be found by solving these tetrahedrons along a path back to the reference probability.

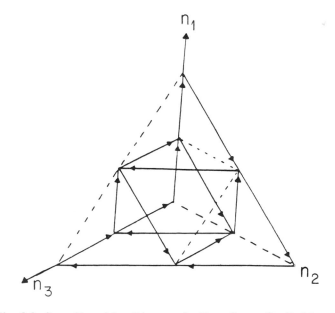

Fig. 3.9: State Transition Diagram for Four Queue Cyclic Network

Note that the state transition diagram of Figure 3.9 is the same as that for an open network of three queues in series, with an upper bound on the *total* number of customers in the queues. This bound takes the form of the plane boundary $n_1+n_2+n_3 \leq N$ in the state transition diagram. As $N \rightarrow \infty$ one has a system of infinite buffer queues and the state transition diagram consists of an infinite number of tetrahedral building blocks.

∇

We have seen that the state transition diagrams of a number of product form networks can be decomposed into elementary building blocks. In fact, this is true of any product form network. One can also look at the reverse case, the construction of state transition diagrams out of elementary building blocks. We will call the process of aggregating building blocks to form a state transition diagram *geometric replication*. The reason is that the same geometric building block is usually replicated to form the overall diagram. Note that in the language of mathematical topology [FIRB] one would call a building block a "cell" and a state transition diagram would be called a "complex". Thus [LAZA 84a]:

Definition: The constructive process of aggregating multi-dimensional cells into a complex is referred to as *geometric replication*.

The geometric replication of cells or building blocks is useful in explaining the nature of the product form solution. It will also be useful later in devising models with this solution.

3.3.3 Building Block Circulatory Structure

In this section the ideas of the last section will be made more precise [WANG 86].

Definition: The *circulatory structure* of a state transition diagram is the pattern of probability flux on the diagram transitions.

Perhaps the most important property of building blocks for product form networks is that they can be embedded into the overall state transition diagram without affecting the flow pattern of the rest of the diagram, except for a renormalization. This property is what allows one to solve building blocks in isolation in order to generate the recursive product form solution.

Definition: An *isolated circulation* consists of the probability flux along a subset of state transition diagram edges for which there is conservation of this flow at each adjacent state when the edges are embedded in the overall state transition diagram. Thus the relative equilibrium probabilities for these states can be solved for in isolation without considering the rest of the state transition diagram.

Definition: A *cyclic flow* is an isolated circulation in the form of a single cycle.

The "circular flows" of the building blocks of the previous section are cyclic flows. We can summarize by saying:

Duality Principle: There is a duality between the existence of local balance and the existence of isolated circulations. That is, the existence of local balance leads to isolated circulations and the presence of isolated circulations leads to local balance.

Proof: The former follows from the usual queueing interpretation of local balance. The latter follows from the definition of an isolated circulation.
∇

In other words, in observing local balance for a pair of transitions at a state, one is observing an isolated circulation passing through the state.

We will now characterize the building blocks:

Definition: A cyclic flow about L edges is said to be of length L.

Now let us recall the examples of the previous section for both open and closed networks. Table 3.1 shows the lengths of the building blocks for the various cases

Table 3.1: # of Queues vs. L		
Closed	Open	L
2	1	2
3	2	3
4	3	4

This suggests the following

Theorem: For cyclic and tandem Markovian product form queueing networks where the state of a queue is described by the number of customers in the queue, a closed cyclic network of L queues corresponds to an L length cyclic flow and an open tandem network of L queues corresponds to a $L+1$ length cyclic flow.

Proof: To prove this for a cyclic network is straightforward. Each queue in a closed path of L queues contributes one transition, so the building block is of length L. For a tandem network of L queues each queue contributes one transition. There is also one additional transition that can be thought of as being due to a "virtual queue" that connects the output to the input. The transition rate of this additional transition is just the arrival rate to the network. This gives building blocks of length $L+1$.

∇

Next, let us characterize the sequence of events in a queueing network that are associated with, or serve to generate, a building block:

Definition a: For a closed network a *relay sequence* is a series of successive events consisting of a customer leaving queue i to enter queue j, followed by a departure from queue j which enters queue k, ... finally followed by an arrival back into queue i. Each of the queues is distinct and they form a closed cyclic path through the queueing network.

The sequence of events associated with a building block is called a relay sequence because a customer arriving at a queue is followed by a customer leaving the same queue. For a FIFO discipline this is not the same customer unless the queue is empty. It is as if customers are carrying a baton around a closed path, as in a relay race. There is an analogous definition for an open network:

Definition b: For an open network a *relay sequence* is a series of successive events corresponding to an arrival into queue i from outside the network, followed by a departure from queue i to queue j, followed by a departure from queue j to queue k, ... followed by a departure from the network. Each of the queues is distinct.

These ideas will now be expanded upon through a series of examples:

Fig. 3.10: Series (Tandem) Open Network

Example: Consider two queues, with exponential servers, in series forming an open network with Poisson arrivals, as in Figure 3.10. As has been mentioned, if the system has an infinite number of buffers, the state transition diagram is comprised of building blocks as in Figure 3.7 but is infinite in extent. The product form solution for this network is

$$p_{n_1 n_2} = \left[\frac{\lambda}{\mu_1}\right]^{n_1} \left[\frac{\lambda}{\mu_2}\right]^{n_2} p_{00}.$$

Suppose now that there is an upper bound on the number of customers in *each* queue: $n_1 \leq n_{1,max}, n_2 \leq n_{2,max}$. This gives rise to the state transition diagram of Figure 3.11 with vertical and horizontal boundaries. This blocking network is now a non-product form network. That is, a (simple) product form solution no longer describes the equilibrium state probabilities.

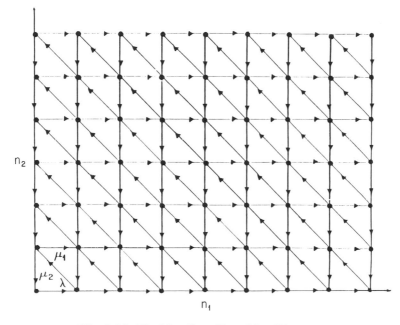

Fig. 3.11: Blocking State Transition Diagram

Why is this so? What has happened is that there are no longer integral building blocks along the upper and right side rectilinear boundary. Put another way, the pairing of transitions for local balance has been disrupted along these boundaries. That is, there are transitions along the boundary that are missing their natural partner for local balance pairing. This deficiency along the boundary disrupts the flow of probability flux throughout the state transition diagram so that there are no longer cyclic flows about the remaining integral building blocks and the product form solution no longer exists. N. van Dijk has written [VAND] that there is a "failure" of local balance in such a case. At this point the only direct way to calculate the state probabilities in most cases is by numerically solving the set of linear global balance equations.

Not every boundary destroys the product form solution. For instance, suppose that we have an upper bound on the *total* number of customers in both queues. As has been mentioned this leads to the state transition diagram like that of Figure 3.4. This queueing network has the same product form solution, except for the normalizing value of p_{00}, as that for a similar infinite buffer network.

What is different, compared to the blocking network, is that the diagonal boundary $n_1+n_2 \leq N$ allows integral building blocks and local balance pairing for all transitions, and thus the product form solution exists.

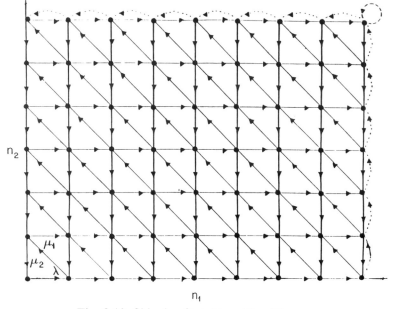

Fig. 3.12: Skipping State Transition Diagram

Finally, consider another modification where if either of the queues is full, customers will instantaneously "skip" over it. The state transition diagram is shown in Figure 3.12. There are now integral building blocks of length two along the upper and right boundaries and integral building blocks of length three throughout the interior. Again there is local balance, cyclic flows and the same product form solution holds.

∇

Example: In this example we will look at the effect of random routing on the building block structure. The example involves a multiprocessor model.

A multiprocessor system [HWAN] consists of a group of processors interconnected with a group of memory modules. A closed Markovian queueing model of a multiprocessor system is shown in Figure 3.13. That is, the service times are exponential. From before, we know that this is a product form network. This model consists of three queueing systems arranged in a cyclic fashion. Two of the systems are comprised of a number of parallel queues and represent, respectively, the CPU's and memory modules. Routing into each of the two banks of parallel queues is random. That is, customers (really jobs) enter the m CPU's with probability $\alpha_1, \alpha_2, \cdots \alpha_m$ and customers enter the n memory modules with probabilities $\beta_1, \beta_2, \cdots \beta_n$. The third system is a single queue,

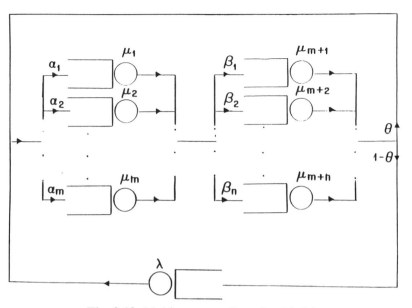

Fig. 3.13: Multiprocessor Queueing Model

departures from which represent submissions of jobs (or interactive commands) to the computer system.

In this multiprocessor system a customer leaving the memory module bank can proceed directly back to the CPU's with probability θ, or it may go to the job submission queue to wait for its next execution with probability $1-\theta$. To simplify the following discussion we will concentrate on the special case when $\theta=0$.

For this reduced system, a "relay sequence" is a sequence of successive events corresponding to a customer entering a CPU queue followed directly by a customer departure from that CPU into a memory module followed directly by a customer departure from that module back to the job submission queue. Again, the departing customers are not necessarily the same as the arriving customers.

We will look at two types of building blocks, the second of which is a decomposed part of the first. The structure of the first building block for this m CPU and n memory module system is shown in Figure 3.14. Assume that "a" is a state $(k_1,k_2, \cdots k_l \cdots k_m,k_{m+1} \cdots k_{m+n})$ where k_l is the # of customers in the lth queue and the job submission queue is non-empty. A customer leaving the job submission queue corresponds to a transition into one of the 1,2 ... m CPU states, i.e., $(k_1,k_2, \cdots k_{l+1}, \cdots k_m,k_{m+1}, \cdots k_{m+n})$. A customer leaving a CPU corresponds to a transition into one of the $m+1$, $m+2$ ... $m+n$ memory module states. Finally a memory module departure corresponds to a transition back to state a.

Thus, in queueing terms, this first building block is generated by *all possible* relay sequences starting from state a. The building block's net flow into and out of the block's states balance when the block is embedded in the state transition diagram. The building block is a subgraph whose relative state probabilities can

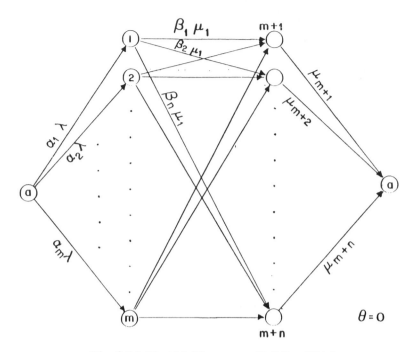

Fig. 3.14: First Multiprocessor Building Block

be solved for in isolation.

Now let us consider the second type of building block. Let p_l be the state probability that the customer is in the CPU queue l if $1 \le l \le m$ and in memory module l if $m+1 \le l \le m+n$. Using flux conservation at a CPU state

$$\alpha_l \lambda p_a = \sum_{j=1}^{n} \beta_j \mu_l p_l = \mu_l p_l, \quad l=1,2,3,...m, \tag{3.45}$$

and at a memory module state

$$\mu_l p_l = \beta_{l-m} \sum_{i=1}^{m} \mu_i p_i, \quad l=m+1,m+2,...m+n. \tag{3.46}$$

Solving these for p_l yields

$$p_l = \frac{\alpha_l \lambda}{\mu_l} p_a, \quad l=1,2,3,...m. \tag{3.47}$$

$$p_l = \frac{\beta_{l-m} \lambda}{\mu_l} p_a, \quad l=m+1,m+2,...m+n. \tag{3.48}$$

The circulatory structure of the building block of Figure 3.14 can be decomposed into an aggregation of smaller cyclic flows that are associated with smaller building blocks. For instance, $\alpha_i \lambda p_a$, the flux from state a to state i, where $1 \leq i \leq m$, can be decomposed as $\sum_{j=1}^{n} \alpha_i \beta_j \lambda p_a$ in which $\alpha_i \beta_j \lambda p_a$ is the portion of flow that corresponds to departures from the ith CPU destined for the jth memory module. The flow corresponding to departures from memory module j can also be decomposed into subflows. These are proportional to $\alpha_1, \alpha_2, \cdots \alpha_m$.

The circulatory structure of the first building block of Figure 3.14 with its splits and joins of flux can thus be decomposed into an aggregation of simple cyclic flows embedded along smaller building blocks. Each such flow corresponds to a *specific* relay sequence originating from state a. One such cyclic flow/building block is shown in Figure 3.15. It illustrates a relay sequence with m=n=3 involving the 2nd CPU and the 1st memory module.

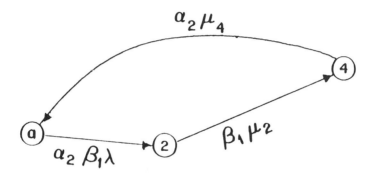

Fig. 3.15: Second Multiprocessor Building Block

These cyclic flow building blocks can also be solved in isolation for the equilibrium probabilities. They are pasted together along edges to form the larger building block of Figure 3.14. They form the most basic decomposition possible of the overall circulatory pattern of the multiprocessor state transition diagram with this state description.

∇

Example: Consider now a different queueing system example. This is a FIFO queue processing two different classes of customers. The server is exponential with rate μ for both classes. The state consists of the class of the job in each queue position. The state transition diagram of this queueing system is shown in Figure 3.16. The numbers associated with each state represent the class of each job in each queueing system position. In the diagram's notation they are served from left to right.

This is a product form network. There are cyclic flows in the state transition diagram. In terms of queueing system events these correspond to a shift register like behavior:

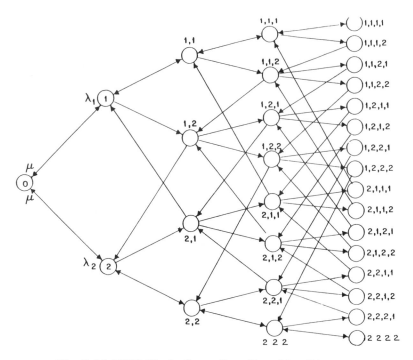

Fig. 3.16: FIFO Single Queue State Transition Diagram

Definition: A "shift register sequence" for a FIFO multiclass product form queue where the state involves the customer class at each queue position is a sequence of successive queueing events corresponding to a class i departure followed by a class i arrival followed by a class j departure (j may $=i$) followed by a class j arrival, and so on, terminating when the queue returns to the original state. This sequence of events generates the state transition diagram building blocks.

For instance, the state transition sequence

$$121 \rightarrow 21 \rightarrow 211 \rightarrow 11 \rightarrow 112 \rightarrow 12 \rightarrow 121$$

corresponds to a cyclic flow of length six. The state transition diagram of this queueing system consists of cyclic flows of different lengths. There is a group of local balance equations for each cyclic flow that can be solved without regard to the rest of the state transition diagram.

Let stage $i=1,2...$ in the diagram consist of the transitions between states with i and i-1 total customers in those states. The cyclic flows exist in a single stage. Moreover, there are cyclic flows of different lengths within the same stage (Table 3.2). This is because the shift register like behavior may return the system to a starting state with a number of shifts less than the queue length (consider 1212).

Table 3.2: Number of Cyclic Flows

Circulation Length	Stage					
	1	2	3	4	5	6
2	2	2	2	2	2	2
4	-	1	-	1	-	1
6	-	-	2	-	-	2
8	-	-	-	3	-	-
10	-	-	-	-	6	-
12	-	-	-	-	-	9
Total	2	3	4	6	8	14

If n_1 is the number of class one customers in the queueing system and n_2 is the number of class two customers in the queueing system, then

$$p(n_1, n_2) = \frac{\lambda_1^{n_1} \lambda_2^{n_2}}{\mu^{n_1 + n_2}} p(0,0).$$

Note that in having the state of the system indicate the queueing order of the two classes one has a more detailed decomposition of the state transition diagram then that when the state indicates only the total number of each class at a queue. The cyclic flows in the state transition diagram of Figure 3.16 are a decomposition of isolated circulations in the diagram where state does not indicate queueing order.

For cyclic and tandem networks of such FIFO queues, the basic geometric building block is also generated by shift register-like sequences. As an example, consider the cyclic two queue system of Figure 3.17. The state transition diagram is shown in Figure 3.18 for four customers. Here μ_1^{II}, for instance, indicates the service rate of the 1st queue for class II customers. To obtain a product form solution, service rates must be class independent. Then

$$p(n_1, n_2) = \left[\frac{\mu_2}{\mu_1} \right]^{n_1 + n_2} p(0,0)$$

where n_1 and n_2 are the number of class I and class II customers in the upper queue, respectively.

Cyclic flows exist between adjacent columns of states in the diagram of Figure 3.18. One such cyclic flow corresponds to the state transition sequence

$$1,122 \rightarrow 11,22 \rightarrow 1,221 \rightarrow 12,21 \rightarrow 2,211 \rightarrow 22,11 \rightarrow 2,112 \rightarrow 21,12 \rightarrow 1,122$$

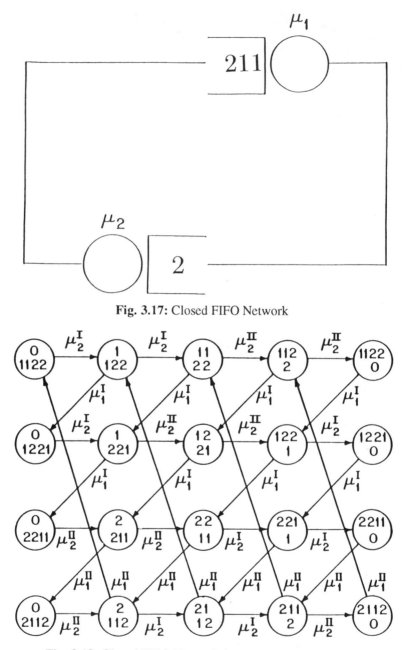

Fig. 3.17: Closed FIFO Network

Fig. 3.18: Closed FIFO Network State Transition Diagram

where the state of the upper queue appears at left.

Note that the sequence starts with a class 1 customer leaving the lower queue followed by a class 1 departure from the upper queue. While this is reminiscent of a relay sequence, one has not returned to the original state since the customer order in both queues is changed. The sequence thus continues cycling until it returns to the original state. It is as if a shift register sequence, in the context of a queueing network, is comprised of successive relay sequences. Let us make precise the sequence of events that generates the state transition diagram building blocks.

Definition: For a closed FIFO product form network in which the state indicates the class of customer at each queue position a "shift register sequence" is a series of successive queueing events beginning with a customer leaving queue i to enter queue j, followed by a departure from queue j that enters queue k... followed by an arrival back into queue i. If the network is not now in the original state, the previous series of events repeats until the network returns to the original state. Again, each of the queues is distinct, and they form a closed cyclic path through the queueing network.

Definition: For an open FIFO network in which the state indicates the class of customer at each queue position, a "shift register sequence" is a series of successive events beginning with an arrival into queue i from outside of the network followed by a departure from queue i to enter queue j, followed by a departure from queue j that enters queue k... followed by a departure from the network. If the network is not now in the original state, the previous series of events repeats until the network returns to the original state. In the sequence, network departures of one class are followed by network arrivals of the same class. Each of the queues is distinct.

Theorem: For cyclic and tandem FIFO product form queueing networks where the state is described by the class of each customer in each queue position and for which there is a consistent (see below) set of local balance equations, a cyclic network of N queues and T customers corresponds to a cyclic flow of length $M \leq NxT$. A tandem network of N queues and T customers corresponds to a cyclic flow of length $M \leq (N+1)xT$.

Proof: The cyclic flow of greatest length occurs when each customer must be shifted completely around the cyclic path of queues (N for a cyclic network and $N+1$ for an tandem network with returns through a virtual queue). For certain states customers may not have to be shifted around the entire path or not all of the customers may have to be shifted to bring the system back to the original state. This accounts for the inequalities in the theorem.
∇

For networks of LCFS queues in which the state indicates the class of customer at each queue position, the basic geometric building block is generated by a relay type sequence. The difference is that, because of the LCFS discipline, the job departing the first queue in a cyclic network is the *same* job that eventually returns. For a tandem network a job of the same class returns.

3.3.4 The Consistency Condition

Recall that the approach adopted in the previous sections calls for decomposing the state transition diagram into its building blocks and for solving the associated local balance equations. In the two-dimensional case, we have noted the existence of distinct paths connecting some arbitrary states with the reference state. This property is not observed in the one-dimensional case since the state transition diagram is a tree and hence a unique path links any node to the reference solution.

Thus the problem of consistency of the set of local balance equations arises as a result of the existence of distinct paths from an arbitrary state to the reference state. If, by taking alternate paths, the equilibrium probabilities for the same state are different, the set of local balance equations is not consistent.

As in the lower-dimensional case the set of global balance equations must be decomposed into a set of local balance equations. Such a decomposition is, however, not unique. In addition, it is not clear whether the resulting set of local balance equations is consistent. If the local balance equations are consistent, however, they are equivalent with the global balance equations. This is because the latter has a unique solution.

To derive the necessary and sufficient conditions for the consistency of a set of local balance equations, we define the following *consistency graph* [LAZA 84C]:

Definition: A consistency graph is an oriented graph, topologically equivalent with the original state transition diagram. In addition, it has the property that the probability at each node is equal to the probability of any adjacent node multiplied with the value of the associated edge connecting the two nodes.

The class of consistency graphs of interest to us derives the algebraic values associated with the arcs directly from the set of local balance equations. Thus, the consistency graph can be seen as an "easy-to-read" graphical representation of the *proposed solution* for the local balance equations.

Example: Consider the state transition diagram of Figure 3.4 for the three queue cyclic Markovian queueing network of Figure 3.3. The consistency graph for this system appears in Figure 3.19. It is the consistency graph for what is, at this point, a *proposed* solution and associated local balance equations. For instance, the relationship between p_{21} and p_{11} is

$$p_{21} = \frac{\mu_1}{\mu_2} p_{11}.$$

Now choose any path (open or closed) on the consistency graph. Associated with it is an algebraic value called the *product* of the edges or simply the product. For example, with the path from state $(0,0)$ to $(0,1)$ to $(1,1)$ to $(1,2)$ is associated

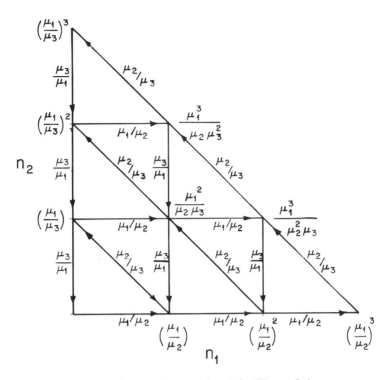

Fig. 3.19: Consistency Graph for Figure 3.4

the product

$$\frac{\mu_1}{\mu_3} \; \frac{\mu_1}{\mu_2} \; \frac{\mu_1}{\mu_3}$$

∇

Note that reversing the orientation of an arc of the consistency graph with value a results in an arc with value a^{-1}. We can now state the following consistency theorem [LAZA 84c]:

Theorem: A system of local balance equations is consistent if and only if any closed path of the consistency graph has the product equal to one.

Proof: Assume that there is a closed path in the consistency graph such that its product is not equal to one. Therefore, for at least one node belonging to this path, there are two distinct paths leading to the reference node, which do not have the same product. Thus two different values for the probability associated with such a node can be obtained. Hence, the local balance equations are not consistent.

The sufficiency part of the theorem can be shown by direct computation. The product of an arbitrary path connecting node $(n_1, n_2, n_3, \cdots n_m)$ with the reference node $(0,0,0, \cdots 0)$ is an algebraic invariant. This is because, by assumption, the product of any closed path is equal to one and reversing the orientation of any edge with value a results in an edge with value a^{-1}. The probability at node $(n_1, n_2, n_3, \cdots n_m)$ is, therefore, equal to the product of the path connecting this node with the reference node times the probability of the reference node.

∇

Hence to determine the consistency of the set of local balance equations, the product of any closed path of the consistency graph has to be computed and compared with the value one. This will be referred to as the *consistency condition*. Naturally, the consistency condition is redundant for some closed paths. The minimum set of products needed to verify the consistency condition is investigated from a topological point of view in [LAZA 84c].

It is easy to verify for Figure 3.19 that the product of any closed path is one and thus there is a consistent set of local balance equations, a building block decomposition, and a product form solution.

3.4 Recursive Solution of Nonproduct Form Networks

3.4.1 Introduction

In terms of applications, there are many important and useful nonproduct form models. In general, these do not have closed form solutions for their equilibrium probabilities. However, quite a few models are amenable to setting up a series of recursive equations to determine their equilibrium probabilities.

The idea of using recursions was first discussed in a systematic way by Herzog, Woo, and Chandy [HERZ], [SAUE 81]. Such recursions can be expressed in matrix form [ROM]. To date such recursions have usually been developed for relatively simple systems with one or two queues. However there has been some work on algorithmically determining such recursions [PARE], [ROBE 89].

In what follows two classes of nonproduct form models for which recursive, or at least decomposed, solutions are possible will be presented.

In [LAZA 87] a class of nonproduct form networks is described whose state transition diagrams can be shown to be equivalent to a lattice tree of simplexes. In this "flow redirection" method the state transition diagram geometry is manipulated by equivalence transformations. Sequential decomposition refers to the related process of solving one subset of states at a time for the equilibrium probabilities:

Definition: A solvable subset of queueing network states is a subset of states for which the equilibrium probabilities can be determined without regard to the equilibrium probabilities of the remaining unknown states. Here probabilities are determined with respect to a reference probability.

Note that the direct solution of linear equations takes time proportional to the cube of the number of equations. If P states can be solved as S subsets, then the computational effect is proportional to $(P/S)^3 *S$ rather than P^3.

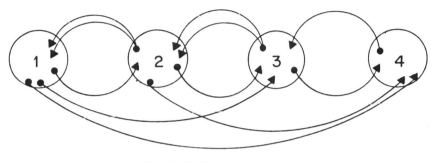

Fig. 3.20: Type A Structure
© 1990 IEEE

Two types of structure that allow the state transition diagram to be decomposed into solvable subsets will now be presented. The first type of structure [ROBE 89], is illustrated in Figure 3.20. Here each circular subset represents a state or a group of states. For the ith subset the rule is that there must be only one state, in the $(i+1)$st subset and with *unknown* probability, from which a transition(s) entering the subset originates. The subsets are solved sequentially, starting from the first subset to the second and so on. There is no restriction on the number of transitions that may leave the ith subset for destinations in the $j=i+1,i+2,...$ subsets. This type of structure will be referred to as Type A structure.

The second type of structure is illustrated in Figure 3.21. Here the first subset consists of a single state. The remaining subsets each consist of a state or a group of states. These are arranged in a tree type of configuration with the flow between subsets from the top of the diagram to the bottom and a return flow from the bottom level back to the top level. The subsets may be solved from the top to the bottom. Transitions may traverse several levels (e.g., dotted line in the figure) as long as the direction of flow is downward. This type of structure will be referred to as Type B structure.

It is possible to write recursive equations for the equilibrium probabilities when each subset in the Type A or Type B structure consists of a single state. In order to understand how recursive solutions are produced it is helpful to take a closer look at the Type A and B structures when one state at a time is decomposed. Figure 3.22 illustrates how a solution is arrived at for the Type A structure. A balance equation is solved for a state whose probability is already known.

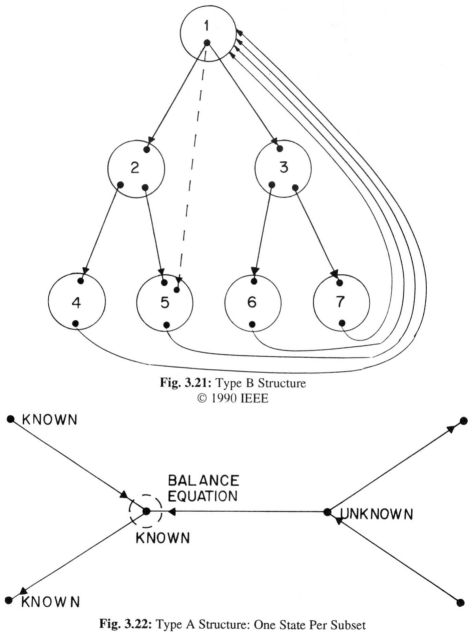

Fig. 3.21: Type B Structure
© 1990 IEEE

Fig. 3.22: Type A Structure: One State Per Subset
© 1989 IEEE

The state must have only one incoming transition from another state with unknown probability. When the balance equation is solved this other state probability becomes known. Often, this other state provides the next balance equation for solution.

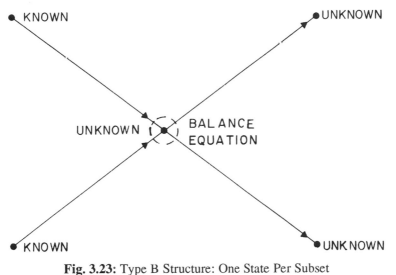

Fig. 3.23: Type B Structure: One State Per Subset
© 1989 IEEE

Figure 3.23 illustrates how a solution is arrived at for the Type B structure. A balance equation is solved for a state with unknown probability where all incoming transitions originate from states with known probabilities.

The types of structures illustrated in Figures 3.22 and 3.23 are particularly suited to software detection [ROBE 89]. However, it is also possible to decompose Type A state transition diagrams by drawing vertical boundaries between adjacent states and equating the flow probability flux from right to left across a boundary with that from left to right. This was done to find a number of the recursions below.

As has been mentioned, models with these recursive structures have a special matrix structure. If we write out the global balance equations in matrix notation $\mathbf{Ax=B}$, then $\mathbf{A}$ can be put in a form where it has nonzero terms in the upper right triangular part and in the first subdiagonal [ROM, pg. 56]. A little thought will show that this format can be solved from the bottom row to the top row for the equilibrium state probabilities, with respect to a reference probability, one equation at a time.

A number of examples of the recursive solution of practical queueing networks follows. They are from [WANG].

3.4.2 Recursive Examples

A. Two Blocking Tandem Queues

This first example concerns two queues in tandem with finite buffers. The first queue has a buffer size of M and the second queue has a buffer size of one. Here δ is the system arrival rate and λ_i and μ are the service rates of the first and second queues, respectively. The service rate of the first queue is state dependent. As with all examples in this correspondence, the arrival process is Poisson and service rates follow an exponential distribution. Recursive equations for this case appear below. The state transition diagram is illustrated in Figure 3.24.

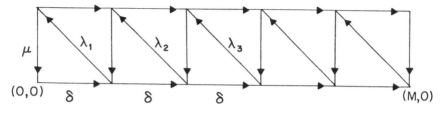

Fig. 3.24: Tandem Queues: Buffer Size 1 for 2nd Queue
© 1990 IEEE

Initialize:

$$p(0,0)=1.0, \tag{3.49}$$
$$p(0,1)=(\delta/\mu)p(0,0).$$

Iterate: $i=1,...,M-1$

$$p(i, 0)=(\delta/\lambda_i)[p(i-1,0)+p(i-1,1)],$$
$$p(i, 1)=(\delta/\mu)[p(i-1,1)+p(i, 0)].$$

Terminate (Right Boundary):

$$p(M, 0)=(\delta/\lambda_M)[p(M-1,0)+p(M-1,1)],$$
$$p(M, 1)=(\delta/\mu)p(M-1,1).$$

These equations have a simple form. The structure is Type A. It should be noted that the equations are not unique. It is also possible to take state-dependent arrival rates into account for this and the remaining examples. Recursive equations for the case when the second queue's buffer is of size two appears below. Here μ_i is the state-dependent service rate of the second queue. The state transition diagram is illustrated in Figure 3.25.

Initialize:

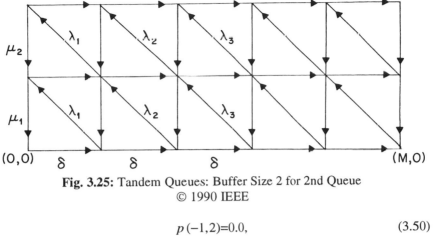

Fig. 3.25: Tandem Queues: Buffer Size 2 for 2nd Queue
© 1990 IEEE

$$p(-1,2)=0.0, \qquad\qquad (3.50)$$
$$p(0,0)=1.0,$$
$$p(0,1)=(\delta/\mu_1)p(0,0).$$

Iterate: $i=1,2,...,M-1$

$$p(i,1)=\delta f(i)/g(i)$$

where

$$f(i)=(\delta+\lambda_i)(\delta+\mu_2)[p(i-1,0)+p(i-1,1)]$$
$$+\delta(\delta+\lambda_i)p(i-2,2)-\lambda_i(\delta+\mu_2)p(i-1,0),$$
$$g(i)=\lambda_i[\mu_1(\delta+\mu_2)+(\delta+\lambda_i)(\delta+\mu_2)-\delta(\delta+\lambda_i)],$$

$$p(i,0)=[1/(\delta+\lambda_i)][\delta p(i-1,0)+\mu_1 p(i,1)],$$
$$p(i-1,2)=[1/(\delta+\mu_2)][\lambda_i p(i,1)+\delta p(i-2,2)].$$

Terminate (Right Boundary):

$$p(M,1)=\delta f'(M)/g'(M)$$

where

$$f'(M)=\lambda_M(\delta+\mu_2)p(M-1,1)+\delta\lambda_M p(M-2,2),$$

$$g'(M) = \lambda^2_M \mu_2 + \lambda_M \mu_1 (\delta + \mu_2),$$

$$p(M, 0) = [1/(\lambda_M)][\delta p(M-1,0) + \mu_1 p(M, 1)],$$
$$p(M-1,2) = [1/(\delta + \mu_2)][\lambda_M p(M, 1) + \delta p(M-2,2)],$$
$$p(M, 2) = (\delta/\mu_2) p(M-1,2).$$

Again, the structure is Type A. The first subset consists of (0,0). Then subsets consist of the states $(i-1,1)$, $(i-1,2)$ and $(i,0)$ for $i=1,2,...M$. At the right boundary the last subset consists of $(M,1)$ and $(M,2)$. It appears that these equations can not be extended to a larger buffer size for the second queue because of the resulting geometry of the state transition diagram. The case where the first queue's buffer is of size one or two and the second queue's buffer is of size M can be handled through a straightforward change of variables. In [ROBE 92] a three queue tandem model with limited buffer space is found to have recursive solutions in certain cases.

B. An ISDN Protocol

Consider a channel that can service either voice or data transmissions [SCHW 87]. The transmitter contains a buffer of size N for data packets. A voice call can seize the channel with rate λ_1 if no data packets are in the buffer. It is serviced at rate μ_1. If a voice transmission has seized the channel, then data packets cannot be serviced but are buffered. Data packets arrive at rate λ_2 and are serviced at rate μ_2. The state transition diagram appears in Figure 3.26. Interestingly the upper row of the diagram has the Type B structure and the lower row has the Type A structure. The recursive equations appear below. We note that the recursive expression for the upper row appears in [SCHW 87].

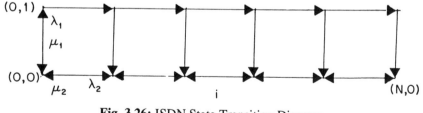

Fig. 3.26: ISDN State Transition Diagram
© 1990 IEEE

Initialize:

$$p(0,0)=1.0, \tag{3.51}$$

$$p(0,1)=[\lambda_1/(\lambda_2+\mu_1)]p(0,0).$$

Iterate: $i=1,...,N-1$

$$p(i,0)=(\lambda_2/\mu_2)[p(i-1,0)+p(i-1,1)],$$

$$p(i,1)=[\lambda_2/(\lambda_2+\mu_1)]p(i-1,1).$$

Terminate (Right Boundary):

$$p(N,0)=(\lambda_2/\mu_2)[p(N-1,0)+p(N-1,1)],$$

$$p(N,1)=(\lambda_2/\mu_1)p(N-1,1).$$

C. A Window Flow Control Protocol

Consider the queueing system of Figure 3.27 which models a window flow protocol [SCHW 87]. Packets are sent from the lower queue (source) through a channel (upper queue) to the destination (W box). The W box only releases packets when all W packets in the closed queueing network are in the W box. This models a reception acknowledgement of the entire group of packets. The state transition diagram appears in Figure 3.28. The structure is Type B. The recursive equations are:

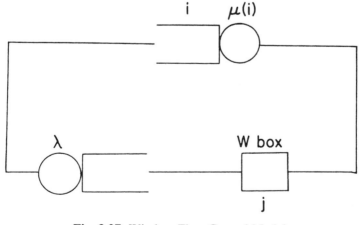

Fig. 3.27: Window Flow Control Model
© 1990 IEEE

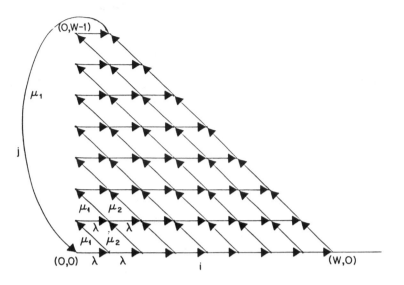

Fig. 3.28: State Transition Diagram of Fig. 3.27
© 1990 IEEE

Initialize:

$$p(0,0)=1.0. \qquad (3.52)$$

Iterate: $i=1,...,W-1$

$$p(i,0)=[\lambda/(\lambda+\mu_i)]p(i-1,0),$$
$$p(W,0)=(\lambda/\mu_W)p(W-1,0).$$

Iterate: $j=1,...,W-1$

$$p(0,j)=(\mu_1/\lambda)p(1,j-1).$$

Iterate: $i=1,...,W-j-1$

$$p(i,j)=[(1/(\mu_i+\lambda)][\lambda p(i-1,j)+\mu_{i+1}p(i+1,j-1)],$$
$$p(W-j,j)=$$
$$(1/\mu_{W-j})[\lambda p(W-j-1,j)+\mu_{W-j+1}p(W-j+1,j-1)].$$

3.4.3 Numerical Implementation of the Equations

Scaling problems are possible with the use of some of these equations. As an example, consider equation (3.49). The first queue may be viewed approximately as an M/M/1 queue. The equilibrium probability at the ith state is equal to λ/μ times the equilibrium probability at the $(i\text{-}1)$th state. Here, λ is the arrival rate, and μ is service rate. Thus, for instance, the probability that there are one hundred packets in the buffer is $(\lambda/\mu)^{100}$ times the probability that the buffer is empty. This can naturally lead to numerical scaling and underflow problems.

One form of automatic scaling is possible in this if it is clear which section of the state transition diagram corresponds to light probability mass and which corresponds to heavy probability mass. This is the case for the linear state transition diagram associated with equation (3.49). The equilibrium probabilities are then calculated in this roughly monotonic order. When the size of equilibrium probability begins to approach the computer register size, it is scaled downward by a factor which brings the equilibrium probability down to the lower limit of machine register size.

3.5 Queueing Networks with Negative Customers

3.5.1 Introduction

An intriguing generalization of the usual product form model involving the concept of "negative" customers has been first developed by E. Gelenbe and co-authors in a series of papers ([GELE 90,91a,91b,91c, ATAL]). In this new queueing paradigm there are two types of customers: positive customers and negative customers. A positive customer acts as the normal sort of customers we have seen so far. A negative customer, however, when it enters a queue will instantly destroy one positive customer and itself. Put another way, a negative and positive customer in a queue instantly cancel each other and disappear from the network.

To be a bit more specific, let us assume that positive customers arrive at the ith queue according to a Poisson process with rate Λ_i and negative customers arrive at the ith queue according to a Poisson process of rate λ_i. A positive customer leaving queue i will enter queue j as a positive customer with probability r_{ij}^+ or will enter queue j as a negative customer with probability r_{ij}^-. A positive customer arriving at a queue increases the queue length by one. A negative customer arriving at a queue will decrease the queue length by one if there is at least one positive customer present in the queue. A negative customer arriving at an empty queue is cleared from the network. Let us also assume that all service times are independently distributed exponential random variables. The service rate of the ith queue is μ_i. Finally the probability that a customer leaving queue i leaves the network is d_i.

Naturally the sum of all the queue departure probabilities should equal one:

$$\sum_j r_{ij}^+ + \sum_j r_{ij}^- + d_i = 1. \qquad 1 \le i \le M. \qquad (3.53)$$

It should be clear that this model only makes sense for open networks. It is impossible to get steady-state results for an analogous closed network with negative customers as the queue lengths go to zero in a finite amount of time. However, it is not so obvious that such a continuous time open model has a product form solution as the underlying equations are nonlinear. In fact, though, as we shall see, it does indeed have a product form solution for the equilibrium state probabilities.

What are the potential applications for this model? One is a resource request model where negative customers are commands to cancel previous requests (positive customers). Another potential application is in real time telecommunications networks where packets are discarded after a certain age [CHAO]. However the application that motivated E. Gelenbe and his co-authors ([ATAL] [GELE 91c]) is that of neural networks [MUEL]. As it is put in [GELE 91a]: "In this analogy each queue represents a neuron. Positive customers moving from one queue to another represent excitation signals, while negative customers going from a queue to another represent inhibition signals. Negative external arrivals can be used to represent the thresholds at each neuron."

The following discussion will closely follow that appearing in [GELE 91a].

3.5.2 Product Form Solutions

In this section the product form solution for queueing networks with negative customers will be presented. Let us state the main result. First the effective utilization is

$$q_i \equiv \frac{\lambda_i^+}{\mu_i + \lambda_i^-}. \qquad (3.54)$$

Here the λ_i^+ and λ_i^- are the solutions of the following nonlinear system of simultaneous traffic equations:

$$\lambda_i^+ = \sum_j q_j \mu_j r_{ji}^+ + \Lambda_i, \qquad (3.55)$$

$$\lambda_i^- = \sum_j q_j \mu_j r_{ji}^- + \lambda_i. \qquad (3.56)$$

Here the variable λ_i^+ is the total mean arrival rate of positive customers to the ith queue. The variable λ_i^- is the total mean arrival rate of negative customers to the ith queue. It can be seen that each equation has two terms: the first corresponding to arrivals into the ith queue from the other queues and the second corresponding arrivals into the ith queue from outside of the network. These equations are intrinsically nonlinear as λ_i^- appears in the denominator of q_j. It should be noted that it is natural for the denominator of q_i to consist of $\mu_i + \lambda_i^-$ as negative arrivals to a queue effectively increase its service rate.

In simple cases it is possible to obtain closed form analytical solutions for these traffic equations. In more complex cases the equations can only be solved iteratively through iteration. That is, one starts with a guess as to the solution and substitutes it into the equations to produce an improved estimate of the solution. After enough successive substitutions, the estimates will converge to the solution. This is a commonly used technique to solve simultaneous sets of nonlinear equations.

What Gelenbe found was that:

Theorem: If (3.55) and (3.56) have a unique nonnegative solution and $q_i < 1$ for each $i = 1,2...M$ then the equilibrium state probabilities are given by

$$p(\underline{n}) = \prod_{i=1}^{M} [1 - q_i] q_i^{n_i}. \tag{3.57}$$

This solution can be seen to be analogous to the product form solution for queues with only positive customers. This is a surprising result as systems of simultaneous nonlinear equations usually do not have a closed form analytic solution. Often one can only solve for special cases of such systems.

To show that the theorem is indeed correct the solution of (3.57) can be shown to satisfy the global balance equations of the states. These equations equate the net flow out of state $\underline{n}$ with the net flow into the same state. The following vectors will be used:

$$\underline{n}_i^+ = (n_1 \cdots n_i + 1 \cdots n_M), \tag{3.58}$$
$$\underline{n}_i^- = (n_1 \cdots n_i - 1 \cdots n_M),$$
$$\underline{n}_{ij}^\pm = (n_1 \cdots n_i + 1 \cdots n_j - 1 \cdots n_M),$$
$$\underline{n}_{ij}^{++} = (n_1 \cdots n_i + 1 \cdots n_j + 1 \cdots n_M).$$

The first two vectors correspond to the external arrival and external departure from a queue, with respect to $p(\underline{n})$, respectively. The third vector, in association with $p(\underline{n})$, corresponds to the transfer of a positive customer from one queue to another. The last vector, in association with $p(\underline{n})$, corresponds to a negative customer leaving one queue for another queue with at least one customer.

The global balance equation for a state in the interior of the state transition diagram, in terms of this notation, is

$$p\,(\underline{n})\sum_i[\Lambda_i+(\lambda_i+\mu_i)\,1[n_i>0]] \tag{3.59}$$

$$=\sum_i\Big[p\,(\underline{n_i^+})\mu_i d_i+p\,(\underline{n_i^-})\Lambda_i\,1[n_i>0]$$

$$+p\,(\underline{n_i^+})\lambda_i$$

$$+\sum_j(p\,(\underline{n_{ij}^\pm})\mu_i r_{ij}^+\,1[n_j>0]+p\,(\underline{n_{ij}^{++}})\mu_i r_{ij}^-$$

$$+p\,(\underline{n_i^+})\mu_i r_{ij}^-\,1[n_j=0])\Big].$$

Here $1[x]$ is an indicator function that is equal to one if the argument x is greater than 0 and equal to zero otherwise.

By substituting (3.57) and then (3.54-3.56) into (3.59) one can show, after several steps, that the product form solution satisfies the global balance equations and thus the theorem is true.

Gelenbe goes on in [GELE 91a] to show that for feedforward networks (networks without feedback paths in the queueing schematic) that the solutions for λ_i^+ and λ_i^- are unique. Moreover this is true for feedback networks with identical q_i.

Example 1: Tandem Network

In this example there are two queues in tandem (series). Positive customers arrive from outside the network to the first queue with Poisson rate Λ_1, and negative customers arrive from outside the network to the first queue with Poisson rate λ_1. Customers leaving the first queue for the second queue are positive with probability r_{12}^+ and negative with probability r_{12}^-. The exponential service rate of the first queue is μ_1 and that of the second queue is μ_2.

For this network we have

$$\lambda_1^+ = \Lambda_1,$$
$$\lambda_2^+ = q_1\mu_1 r_{12}^+,$$
$$\lambda_1^- = \lambda_1,$$
$$\lambda_2^- = q_1\mu_1 r_{12}^-.$$

From this it is easy to see that

$$q_1 = \frac{\Lambda_1}{\mu_1 + \lambda_1}.$$

Moreover

$$q_2 = \frac{q_1 \mu_1 r_{12}^+}{\mu_2 + q_1 \mu_1 r_{12}^-}.$$

This can be solved as

$$q_2 = \frac{\Lambda_1 \mu_1 r_{12}^+}{\mu_1 \mu_2 + \lambda_1 \mu_2 + \Lambda_1 \mu_1 r_{12}^-}.$$

Note that if only positive customers arrive at the first queue, this reduces to

$$q_1 = \frac{\Lambda_1}{\mu_1},$$

$$q_2 = \frac{\Lambda_1 r_{12}^+}{\mu_2 + \Lambda_1 r_{12}^-}.$$

Example 2: Pair of Feedback Queues

In this example there are two parallel queues. Each has an external positive Poisson arrival stream of rates Λ_1 and Λ_2, respectively. The service rates of the queues are μ_1 and μ_2, respectively. Customers departing queue 1 will enter queue 2 as a positive customer with probability r_{12}^+ will enter queue 2 as a negative customer with probability r_{12}^- and leave the network with proability d_1. Analogous probabilities hold for customers leaving queue 2.

Let us write out the traffic equations for this network. One first has

$$\lambda_1^! = \Lambda_1 + q_2 \mu_2 r_{21}^+$$

Or

$$\lambda_1^+ = \Lambda_1 + \frac{\lambda_2^+ \mu_2 r_{21}^+}{\mu_2 + \lambda_2^-}.$$

Similarly

$$\lambda_2^+ = \Lambda_2 + \frac{\lambda_1^+ \mu_1 r_{12}^+}{\mu_1 + \lambda_1^-}.$$

Also

$$\lambda_1^- = q_2 \mu_2 r_{21}^-$$

Or

$$\lambda_1^- = \frac{\lambda_2^+ \mu_2 r_{21}^-}{\mu_2 + \lambda_2^-}.$$

And similarly

$$\lambda_2^- = \frac{\lambda_1^+ \mu_1 r_{12}^-}{\mu_1 + \lambda_1^-}.$$

These four implicit equations cannot be easily solved in closed form. Instead, they can be solved iteratively. One starts with an (incorrect) guess as to the solution. Then one continually substitutes the current estimate of the solutions for λ_1^+, λ_1^-, λ_2^+, λ_2^- into the right side of the previous four equations to generate a new estimate at the left side. A program to do this will eventually converge to the correct values.

For instance, suppose that

$$\Lambda_1=1, \quad \Lambda_2=3, \quad r_{12}^+=0.5, \quad r_{21}^+=0.4,$$
$$\mu_1=2, \quad \mu_2=5, \quad r_{12}^-=0.3, \quad r_{21}^-=0.3.$$

After iteration these will converge to

$$\lambda_1^+=2.384, \quad \lambda_2^+=3.785,$$
$$\lambda_1^-=1.038, \quad \lambda_2^-=0.4708.$$

And

$$q_1=.7847, \quad q_2=0.6918.$$

Naturally this means that

$$p(n_1,n_2) = 0.2153 \times (0.7847)^{n_1} \times 0.3082 \times (0.6918)^{n_2}$$

3.5.3 The Chao/Pinedo Model

3.5.3.1 Introduction

A somewhat different approach to Gelenbe's was recently proposed by X. Chao and M. Pinedo [CHAO]. In this model each customer in a queue position receives some service effort (possibly zero effort). An arriving negative customer is thought of as a "signal" that causes a customer in a queue position to complete service with a certain probability and leave for either another queue or for a network departure. The resulting traffic equations are linear, as will be discussed below. However, they reduce, in some sense, to Gelenbe's nonlinear equations if one assumes that an arriving negative customer simply cancels out a positive customer. In the following a "G-network" refers to a general network allowing negative customers.

3.5.3.2 The Model

By X. Chao and M. Pinedo

Consider a G-network with J nodes (queues) and I different classes of customers. Regular customers of class i arrive at node j according to a Poisson process of rate Λ_{ji}. When a regular class i customer completes its service at node j, it will go to node k as a regular class h customer with probability $r^+_{ji;kh}$, and as a negative arrival with probability $r^-_{ji;k}$. It leaves the system with probability $r_{ji;0}$ (similar to the d_i of the previous section). These transition probabilities constitute a Markov chain on the state space $\{0,1...J\}\times\{1,...J\}\times\{+,-\}$. Hence

$$\sum_{k,h} r^+_{ji;kh} + \sum_k r^-_{ji;k} + r_{ji;0} = 1, \qquad j=1,...J, i=1,...I. \tag{3.60}$$

In addition to the regular customers, negative arrivals enter node j with rate λ^-_j. Whenever a negative signal arrives at node j, either from outside the system or from another node, it induces a service completion. In other words, when a signal arrives at node j, a regular customer departs at once. It will either leave the system, or it will go to another node. When the customers go to another node, they either go as a regular customer or as a signal, with probabilities defined above. From the point of view of node j, the negative arrival process is again equivalent to a secondary servicing process. It is clear that such a network is a straightforward generalization of the class of networks described in [KELL] by Kelly in 1979.

In this section we assume that node j operates in the manner described in section 2.1 of [CHAO]:

(i)   Each customer requires an amount of service at node j which is exponentially distributed with rate μ_j.

(ii) The total service effort at node j is state dependent; when there are n_j customers at node j the service rate is $\Phi_j(n_j)\mu_j$.

(iii) A proportion $\gamma_j(l,n_j)$ of the total service effort at node j is directed to the customer in position l, $l=1,...n_j$; when its service is completed, customers in positions $l+1,l+2,\cdots n_j$ move into positions $l,l+1,...,n_j-1$, respectively.

(iv) When a regular customer arrives at node j it moves into position $l,l=1,2,...n_j+1$, with probability $\alpha_j(l,n_j+1)$; customers previously in position $l,l+1,...,n_j$ move into positions $l+1,l+2,...,n_j+1$, respectively.

(v) When a signal arrives at node j, it results in the service completion of the customer in position l with probability $\beta_j(l,n_j)$; with its departure, customers in positions $l+1,l+2,\cdots n_j$ move into positions $l,l+1,\cdots n_j-1$, respectively.

As noted in [KELL], this type of service discipline, determined by the $\gamma_j(l,n_j)$, $\alpha_j(l,n_j+1)$,   and $\beta_j(l,n_j)$ is   very   general.   For   example,   if $\Phi_j(n)=\min(K,n)$, and

$$\gamma_j(l,n) = \begin{cases} 1/n, & l=1,2,...n\,;\, n=1,...K\,; \\ 1/K, & l=1,2,...K\,;\, n=K+1,K+2...\,; \\ 0 & otherwise, \end{cases} \qquad (3.61)$$

and

$$\alpha_j(l,n) = \begin{cases} 1, & l=n, \\ 0 & otherwise\,; \end{cases} \qquad (3.62a)$$

$$\beta_j(l,n_j) = \frac{1}{n}, \qquad l=1,2,...n\,;\, n=1,2..., \qquad (3.62b)$$

then node j behaves as a K server queue in which customers are served in the order of their arrivals and a negative arrival selects a customer randomly to possibly leave the network.

If $c_j(l)$ denotes the class of the customer in position l in queue j, then the vector

$$c_j = (c_j(1),c_j(2),...c_j(n_j)) \qquad (3.63)$$

describes the state of node j and

$$C = (c_1,c_2,\cdots c_J) \qquad (3.64)$$

describes the state of the entire system.

If λ_{ji}^+ denotes the total arrival rate of regular customers of class i at node j, and λ_j^- denotes the total arrival rate of negative signals at node j, then these arrival rates satisfy the following traffic equations:

$$\lambda_{ji}^+ = \Lambda_{ji} + \sum_{k=1}^{J}\sum_{h=1}^{I}\lambda_{kh}^+ r_{kh\,;ji}^+, \qquad (3.65)$$

$$\lambda_j^- = \lambda_j + \sum_{k=1}^{J} \sum_{h=1}^{I} \lambda_{kh}^+ r_{kh;j}^-. \tag{3.66}$$

Define

$$a_j = \sum_{i=1}^{I} \lambda_{ji}^+ \tag{3.67}$$

as the total positive arrival rate (work load) at node j. Since the negative arrival process at node j represents a secondary servicing process, the total service rate at node j is the sum of two servicing processes, i.e., $\lambda_j^- + \mu_j \Phi_j(n_j)$ when there are n_j customers present. Our main result in this discussion is contained in the following theorem.

Theorem: The equilibrium distribution of the open network of queues described above is

$$\pi(C) = \prod_{j=1}^{J} \pi_j(c_j) \tag{3.68}$$

where

$$\pi_j(c_j) = b_j \prod_{l=1}^{n_j} \frac{\lambda_{jc_j(l)}^+}{\lambda_j^- + \mu_j \Phi_j(l)} \tag{3.69}$$

and

$$b_j^{-1} = \sum_{n=0}^{\infty} \frac{a_j^n}{\prod_{l=1}^{n} (\lambda_j^- + \mu_j \Phi_j(l))}. \tag{3.70}$$

From this theorem we conclude that the probability that the system is in state n is $\pi(n)$, where $n = (n_1, n_2, ...n_J)$ and n_j is the number of jobs at node j, $j = 1...J$,

$$\pi(n) = \prod_{j=1}^{J} \pi_j(n_j) \tag{3.71}$$

where

$$\pi_j(n_j) = b_j \frac{a_j^{n_j}}{\prod_{l=1}^{n_j}(\lambda_j^- + \mu_j \Phi_j(l))}. \tag{3.72}$$

In particular, if all customers are of one class and each node consists of a single server, the equilibrium distribution of the G-network is

$$\pi(n) = \prod_{j=1}^{J}(1 - \frac{\lambda_j^+}{\lambda_j^- + \mu_j})(\frac{\lambda_j^+}{\lambda_j^- + \mu_j})^{n_j} \tag{3.73}$$

Corollary: When a signal or a customer of any class arrives at queue j, the probability that it finds queue j in state c_j is $\pi_j(c_j)$. The probability that it finds n_j customers in queue j, $j=1,...J$ is given by equation (3.71). This is a generalization of a well-known result for conventional Jackson networks.

Note the difference between our definition of negative arrivals and Gelenbe's definition. In [GELE 91a] a negative arrival kills a customer, i.e., the service of a customer is completed at once and the customer leaves the system, and the probability the customer goes to another node is zero. It is not hard to see that in the special case of a single exponential server at each node the two definitions can be transformed into one another by changing the transition rates. For example, consider a node, say node j, with service rate μ_j, an arrival rate of regular customers λ_j^+, and an arrival rate of negative signals λ_j^-. According to our definition, the regular servicing process as well as the secondary servicing process may result in the service completion of a customer. This customer leaves the system with probability r_{j0} and goes to another node (say node k) with probability r_{jk}. According to Gelenbe's definition a negative arrival kills a customer with probability one (that is, it leaves the system). Let r_{j0}^G denote the probability that a customer upon leaving node j departs the system according to Gelenbe's definition, and let r_{jk}^G denote the probability that a customer upon leaving node j goes to node k, $k=1...J$. We then have

$$r_{j0}(\lambda_j^- + \mu_j) = \lambda_j^- + \mu r_{j0}^G, \tag{3.74a}$$

$$r_{jk}(\lambda_j^- + \mu_j) = \mu_j r_{jk}^G. \tag{3.74b}$$

So

$$r_{j0} = \frac{\lambda_j^- + \mu_j r_{j0}^G}{\lambda_j^- + \mu_j}, \tag{3.75a}$$

$$r_{jk} = \frac{\mu_j r_{jk}^G}{\lambda_j^- + \mu_j}.$$

(3.75b)

For instance, consider the single class, single server G-network model of
Gelenbe [GELE 91a] and substitute (3.74) and (3.75) into (3.65) and (3.66). We
then obtain the following nonlinear traffic equations

$$\lambda_j^+ = \Lambda_j + \sum_{k=1}^{J} \lambda_k^+ \frac{\mu_j}{\lambda_j^- + \mu_j} r_{kj}^+,$$

(3.76)

$$\lambda_j^- = \lambda_j + \sum_{k=1}^{J} \lambda_k^+ \frac{\mu_j}{\lambda_j^- + \mu_j} r_{kj}^-,$$

(3.77)

and the equilibrium distribution is given by (3.73). This is consistent with the
result obtained in [GELE 91a]. The analysis above also explains why the model
of [GELE 91a] has nonlinear traffic equations. If the effect of negative arrivals is
defined as in assumption (v), the traffic equations are linear. If a negative arrival
just kills another customer, then this action is not "linear" in comparison with
the effects of the regular servicing process; this results in a nonlinear traffic equation.

Example 1: Tandem Network

This example is similar to Example 1 of the previous section dealing with
Gelenbe's original model. There is a single class of customers in this example.
However assume now that there are c_i servers at station i, $i=1,2$. Let λ_i^+ be the
total arrival rate of positive customers at station i, and let λ_i^- be the total negative
arrival rate at station i. Then they satisfy the following traffic equations:

$$\lambda_1^+ = \Lambda_1 \quad \lambda_1^- = \lambda_1,$$
$$\lambda_2^+ = \lambda_1^+ r_{12}^+ = \Lambda_1 r_{12}^+,$$
$$\lambda_2^- = \lambda_1^+ r_{12}^- = \Lambda_1 r_{12}^-.$$

Furthermore, $\Phi_j(n_j) = \min[n, c_j]$. We have

$$\pi(n) = \pi_1(n_1)\pi_2(n_2)$$

where

$$\pi_1(n_1) = b_1 \prod_{l=1}^{n_1} \frac{\Lambda_1}{\lambda_1 + \min[l, c_1]\mu_1},$$

$$\pi_2(n_2) = b_2 \prod_{l=1}^{n_2} \frac{\Lambda_1 r_{12}^+}{\Lambda_1 r_{12}^- + \min[l, c_2]\mu_2},$$

and

$$b_1^{-1} = \sum_{n=0}^{\infty} \prod_{l=1}^{n} \frac{\Lambda_1}{\lambda_1 + \min[l, c_1]\mu_1},$$

$$b_2^{-1} = \sum_{n=0}^{\infty} \prod_{l=1}^{n} \frac{\Lambda_1 r_{12}^+}{\Lambda_1 r_{12}^- + \min[l, c_2]\mu_2}.$$

In particular, when $c_1 = c_2 = 1$,

$$\pi_1(n_1) = (1 - \frac{\Lambda_1}{\lambda_1 + \mu_1})(\frac{\Lambda_1}{\lambda_1 + \mu_1})^{n_1},$$

$$\pi_2(n_2) = (1 - \frac{\Lambda_1 r_{12}^+}{\Lambda_1 r_{12}^- + \mu_2})(\frac{\Lambda_1 r_{12}^+}{\Lambda_1 r_{12}^- + \mu_2})^{n_2}.$$

Example 2: Pair of Feedback Queues

This example is similar to Example 2 of the previous section dealing with Gelenbe's original model. However, now there are c_i servers at node i. Let λ_i^+ be the total arrival rate of positive customers at node i, $i=1,2$. Note that there is a single class of customers in this example. Then

$$\lambda_1^+ = \Lambda_1 + \lambda_2^+ r_{21}^+,$$
$$\lambda_2^+ = \Lambda_2 + \lambda_1^+ r_{12}^+,$$
$$\lambda_1^- = \lambda_2^+ r_{21}^-,$$
$$\lambda_2^- = \lambda_1^+ r_{12}^-.$$

Solving for λ_1^+ and λ_2^+,

$$\lambda_1^+ = \frac{\Lambda_1 + \Lambda_2 r_{21}^+}{1 - r_{12}^+ r_{21}^+},$$

$$\lambda_2^+ = \frac{\Lambda_2 + \Lambda_1 r_{12}^+}{1 - r_{21}^+ r_{12}^+},$$

hence

$$\lambda_1^- = \frac{\Lambda_1 + \Lambda_2 r_{21}^+}{1 - r_{12}^+ r_{21}^+} r_{21}^-,$$

$$\lambda_2^- = \frac{\Lambda_2 + \Lambda_1 r_{12}^+}{1 - r_{21}^+ r_{12}^+} r_{12}^-.$$

Therefore, the stationary distribution is

$$\pi(n) = \pi_1(n_1)\pi_2(n_2)$$

$$\pi_1(n_1) = b_1 \prod_{l=1}^{n_1} \frac{\lambda_1^+}{\lambda_1^- + \min[l,c_1]\mu_1},$$

$$\pi_2(n_2) = b_2 \prod_{l=1}^{n_2} \frac{\lambda_2^+}{\lambda_2^- + \min[l,c_2]\mu_2}.$$

In particular, when $c_1 = c_2 = 1$,

$$\pi_1(n_1) = \left[1 - \frac{\dfrac{\Lambda_1 + \Lambda_2 r_{21}^+}{1 - r_{12}^+ r_{21}^+}}{\dfrac{\Lambda_1 + \Lambda_2 r_{21}^+}{1 - r_{12}^+ r_{21}^+} r_{21}^- + \mu_1}\right] \left[\frac{\dfrac{\Lambda_1 + \Lambda_2 r_{21}^+}{1 - r_{12}^+ r_{21}^+}}{\dfrac{\Lambda_1 + \Lambda_2 r_{21}^+}{1 - r_{12}^+ r_{21}^+} r_{21}^- + \mu_1}\right]^{n_1},$$

$$\pi_2(n_2) = \left[1 - \frac{\dfrac{\Lambda_2 + \Lambda_1 r_{12}^+}{1 - r_{21}^+ r_{12}^+}}{\dfrac{\Lambda_2 + \Lambda_1 r_{12}^+}{1 - r_{21}^+ r_{12}^+} r_{12}^- + \mu_2}\right] \left[\frac{\dfrac{\Lambda_2 + \Lambda_1 r_{12}^+}{1 - r_{21}^+ r_{12}^+}}{\dfrac{\Lambda_2 + \Lambda_1 r_{12}^+}{1 - r_{21}^+ r_{12}^+} r_{12}^- + \mu_2}\right]^{n_2}.$$

To Look Further

At about the time that the BCMP paper was published a number of related works also appeared. These were written by F. P. Kelly [KELL 75] [KELL 76] and by M. Reiser and H. Kobayashi [REIS 75]. The relation between these papers is discussed in [CONW 89b]. Product form solutions for multiserver service centers with concurrent classes of customers appear in [CROS]. Work on product form solutions for batch arrivals, batch servicing, and batch routings appears in [HEND 90] and [BOUC].

A property of certain models is "insensitivity". That is, analytical expressions for quantities such as the state probabilities may depend only on the means of relevant distributions (i.e., arrival and service distributions) rather than on the distributions themselves. An example of this involving blocking probabilities in circuit switched networks appears in [BURM]. The question of insensitivity is also examined in [SURI 83]. General formulas to determine the impact of changing model parameters appear in [LIU]. Insensitivity in discrete time queues with a "moving" server is examined in [HEND 92]. A moving server is one that moves from queue to queue. The concept arises in certain communication network applications.

The consistency graph has been used by R. Lee Hamilton and E. J. Coyle to synthesize multihop radio networks with product form solution [HAMI 86a, 86b, 89]. This was previously not possible.

A recent generalized recursive approach to solving for the equilibrium state probabilities appears in [YANG]. This work also provides references to related work.

Problems

3.1 The joint state equilibrium probabilities for an open two queue network are given in the following table. Here at most two customers may enter the network at one time

i	j	$p(i,j)$
0	0	0.05
0	1	0.15
1	0	0.15
1	1	0.25
0	2	0.20
2	0	0.20

Here i is the number of customers in Q1 and j is the number of customers in Q2.

a) For Q1 calculate the marginal state probabilities $p(0)$, $p(1)$ and $p(2)$.

b) If the service rate of Q1 is μ, calculate the mean throughput and utilization.

3.2 For the queueing network of Figure 3.2 solve the traffic equations for the average throughput θ_1 and θ_2.

3.3 Consider a two-queue Markovian tandem network with feedback. Let the first queue (nearest the network input) be Q1 and the second queue be Q2. Let $r_{12}=1-\alpha$, $r_{11}=\alpha$, $r_{22}=\beta$, $r_{21}=\gamma$, and let the probability that a customer leaving Q2 departs the system be δ. Naturally

$$\beta + \gamma + \delta = 1.0$$

a) Draw and label the queueing network.

b) Solve the traffic equations for this network to determine θ_1 and θ_2.

3.4 For an M queue tandem Markovian network with service rates $\mu_1=\mu_2=\mu_3=\mu$ and arrival rate λ calculate the mean delay a customer experiences in passing through the network.

3.5 Consider a single queue where $r_{11}=\alpha$ so that a customer leaving the queue either joins the queue again with probability α or leaves the system with probability $1-\alpha$. The arrival rate is λ, and the service rate is μ. Calculate the mean delay a customer experiences in passing through this system. Hint: Make use of Little's Law, the traffic equations, and a useful summation from the appendix.

3.6 Find the mean delay for a customer to move through the system of Figure 3.2. Hint: Make use of the result of the previous problem.

3.7 [KOBA] Consider a state-dependent open queueing network as in section 3.2.2. The service rate of the ith queue, $\mu_i(n_i)$, is a function of the number of customers in the ith queue, n_i. The arrival rate is as it is in 3.2.2 (state independent). Write out the global balance equation for state $\underline{n}$. Show that the product form solution is

$$p(\underline{n}) = p(0) \prod_{i=1}^{M} \frac{\theta_i^{n_i}}{\prod_{n=1}^{n_i} \mu_i(n)}.$$

3.8 [KOBA] Consider a state-dependent closed queueing network as in section 3.2.4. The service rate of the ith queue, $\mu_i(n_i)$, is a function of the number of customers in the ith queue, n_i. Write out the global balance equation for state $\underline{n}$. Show that the product form solution is

$$p(\underline{n}) = \frac{1}{G(N)} \prod_{i=1}^{M} \frac{\theta_i^{n_i}}{\prod_{n=1}^{n_i} \mu_i(n)}.$$

3.9 Plot the marginal queue length distributions for Q1 and Q2 in Figure 3.2. For this plot let $\lambda=1.0$ and $\mu_1=\mu_2=4.0$.

3.10 Find the constraints on λ in Figure 3.2 for the two infinite buffers queues to be stable (arrival rate less than service rate).

3.11 a) Consider a three queue cyclic (closed) Markovian queueing network as in Figure 3.3. Let $N=3$. Compute numerically $G(N)$ if $\mu_1=1.0$, $\mu_2=2.0$, and $\mu_3=3.0$. See equations (3.35) and (3.36).

b) Compute the joint state probability $p(1,1)$.

c) Compute the marginal state probabilities $p_1(1)$ and $p_2(1)$.

d) Is $p(1,1)=p_1(1)p_2(1)$? Why or why not?

3.12 Verify that the local balance pairings of Figure 3.5 follow the rule in section 3.2.3 on Local Balance.

3.13 Consider a three-queue Markovian cyclic network with two customers. The state transition diagram appears in Figure 3.4. By identifying and solving the appropriate local balance equations along a path back to p_{00}, solve for p_{12}.

3.14 For an open two-queue Markovian series network, as in Figure 3.10, write the local balance equations involving the (n_1, n_2) state. Verify that the product form solution

$$p_{n_1 n_2} = \left[\frac{\lambda}{\mu_1} \right]^{n_1} \left[\frac{\lambda}{\mu_2} \right]^{n_2} p_{00}$$

satisfies these equations. Do the same for the global balance equations for the (n_1, n_2) state.

3.15 Consider an open network of two finite buffer Markovian queues in series (tandem). Suppose that when either queue is full, the arrival and service process at the other queue will be stopped until the full queue releases a customer. Draw the state transition diagram. Verify that every transition belongs to a local balance pair and that the product form solution exists.

3.16 Consider an arbitrarily configured network of Markovian finite buffer queues with skipping. That is, when a queue is full, arriving customers skip over it. Does this network have a product form solution? Hint: Develop an argument based on state-dependent service rates.

3.17 For Figure 3.14 we have

$$\alpha_l \lambda p_a = \sum_{j=1}^{n} \beta_j \mu_l p_l = \mu_l p_l, \qquad l = 1, 2, 3, \ldots m,$$

$$\mu_l p_l = \beta_{l-m} \sum_{i=1}^{m} \mu_i p_i, \qquad l = m+1, m+2, \ldots m+n.$$

Solve for p_l, $l = 1, 2, 3 \ldots m, m+1, \ldots m+n$. Explain how you arrive at your results.

3.18 Consider the FIFO queueing system associated with the state transition diagram of Figure 3.16. Using the consistency condition, prove that one can not have distinct service rates for the two different classes of customers, i.e., μ_1 and μ_2, and still maintain consistency. Be sure to generate the consistency graph directly from the proposed local balance equations. Note: It is true in general for FIFO networks that the service rate must not be class dependent [BASK 75].

3.19 Consider a packet multiplexer. There are two input lines, one for class 1 packets and one for class 2 packets. There are also separate buffers for class 1 and class 2 packets. Packets from both buffers enter a joint service facility leading to a single output line. The service facility contains two servers, one for each class of traffic.

There is a maximum of three class 1 and three class 2 packets in the multiplexer (1 in the server, 2 in the buffer). If there are only packets of a single class present, that class's server operates at rate μ (has full use of the output channel). If there are packets of both classes present, each server operates at rate $\mu/2$ (has half the output capacity).

a) Draw a diagram of the multiplexer.

b) Draw and label the state transition diagram for up to three packets of each class. Let i, the total number of class 1 packets in the multiplexer, be the horizontal coordinate. Similarly, let j, the total number of class two packets in the multiplexer, be the vertical coordinate. Here "total number" refers to the packets in both buffer and server.

c) Is there local balance (and by implication a product form solution) for this system? *Assume* local balance holds between horizontally and vertically adjacent states. Let $(0,0)$ be the reference state. Can you find a state for which you can recursively compute its state probability by using the local balance equations along two different paths from the reference state and come up with two different solutions? If you can, specify the state, the paths and the different solutions. State also the implication of this.

d) What would happen if each server has fixed service rate $\mu/2$, independent of multiplexer state (sort of like time division multiplexing).

3.20 Consider a queue where the service time is an Erlang distributed random variable consisting of N stages of independent exponential servers. Draw the state transition diagram in two-dimensional form. That is, let the horizontal axis correspond to the number of customers in the queue (not including the one in service), and let the vertical axis correspond to the stage of service in which the customer being serviced is. Let $(0,0)$ correspond to an empty queueing system. Does this state transition diagram have Type A or Type B structure? Write a set of recursive equations for the equilibrium state probabilities.

3.21 Consider a queue that processes two classes of customers according to a priority discipline. That is, class 1 customers are always serviced before class 2 customers. If a class 2 customer is being serviced and a class 1 customer arrives, the class 1 customer immediately receives service and the class 2 customer will have to start service all over again.

Draw the state transition diagram. Let the horizontal axis correspond to the number of class 1 customers in the queueing system and let vertical axis correspond to the number of class 2 customers in the queueing system. Is the structure of the state transition diagram Type A or Type B? Identify the subsets of states that can be solved for individually.

3.22 Consider the state transition diagram of an ISDN protocol appearing in Figure 3.26. Find a *different* recursive equation from the one used in the book for the lower row equilibrium state probabilities. Make use of an equation like that described in Figure 3.22.

3.23 Consider the tandem queueing system of Figure 3.24 and its associated recursion in the text (equation (3.49)). This recursion was arrived at by drawing vertical and diagonal boundaries through the diagram and equating the flow of probability flux moving in each direction across the boundaries. Arrive at a different set of recursions for this system by utilizing equations like those illustrated in Figure 3.22.

3.24 A communication node has a transmission queue, Q1, and a local traffic source buffer, Q2. Normally packet arrivals to the node enter Q1. If Q1 empties, 2 packets at a time may enter Q1 from Q2 (which is always full) at rate γ. That is, after an exponentially distributed delay of $1/\gamma$, two packets are delivered from Q2 to Q1. Q1 will then accept arrivals from the node input, at rate λ, until the next time it empties. Packets arriving to the node while Q1 is waiting for Q2 arrivals are lost.

a) Draw and label the state transition diagram for Q1.

b) Express $p(3), p(2), p(1)$ as a function of $p(0)$ and the queueing parameters.

c) On average how many packets are lost while Q1 is waiting for arrivals from Q2?

3.25 Consider two M/M/1 queues in parallel, Q1 and Q2. The network arrival rate is λ, and the individual queue arrival rates are $p_1\lambda$ and $p_2\lambda$, respectively. Naturally $p_1+p_2=1$. The service rates are μ_1 and μ_2, respectively.

a) Write an expression for the average delay of packets in the system. Average delay for the whole system must be averaged over both queues.

b) Find the optimal choice for p_1 for minimal avg. queue length under light load:

$$\frac{\lambda}{\mu_1}\ll 1 \qquad \frac{\lambda}{\mu_2}\ll 1.$$

3.26 Consider a single queue operating under the Gelenbe model of negative customers discussed in the text. Let positive customers enter the queue at rate Λ. Let customers leaving the queue reenter the queue as a negative customer with probability r_{11}^-. Otherwise, customers depart the queue with probability d_1. Naturally

$$r_{11}^- + d_1 = 1.0.$$

a) Draw the queueing system.

b) Draw and label the state transition diagram.

c) Find an expression for the equilibrium state probabilities.

d) Verify the global balance equations at states 0, 1 and 2, if $\Lambda=3.0$, $\mu=5.0$, $r_{11}^-=0.3$, and $d_1=0.7$.

3.27 Consider a two-queue tandem Markovian network with negative arrivals following a Gelenbe type model. Specifically, positive customers arrive from outside the network for Q1 with Poisson rate Λ_1. Negative customers arrive from outside the network for Q1 with rate λ_1 and for Q2 with rate λ_2. Customers leaving the first queue for the second queue are positive with probability r_{12}^+ and are negative with probability r_{12}^-. The exponential service rate of the first queue is μ_1 and that of the second queue is μ_2.

a) Draw the queueing system diagram.

b) Find the traffic equations, q_1 and q_2, and the product form solution.

c) Find q_1 and q_2 if there are no negative customers. In other words, show that the model reduces to a standard Markovian product form network.

3.28 Consider two tandem queues following a Gelenbe type model. Specifically, positive arrivals to the first queue, Q1, follow a Poisson process with rate Λ_1. Only positive customers leave Q1 for Q2. Customers leaving Q2 for Q1 may be negative with probability r_{21}^-. Customers depart Q2 and leave the system with probability d_2. Naturally

$$r_{21}^- + d_2 = 1.0.$$

The service rate of Q1 is μ_1, and the service rate of Q2 is μ_2.

a) Draw the queueing system.

b) Find the traffic equations and closed form solutions for q_1 and q_2.

c) As has been mentioned in the text for more complex models iteration can be used to solve for q_1 and q_2. Try this for this model. That is, make use of the implicit equations for q_1 as a function of q_2 and for q_2 as a function of q_1. Pick initial guesses for q_1 and q_2 and substitute these into the equations to provide new estimates of q_1 and q_2. Continue the substitutions until the results converge. Compare with the closed form solution of part (b). For this problem let $\Lambda_1=2.0$, $\mu_1=2.5$, $\mu_2=3.0$, and $r_{21}^-=0.3$.

3.29 Consider a single queue operating under the Chao/Pinedo model of negative customers discussed in the text. Let positive customers enter the queue at rate Λ. Let customers leaving the queue reenter the queue as a negative customer with probability $r_{1;1}^-$. Otherwise, customers depart the queue with probability $r_{1;0}$. There is a single customer class. Naturally

$$r_{1;1}^- + r_{1;0} = 1.0.$$

a) Draw the queueing system schematic.

b) Find an expression for the equilibrium state probabilities.

3.30 Consider a two-queue tandem Markovian network with negative arrivals following a Chao/Pinedo type model. Specifically, positive customers arrive from outside the network for Q1 with Poisson rate Λ_1. Negative customers arrive from outside the network for Q1 with rate λ_1 and for Q2 with rate λ_2. Customers leaving the first queue for the second queue are positive with probability $r_{1;2}^+$ and are negative with probability $r_{1;2}^-$. The exponential service rate of the first queue is μ_1 and that of the second queue is μ_2. There is a single customer class.

a) Draw the queueing system schematic.

b) Find the traffic equations and an expression for the equilibrium state probabilities.

c) If there are no negative customers, show that the result of (b) reduces to that for a purely positive customer network.

3.31 Consider two tandem queues following a Chao/Pinedo type model. Specifically, positive arrivals to the first queue, Q1, follow a Poisson process with rate Λ_1. Only positive customers leave Q1 for Q2. Customers leaving Q2 for Q1 may be negative with probability $r_{2;1}^-$. Customers depart Q2 and leave the system with probability $r_{2;0}$. There is a single customer class. Naturally:

$$r_{2;1}^- + r_{2;0} = 1.0.$$

The service rate of Q1 is μ_1, and the service rate of Q2 is μ_2.

a) Draw the queueing system schematic.

b) Find the traffic equations and expressions for the equilibrium state probabilities.

Chapter 4: Numerical Solution

of Models

4.1 Introduction

The ultimate goal of much performance analysis is to generate numerical performance results for a particular system under specific conditions. These can take the form of tables or a set of performance curves. There are a variety of numerical techniques available, ranging from the brute force solution of the state equations to clever algorithms to Monte Carlo type simulation. With all these techniques the *cost* of solution in terms of solution time and computer memory requirements are important considerations. The cost of solving a system of even moderate complexity may be prohibitively expensive for a particular technique and computer installation. There is a tradeoff between our ability to model and our ability to solve such models. There is thus an advantage to simplified models that capture the most important aspects of the system in question.

The algorithms that are available deal with closed product form networks. The algorithms generally compute performance measures for the network with N customers based on previous results for N-1 customers. One approach, due independently to J. Buzen [BUZE] and to M. Reiser and H. Kobayashi [REIS 73], is called the *convolution algorithm*. It is basically a clever recursion for the normalization constants for increasing populations of customers. What makes this useful is that a variety of performance measures can be written as functions of these normalization constants.

Another approach, due to M. Reiser and S. Lavenberg [REIS 80,81,82], is *mean value analysis*. Just as the convolution algorithm avoids the need to calculate equilibrium state probabilities, mean value analysis avoids even the calculation of the normalization constants. It is based on a number of recursions arising from fundamental concepts of queueing theory.

A very different approach is the PANACEA technique of J. McKenna, D. Mitra and K. G. Ramakrishnan for large Markovian queueing networks. Here the problem of calculating the normalization constant is transformed into a problem in integration whose solution takes the form of asymptotic power series.

Of course, many practical Markovian models do not have a product form solution. Certain limited models are known to have recursive solutions for the equilibrium state probabilities. If this is not so one can, in theory, solve the global balance equations in much the same way one solves Kirchoff's nodal equations for DC electric circuits. The difficulty is that for even for relatively simple systems the state space can be enormous. In such cases a Monte Carlo simulation may provide results of sufficient accuracy for most purposes.

To simplify the solution of large product form queueing networks one can reduce sub-networks into equivalent queues. This idea was first proposed by K. M. Chandy, U. Herzog and L. Woo. The concept is analogous to that of an equivalent circuit in electric circuit theory (Norton's theorem).

In this chapter we will describe these techniques.

4.2 Closed Queueing Networks: Convolution Algorithm

4.2.1 Lost in the State Space

For a closed, product form, queueing network we know that

$$p(\underline{n}) = \frac{1}{G(N)} \, f_1(n_1)f_2(n_2).....f_M(n_M) \tag{4.1}$$

where

State Independent Service Rate:

$$f_i(n_i) = \left[\frac{\theta_i}{\mu_i}\right]^{n_i}, \tag{4.2}$$

State Dependent Service Rate:

$$f_i(n_i) = \frac{\theta_i^{n_i}}{\prod_{k=1}^{n_i}\mu_i(k)}, \tag{4.3}$$

and the θ_i's are the (nonunique) solutions to the traffic equations:

$$\theta_i = \sum_{k=1}^{M}\theta_k r_{ki}, \qquad i=1,2,...M. \tag{4.4}$$

The difficulty in evaluating $p(\underline{n})$ lies in the determination of $G(N)$. It does not factor out as $p(0)$ does for open networks. We will write a direct expression for it based on the fact that the sum of the probabilities of all states must be one. Use will be made of the notation from Buzen's 1973 paper [BUZE]. The legitimate state space for N customers and M queues is

$$S(N,M) = \qquad (4.5)$$

$$\left\{ \underline{n} = (n_1, n_2 \cdots n_M) \mid \sum_{i=1}^{M} n_i = N \text{ and } n_i \geq 0, \ i = 1, 2, \ldots M \right\}.$$

Then one can write [BRUE 80]:

$$\sum_{\underline{n} \, \varepsilon \, S(N,M)} p(\underline{n}) = 1 = \frac{1}{G(N)} \sum_{\underline{n} \, \varepsilon \, S(N,M)} \prod_{i=1}^{M} f_i(n_i) \qquad (4.6)$$

or

$$\boxed{G(N) = \sum_{\underline{n} \, \varepsilon \, S(N,M)} \prod_{i=1}^{M} f_i(n_i). \qquad (4.7)}$$

The constraint that the sum of the customers must be equal to N defines a hyperplane in the much larger state space of an open network. Still, the remaining state space can be enormous. Consider a single-class network. There are N indistinguishable objects (customers) that must be placed in M queues. There are altogether

$$\begin{bmatrix} M+N-1 \\ N \end{bmatrix} \ states.$$

If, for instance, there are ten queues and twenty-five customers, this is

$$\frac{34!}{25! \, 9!} = 52{,}451{,}256 \ states \ or \ terms.$$

Each term requires the multiplication of ten constants so that the total number of multiplications is over half a billion! The associated floating point summation of terms would also require special care [ROBE 88a].

Buzen's 1973 paper showed that the normalization constant can be evaluated without recourse to the previous direct sum. This technique is called the convolution algorithm. Moreover, many important performance measures are simple functions of such normalization constants. This avoids the need to deal directly with an enormous state space. The algorithm will now be discussed.

4.2.2 Convolution Algorithm: Single Customer Class

For the case of state dependent servers we will follow the treatment in Bruell's Ph.D. dissertation [BRUE 78] (later condensed into the very readable [BRUE 80]). Let $g(n,m)$ be the normalization constant for n customers and the first m of the M queues. The reason that small letters are now used is that a series of normalization constants for increasing population sizes will be calculated up to and including $G(N)$. By definition;

$$g(n,m) = \sum_{\underline{n} \, \varepsilon \, S(n,m)} \prod_{i=1}^{m} f_i(n_i). \tag{4.8}$$

This can be expanded as

$$g(n,m) = \sum_{k=0}^{n} \left[\sum_{\substack{\underline{n} \, \varepsilon \, S(n,m) \\ n_m = k}} \prod_{i=1}^{m} f_i(n_i) \right]. \tag{4.9}$$

Factoring one has

$$g(n,m) = \sum_{k=0}^{n} f_m(k) \left[\sum_{\underline{n} \varepsilon S(n-k,m-1)} \prod_{i=1}^{m-1} f_i(n_i) \right]. \tag{4.10}$$

This expression in brackets can be identified, so

$$g(n,m) = \sum_{k=0}^{n} f_m(k) g(n-k,m-1). \tag{4.11}$$

This convolution-like expression accounts for the name "convolution" algorithm. From the direct definition of $g(n,m)$ the initial conditions for the algorithm are

$$g(n, 1) = f_1(n), \qquad n = 0, 1, \ldots N,$$
$$g(0, m) = 1, \qquad m = 1, 2, \ldots M.$$

Table 4.1 illustrates the operation of the algorithm. Entries can be thought of as being calculated in this tabular form although a little thought will show that only about a column's worth or $N+1$ memory locations are needed ([BRUE 80]). Entries are calculated from top to bottom, left to right. Each entry is a function of the entries in the column to its left that are at its level or above.

The payoff from the use of this algorithm comes in the last column. A comparison of the direct expression for $G(N)$ and $g(n,m)$ shows that the entries of the last column, $g(n,M)$, equals $G(n)$. Thus the normalization constants for increasing numbers of customers are calculated as the algorithm proceeds.

For the case where the queues are not state dependent we will follow Buzen's original paper and [SCHW 87]. From before

$$g(n,m) = \sum_{\underline{n} \, \varepsilon \, S(n,m)} \prod_{i=1}^{m} f_i(n_i) \tag{4.12}$$

where

$$f_i(n_i) = \left[\frac{\theta_i}{\mu_i} \right]^{n_i}. \tag{4.13}$$

This expression for $g(n,m)$ can be decomposed as

$$g(n,m) = \tag{4.14}$$

$$\sum_{\substack{\underline{n} \, \varepsilon \, S(n,m) \\ n_m = 0}} \prod_{i=1}^{m-1} f_i(n_i) + \sum_{\substack{\underline{n} \, \varepsilon \, S(n,m) \\ n_m > 0}} \prod_{i=1}^{m} f_i(n_i).$$

The first term here is just the normalization constant for the network without the mth queue or simply $g(n,m-1)$. For the second term if one factors out $f_m(1) = \dfrac{\theta_m}{\mu_m}$ it can be shown that what remains is the normalization constant for m queues with n-1 customers or simply $g(n-1,m)$. Thus

$$\boxed{g(n,m) = g(n,m-1) + f_m(1)g(n-1,m). \qquad (4.15)}$$

This recursion is simpler than the one for the state-dependent case. The computation can also be illustrated in a tabular form. From table 4.2 it can be seen that each entry is a function of the entry to the left and the entry above. Again, only $N+1$ memory locations are actually needed to implement the algorithm.

What can be said of the computational complexity in time of this algorithm? It involves approximately MN additions and MN multiplications. For the state-dependent recursion of Table 4.1 there are approximately $N^2/2$ additions and $N^2/2$ multiplications per column. The entire table thus requires approximately $MN^2/2$ additions and $MN^2/2$ multiplications. For the previous example of ten

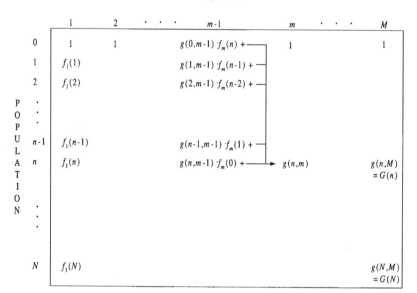

Table 4.1

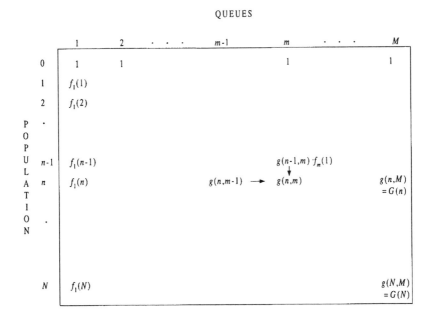

Table 4.2

queues and twenty-five customers this results in 3,125 multiplications rather than the half a billion mentioned before!

4.2.3 Performance Measures from Normalization Constants

Performance measures can be expressed as functions of the equilibrium state probabilities. Unfortunately this approach can lead to the same problems of excessive computation that were encountered in the calculation of the normalization constant. Fortunately, a number of important performance measures can be computed as functions of the various normalization constants, which are the products of the convolution algorithm. In this section it will be shown how this can be done.

Marginal Distribution of Queue Length

Following Buzen's 1973 paper, let

$$p(n_i = n) = \sum_{\substack{n \, \varepsilon \, S(N,M) \\ n_i = n}} p(\underline{n}). \tag{4.16}$$

Consider the state-independent server first. To arrive at the marginal distribution it will be easier to first calculate

$$p(n_i \geq n) = \sum_{\substack{n \, \varepsilon \, S(N,M) \\ n_i \geq n}} p(\underline{n}) \tag{4.17}$$

or

$$p(n_i \geq n) = \sum_{\substack{n \, \varepsilon \, S(N,M) \\ n_i \geq n}} \frac{1}{G(N)} \prod_{j=1}^{M} f_j(n_j) \tag{4.18}$$

where from before

$$f_j(n_j) = \left[\frac{\theta_j}{\mu_j} \right]^{n_j} \tag{4.19}$$

and the θ_j's satisfy the traffic equations:

$$\theta_j = \sum_{k=1}^{M} \theta_k r_{kj}, \qquad j = 1, 2, \ldots M. \tag{4.20}$$

Now every term has a factor $\left[\dfrac{\theta_i}{\mu_i}\right]^n$, and this can be factored out:

$$p\,(n_i \geq n) = \left[\frac{\theta_i}{\mu_i}\right]^n \frac{1}{G(N)} \sum_{n \,\varepsilon\, S(N-n,M)} \prod_{j=1}^{M} f_j(n_j). \tag{4.21}$$

This summation is simply the normalization constant with $N\text{-}n$ customers. Thus

$$p\,(n_i \geq n) = \left[\frac{\theta_i}{\mu_i}\right]^n \frac{G(N-n)}{G(N)}. \tag{4.22}$$

Now the key to calculating the marginal distribution is to recognize that

$$p\,(n_i = n) = p\,(n_i \geq n) - p(n_i \geq n+1), \tag{4.23}$$

so

$$p\,(n_i = n) = \left[\frac{\theta_i}{\mu_i}\right]^n \frac{1}{G(N)} \left[G(N-n) - \left[\frac{\theta_i}{\mu_i}\right] G(N-n-1) \right]. \tag{4.24}$$

Once $G(1)$, $G(2)$, ... $G(N)$ are computed the marginal distributions can be found. Note that $p\,(n_i = n)$ depends on the nonunique θ_i. Thus the (also nonunique) G's work in conjunction with θ_i in the previous equation to produce the actual value of $p(n_i = n)$.

Perhaps the most useful statistic that can be derived from the distribution is its mean. A little thought will show that

$$\bar{n}_i(N) = \sum_{n=1}^{N} np(n_i = n) = \sum_{n=1}^{N} p(n_i \geq n), \tag{4.25}$$

so substituting for $p\,(n_i \geq n)$

$$\bar{n}_i(N) = \sum_{n=1}^{N} \left[\frac{\theta_i}{\mu_i}\right]^n \frac{G(N-n)}{G(N)}. \tag{4.26}$$

What of the case of state-dependent servers? One can start with

$$p(n_M = n) = \sum_{\substack{\underline{n} \in S(N,M) \\ n_M = n}} p(\underline{n}),$$ (4.27)

$$p(n_M = n) = \sum_{\substack{\underline{n} \in S(N,M) \\ n_M = n}} \frac{1}{G(N)} \prod_{j=1}^{M} f_j(n_j)$$ (4.28)

where from before

$$f_j(n_j) = \frac{\theta_j^{n_j}}{\prod\limits_{n=1}^{n_j} \mu_j(n)}$$ (4.29)

and the θ_j's are the solutions of the traffic equations. Each term in the above summation has a term $f_M(n)$, which can be factored out. This leaves

$$p(n_M = n) = \frac{f_M(n)}{G(N)} \sum_{\substack{\underline{n} \in S(N,M) \\ n_M = n}} \prod_{\substack{j=1 \\ j \neq M}}^{M} f_j(n_j)$$ (4.30)

or

$$\boxed{p(n_M = n) = \frac{f_M(n)}{G(N)} g(N-n, M-\{M\});}$$ (4.31)

$$g(n, m-\{i\}) = \sum_{\substack{\underline{n} \in S(N,M) \\ n_i = N-n}} \prod_{\substack{j=1 \\ j \neq i}}^{m} f_j(n_j)$$ (4.32)

is an auxiliary function defined in [BRUE 80]. The notation $-\{i\}$ indicates that the ith queue has been removed. The function $g(N-n, M-\{M\})$ can be calculated using the convolution algorithm where the queue of interest, M, is associated with the last column. This is possible as $g(n, M - \{M\}) = g(n, M-1)$. In other words the normalization constants that are generated after processing the first M-1 columns are the same as those generated for the network of all M queues with the Mth queue removed. If one wants the marginal distribution for a different queue, the queue must be reordered so that it is associated with the last column [BUZE

73]. Bruell and Balbo provide an alternative to this expensive procedure [BRUE 80]. Since

$$\sum_{n=0}^{N} p(n_i = n) = 1 \tag{4.33}$$

or

$$\sum_{n=0}^{N} \frac{f_i(n)}{G(N)} g(N-n, M-\{i\}) = 1. \tag{4.34}$$

This can be rewritten as

$$\sum_{n=0}^{N} f_i(n) g(N-n, M-\{i\}) = G(N) \tag{4.35}$$

or

$$g(N, M - \{i\}) = G(N) - \sum_{n=1}^{N} f_i(n) g(N-n, M-\{i\}) \quad (4.36)$$

where the initial condition is

$$g(0, M - \{i\}) = 1.0 = G(0).$$

This equation can be used to recursively compute $g(N, M - \{i\})$. One then has

$$p(n_i = n) = \frac{f_i(n)}{G(N)} g(N-n, M-\{i\}) \tag{4.37}$$

where $g(n, m - \{i\})$ is defined as before. This calculation can be carried out for the ith queue without having to order the queues.

What of the expected number of customers for this state-dependent case? In general it must be calculated from its definition [BRUE 80]:

$$\bar{n}_i(N) = \sum_{n=1}^{N} np(n_i = n). \qquad (4.38)$$

On the other hand, the computation of the Mth queue turns out to be similar to that of the normalization constant, $G(N)$. One starts out with the definition of the expected number of customers

$$\bar{n}_M(N) = \sum_{n=1}^{N} np(n_M = n) \qquad (4.39)$$

and substitutes in the previous expression for the marginal probability:

$$\bar{n}_M(N) = \sum_{n=1}^{N} n \frac{f_M(n)}{G(N)} g(N-n, M-\{M\}), \qquad (4.40)$$

or since $g(n, M-\{M\}) = g(n, M-1)$ one has

$$\bar{n}_M(N) = \frac{1}{G(N)} \sum_{n=1}^{N} n f_M(n) g(N-n, M-1). \qquad (4.41)$$

This equation is very similar to the previous convolution expression for $g(n,m)$. Bruell and Balbo [BRUE 80] suggest performing the calculation of these two quantities together to minimize the amount of calculation.

Mean Throughput

The mean throughput for state-dependent servers is defined from first principles to be

$$\bar{Y}_i(N) = \sum_{n=1}^{N} p(n_i = n) \mu_i(n) \qquad (4.42)$$

where $\mu_i(n)$ is the state-dependent service rate of the ith queue. The dependence of $\bar{Y}_i(N)$ on the number of customers in the system, N, is explicitly shown. Using the previous expression for $p(n_i = n)$,

$$\bar{Y}_i(N) = \sum_{n=1}^{N} \frac{f_i(n)}{G(N)} g(N-n, M-\{i\}) \mu_i(n). \qquad (4.43)$$

Following [BRUE 80] from the definition of $f_i(n)$ one has

$$\overline{Y}_i(N) = \sum_{n=1}^{N} \frac{\theta_i}{\mu_i(n)} \frac{f_i(n-1)}{G(N)} g(N-n, M-\{i\}) \mu_i(n).$$ (4.44)

Now this can simplified and rearranged to

$$\overline{Y}_i(N) = \frac{\theta_i}{G(N)} \sum_{n=1}^{N} f_i(n-1) g(N-n, M-\{i\})$$ (4.45)

or with a change of variables

$$\overline{Y}_i(N) = \frac{\theta_i}{G(N)} \sum_{n=0}^{N-1} f_i(n) g(N-n-1, M-\{i\})$$ (4.46)

This last summation can be identified as being simply $G(N-1)$ so

$$\boxed{\overline{Y}_i(N) = \theta_i \frac{G(N-1)}{G(N)}.\qquad (4.47)}$$

Again, the nonunique θ_i and the nonunique normalization constants combine to form the actual mean throughput.

Utilization

For a load dependent queue one can clearly write

$$U_i(N) = 1 - p(n_i = 0).$$ (4.48)

However, we can find $p(n_i = 0)$ from the previous expression for the marginal distribution [BRUE 80]:

$$\boxed{U_i(N) = 1 - \frac{g(N, M-\{i\})}{G(N)}.\qquad (4.49)}$$

The numerator can be computed using the previous recursive expression for the auxiliary function. If M is the last queue in the ordering this becomes simply

$$U_M(N) = 1 - \frac{g(N,M-1)}{G(N)}.$$
(4.50)

A more specific expression can be found when the service rate is state independent. From [BUZE 76]

$$U_i(N) = \frac{1}{\mu_i} \times \overline{Y}_i(N)$$
(4.51)

it is evident that

$$\overline{Y}_i(N) = U_i(N) \times \mu_i.$$
(4.52)

Using the previous expression for throughput,

$$U_i(N) = \frac{\theta_i}{\mu_i} \frac{G(N-1)}{G(N)}.$$
(4.53)

Examples of the Use of the Convolution Algorithm

Example 1:

A Convolution Algorithm Example: State Independent Servers

In this example we will consider a cyclic queueing system of three queues. The service rates are $\mu_1 = \mu_2 = \mu_3 = \mu$. The mean throughput of each queue is equal and $\theta_1 = \theta_2 = \theta_3 = 1.0$ as the queues are in series. This example will be solved again, later, using the Mean Value Analysis algorithm. Table 4.3 can be filled in as shown below:

Table 4.3			
Stations			
Loads	1	2	3
0	1.0	1.0	1.0
1	$f_1(1)=\dfrac{1}{\mu}$	$\dfrac{1}{\mu}+1\times\dfrac{1}{\mu}=$ $\dfrac{2}{\mu}$	$\dfrac{2}{\mu}+1\times\dfrac{1}{\mu}=$ $\dfrac{3}{\mu}$
2	$f_1(2)=\dfrac{1}{\mu^2}$	$\dfrac{1}{\mu^2}+\dfrac{2}{\mu}\times\dfrac{1}{\mu}=$ $\dfrac{3}{\mu^2}$	$\dfrac{3}{\mu^2}+\dfrac{3}{\mu}\times\dfrac{1}{\mu}=$ $\dfrac{6}{\mu^2}$
3	$f_1(3)=\dfrac{1}{\mu^3}$	$\dfrac{1}{\mu^3}+\dfrac{3}{\mu^2}\times\dfrac{1}{\mu}=$ $\dfrac{4}{\mu^3}$	$\dfrac{4}{\mu^3}+\dfrac{6}{\mu^2}\times\dfrac{1}{\mu}=$ $\dfrac{10}{\mu^3}$
4	$f_1(4)=\dfrac{1}{\mu^4}$	$\dfrac{1}{\mu^4}+\dfrac{4}{\mu^3}\times\dfrac{1}{\mu}=$ $\dfrac{5}{\mu^4}$	$\dfrac{5}{\mu^4}+\dfrac{10}{\mu^3}\times\dfrac{1}{\mu}=$ $\dfrac{15}{\mu^4}$
5	$f_1(5)=\dfrac{1}{\mu^5}$	$\dfrac{1}{\mu^5}+\dfrac{5}{\mu^4}\times\dfrac{1}{\mu}=$ $\dfrac{6}{\mu^5}$	$\dfrac{6}{\mu^5}+\dfrac{15}{\mu^4}\times\dfrac{1}{\mu}=$ $\dfrac{21}{\mu^5}$
—			

From the right most column in the table we can read

$$G(1) = \frac{3}{\mu},$$

$$G(2) = \frac{6}{\mu^2},$$

$$G(3) = \frac{10}{\mu^3},$$

$$G(4) = \frac{15}{\mu^4},$$

$$G(5) = \frac{21}{\mu^5},$$

so we can infer that

$$G(N) = \frac{(N+1)(N+2)}{2\mu^N}.$$

Some performance measures that will also be computed later using the Mean Value Analysis algorithm will now be calculated. The mean throughput of the network as a function of the number of customers will first be calculated. The expression for this is

$$\overline{Y}_i(N) = \theta_i \frac{G(N-1)}{G(N)},$$

so

$$\overline{Y}_i(1) = \theta_i \frac{G(0)}{G(1)} = 1 \times \frac{1.0}{3/\mu} = \frac{\mu}{3},$$

$$\overline{Y}_i(2) = \theta_i \frac{G(1)}{G(2)} = 1 \times \frac{3/\mu}{6/\mu^2} = \frac{2\mu}{4},$$

$$\overline{Y}_i(3) = \theta_i \frac{G(2)}{G(3)} = 1 \times \frac{6/\mu^2}{10/\mu^3} = \frac{3\mu}{5},$$

$$\bar{Y}_i(4) = \theta_i \frac{G(3)}{G(4)} = 1 \times \frac{10/\mu^3}{15/\mu^4} = \frac{4\mu}{6},$$

$$\bar{Y}_i(5) = \theta_i \frac{G(4)}{G(5)} = 1 \times \frac{15/\mu^4}{21/\mu^5} = \frac{5\mu}{7}$$

or in general:

$$\bar{Y}_i(N) = \frac{N\mu}{N+2}.$$

The mean throughput can be seen to approach a value of μ, the service rate of the queues.

The mean number of customers in each queue can be calculated using

$$\bar{n}_i(N) = \sum_{n=1}^{N} \left[\frac{\theta_i}{\mu_i}\right]^n \frac{G(N-n)}{G(N)}.$$

Thus for one customer:

$$\bar{n}_i(1) = \left[\frac{\theta_i}{\mu_i}\right] \frac{G(0)}{G(1)} = \frac{1}{\mu} \frac{1.0}{3/\mu} = 0.333.$$

For two customers:

$$\bar{n}_i(2) = \left[\frac{\theta_i}{\mu_i}\right] \frac{G(1)}{G(2)} + \left[\frac{\theta_i}{\mu_i}\right]^2 \frac{G(0)}{G(2)}$$

$$= \frac{1}{\mu} \frac{3/\mu}{6/\mu^2} + \frac{1}{\mu^2} \frac{1.0}{6/\mu^2}$$

$$= 0.5 + 0.166 = 0.666.$$

And finally for three customers:

$$\bar{n}_i(3) =$$

$$= \left[\frac{\theta_i}{\mu_i}\right] \frac{G(2)}{G(3)} + \left[\frac{\theta_i}{\mu_i}\right]^2 \frac{G(1)}{G(3)} + \left[\frac{\theta_i}{\mu_i}\right]^3 \frac{G(0)}{G(3)}$$

$$= \frac{1}{\mu} \frac{6/\mu^2}{10/\mu^3} + \frac{1}{\mu^2} \frac{3/\mu}{10/\mu^3} + \frac{1}{\mu^3} \frac{1.0}{10/\mu^3}$$

$$= 0.6 + 0.3 + 0.1 = 1.0.$$

So we may deduce that

$$\bar{n}_i(N) = 0.33 \times N.$$

Finally, the mean waiting time spent in each queue can be calculated from Little's Law as

$$\bar{\tau}_i(N) = \frac{\bar{n}_i(N)}{\bar{Y}_i(N)} = \frac{.33 \times N}{N\mu/N+2} = \frac{0.33(N+2)}{\mu}.$$

Example 2:

A Convolution Algorithm Numerical Example: State Independent Servers

Consider a cyclic queueing system of three queues. The service rates are $\mu_1 = 2.0$, $\mu_2 = 3.0$ and $\mu_3 = 6.0$. Naturally, the mean throughput of each queue is equal and $\theta_1 = \theta_2 = \theta_3 = 1.0$. The normalization constants are shown below in table 4.4.

Table 4.4			
Stations			
Loads	1	2	3
0	1.0	1.0	1.0
1	$f_1(1)= \frac{1}{2} = 0.5$	$0.5+1\times\frac{1}{3} =$ 0.833	$0.833+1\times\frac{1}{6} =$ 1.0
2	$f_1(2)= \frac{1}{4} = 0.25$	$0.25+0.833\times\frac{1}{3} =$ 0.527	$0.527+1\times\frac{1}{6} =$ 0.694
3	$f_1(3)= \frac{1}{8} = 0.125$	$0.125+0.527\times\frac{1}{3} =$ 0.300	$0.3+0.694\times\frac{1}{6} =$ 0.416
4	$f_1(4)= \frac{1}{16} = 0.0625$	$0.0625+.3\times\frac{1}{3} =$ 0.1625	$0.1625+0.416\times\frac{1}{6} =$ 0.232
5	$f_1(5)= \frac{1}{32} = 0.03125$	$0.03125+0.1625\times\frac{1}{3} =$ 0.0854	$0.0854+0.232\times\frac{1}{6} =$ 0.124
–			

From the right most column in the table we can read

$$G(1) = 1.0,$$

$$G(2) = 0.694,$$

$$G(3) = 0.416,$$

$$G(4) = 0.232,$$

$$G(5) = 0.124.$$

We are now in a position to determine a number of performance measures from the normalization constants in table 4.4. The mean throughput of the network as a function of the number of customers will first be calculated. The expression for this is

$$\overline{Y}_i(N) = \theta_i \frac{G(N-1)}{G(N)},$$

so:

$$\overline{Y}_i(1) = \theta_i \frac{G(0)}{G(1)} = 1 \times \frac{1.0}{1.0} = 1.00,$$

$$\overline{Y}_i(2) = \theta_i \frac{G(1)}{G(2)} = 1 \times \frac{1.0}{0.694} = 1.44,$$

$$\overline{Y}_i(3) = \theta_i \frac{G(2)}{G(3)} = 1 \times \frac{0.694}{0.416} = 1.67,$$

$$\overline{Y}_i(4) = \theta_i \frac{G(3)}{G(4)} = 1 \times \frac{0.416}{0.232} = 1.79,$$

$$\overline{Y}_i(5) = \theta_i \frac{G(4)}{G(5)} = 1 \times \frac{0.232}{0.124} = 1.87.$$

As the number of customers increases the throughput levels off at just under a value of two, as this is the service rate of the slowest queue.

We will first calculate the marginal queue length distribution when there are four customers in the cyclic queueing network using the equation for state-independent servers:

$$p(n_i = n) = \left[\frac{\theta_i}{\mu_i}\right]^n \frac{1}{G(N)} \left[G(N-n) - \frac{\theta_i}{\mu_i} G(N-n-1)\right].$$

Using this equation we have

$$p(n_3 = 0) = \left[\frac{\theta_3}{\mu_3}\right]^0 \frac{1}{G(4)} \left[G(4) - \frac{\theta_3}{\mu_3} \times G(3)\right]$$

$$= \left[\frac{1}{6}\right]^0 \frac{1}{0.232} \left[0.232 - \frac{1}{6} \times 0.416\right]$$

$$= 0.700,$$

$$p(n_3 = 1) = \left[\frac{\theta_3}{\mu_3}\right]^1 \frac{1}{G(4)} \left[G(3) - \frac{\theta_3}{\mu_3} \times G(2)\right]$$

$$= \left[\frac{1}{6}\right]^1 \frac{1}{0.232} \left[0.416 - \frac{1}{6} \times 0.694\right]$$

$$= 0.216,$$

$$p(n_3 = 2) = \left[\frac{\theta_3}{\mu_3}\right]^2 \frac{1}{G(4)} \left[G(2) - \frac{\theta_3}{\mu_3} \times G(1)\right]$$

$$= \left[\frac{1}{6}\right]^2 \frac{1}{0.232} \left[0.694 - \frac{1}{6} \times 1.0\right]$$

$$= 0.0631,$$

$$p\,(n_3 = 3) = \left[\frac{\theta_3}{\mu_3}\right]^3 \frac{1}{G\,(4)}\left[G\,(1) - \frac{\theta_3}{\mu_3}\times G\,(0)\right]$$

$$= \left[\frac{1}{6}\right]^3 \frac{1}{0.232}\left[1.0 - \frac{1}{6}\times 1.0\right]$$

$$= 0.0166,$$

$$p\,(n_3 = 4) = \left[\frac{\theta_3}{\mu_3}\right]^4 \frac{1}{G\,(4)}\left[G\,(0) - \frac{\theta_3}{\mu_3}\times G\,(-1)\right]$$

$$= \left[\frac{1}{6}\right]^4 \frac{1}{0.232}\left[1.0 - \frac{1}{6}\times 0.0\right]$$

$$= 0.0033.$$

We can also use the equation that was developed for the case of state-dependent servers for this state-independent server example. After all, state-independent servers are a specialization of state-dependent servers. We first need the auxiliary function:

$$g\,(N,M-\{i\}) = G\,(N) - \sum_{n=1}^{N} f_i(n)g\,(N-n,M-\{i\}).$$

We will assume that the i=3rd queue is removed from the network. The specific values, when there are four customers, are

$$g\,(0,M-\{3\}) = 1.0,$$

$$g\,(1,M-\{3\}) = G\,(1) - f_3(1)g\,(0,M-\{3\})$$

$$= 1.0 - \frac{1}{6} \times 1.0 = \frac{5}{6},$$

$$g\,(2,M-\{3\}) = G\,(2) \cdots f_3(1)g\,(1,M-\{3\}) - f_3(2)g\,(0,M-\{3\})$$

$$= 0.694 - \frac{1}{6}\,\frac{5}{6} - \frac{1}{36}\,1.0 = 0.527$$

$$g\,(3,M-\{3\}) =$$

$$= G\,(3) - f_3(1)g\,(2,M-\{3\}) - f_3(2)g\,(1,M-\{3\}) - f_3(3)g\,(0,M-\{3\})$$

$$= 0.416 - \frac{1}{6}\,0.527 - \frac{1}{36}\,\frac{5}{6} - \frac{1}{216}\,1.0 = 0.3$$

$$g\,(4,M-\{3\}) = G\,(4) - f_3(1)g\,(3,M-\{3\})$$

$$- f_3(2)g\,(2,M-\{3\}) - f_3(3)g\,(1,M-\{3\}) - f_3(4)g\,(0,M-\{3\})$$

$$= 0.232 - \frac{1}{6}\,0.3 - \frac{1}{36}\,0.527 - \frac{1}{216}\,\frac{5}{6} - \frac{1}{1296}\,1 = 0.1625.$$

These values can be seen to match the entries in the middle column of the normalization constants table 4.3. This is because the results from the columns for stations 1 and 2 are equivalent to results for the network of three queues without the third queue. That is $g(n, M - \{M\}) = g(n, M - 1)$. With these values of the normalization constants calculated we can now use the expression

$$p(n_i = n) = \frac{f_i(n)}{G\,(N)}\,g\,(N-n,M-\{i\})$$

to calculate the marginal state probabilities for the third queue. That is:

$$p(n_3 = 0) = \frac{f_3(0)}{G\,(4)}\,g\,(4,M-\{3\}) = \frac{1.0}{0.232}\,0.1625 = 0.700,$$

$$p(n_3 = 1) = \frac{f_3(1)}{G\,(4)}\,g\,(3,M-\{3\}) = \frac{1/6}{0.232}\,0.3 = 0.216,$$

$$p(n_3 = 2) = \frac{f_3(2)}{G(4)} g(2, M-\{3\}) = \frac{1/36}{0.232} 0.527 = 0.0631,$$

$$p(n_3 = 3) = \frac{f_3(3)}{G(4)} g(1, M-\{3\}) = \frac{1/216}{0.232} \frac{5}{6} = 0.0166,$$

$$p(n_3 = 4) = \frac{f_3(4)}{G(4)} g(0, M-\{3\}) = \frac{1/1296}{0.232} 1.0 = 0.0033.$$

A quick check will show that these marginal state probabilities do indeed sum to one.

The utilization of each queue may be calculated using

$$U_i(N) = \frac{\theta_i}{\mu_i} \frac{G(N-1)}{G(N)}.$$

Specifically for four customers:

$$U_1(4) = \frac{1}{2} \frac{0.416}{0.232} = 0.90,$$

$$U_2(4) = \frac{1}{3} \frac{0.416}{0.232} = 0.60,$$

$$U_3(4) = \frac{1}{6} \frac{0.416}{0.232} = 0.30.$$

As one might expect, the queue with the slowest service rate has the highest utilization and the queue with the fastest service rate has the lowest utilization.

The average number of customers in each queue can be calculated using

$$\bar{n}_i(N) = \sum_{n=1}^{N} \left[\frac{\theta_i}{\mu_i} \right]^n \frac{G(N-n)}{G(N)}.$$

We will calculate this quantity for the third queue (and with four customers in the network):

$$\bar{n}_3(4)$$

$$= \left[\frac{\theta_3}{\mu_3}\right]^1 \frac{G(3)}{G(4)} + \left[\frac{\theta_3}{\mu_3}\right]^2 \frac{G(2)}{G(4)} + \left[\frac{\theta_3}{\mu_3}\right]^3 \frac{G(1)}{G(4)} + \left[\frac{\theta_3}{\mu_3}\right]^4 \frac{G(0)}{G(4)}$$

$$= \frac{1}{6} \frac{0.416}{0.232} + \frac{1}{36} \frac{0.694}{0.232} + \frac{1}{216} \frac{1.0}{0.232} + \frac{1}{1296} \frac{1.0}{0.232}$$

$$= 0.299 + 0.083 + 0.02 + 0.0033$$

$$= 0.405.$$

As a check, the average number of customers in queue three can be calculated using the marginal probability distribution that we have already calculated for the third queue:

$$\bar{n}_3(4) = 1 \times 0.216 + 2 \times 0.0631 + 3 \times 0.0166 + 4 \times 0.0033 = 0.405.$$

The results can be seen to match.

Finally, the mean waiting time spent in the third queue can be calculated from Little's Law as

$$\bar{\tau}_3(4) = \frac{\bar{n}_3(4)}{\bar{Y}_3(4)} = \frac{0.405}{1.79} = 0.226.$$

The mean delay in the third queue consists of the mean service time of 0.166 seconds and the mean queueing time of 0.0593 seconds for a total mean waiting time of 0.226 seconds.

Example 3:

A Convolution Algorithm State Independent Example: Random Routing

Consider a cyclic queueing system of three queues. Here Q1 is followed by two queues in parallel, Q2 and Q3. Departures from Q2 and Q3 return to Q1 (see Figure 4.1). The service rates are $\mu_1 = 2.0$, $\mu_2 = 3.0$, and $\mu_3 = 4.0$. The probability of a Q1 departure entering Q2 is 1/3 (and entering Q3 is 2/3). Naturally, $\theta_1 = 1.0$, $\theta_2 = 1/3$ and $\theta_3 = 2/3$. The normalization constants are in Table 4.4.

Table 4.4		
Stations		
Loads	1	2	3
0	1.0	1.0	1.0
1	$f_1(1) = \dfrac{1}{2} = 0.5$	$0.5 + 1 \times \dfrac{0.3333}{3} =$ 0.6111	$0.6111 + 1 \times \dfrac{0.6666}{4} =$ 0.7777
2	$f_1(2) = \dfrac{1}{4} = 0.25$	$0.25 + 0.6111 \times \dfrac{1}{9} =$ 0.3179	$0.3179 + 0.7777 \times \dfrac{1}{6} =$ 0.4475
3	$f_1(3) = \dfrac{1}{8} = 0.125$	$0.125 + 0.3179 \times \dfrac{1}{9} =$ 0.1603	$0.1603 + 0.4475 \times \dfrac{1}{6} =$ 0.2349
4	$f_1(4) = \dfrac{1}{16} = 0.0625$	$0.0625 + 0.1603 \times \dfrac{1}{9} =$ 0.08031	$0.08031 + 0.2349 \times \dfrac{1}{6} =$ 0.1195

From the right most column in the table we can read

$$G(1) = 0.7777,$$

$$G(2) = 0.4475,$$

$$G(3) = 0.2349,$$

$$G(4) = 0.1195.$$

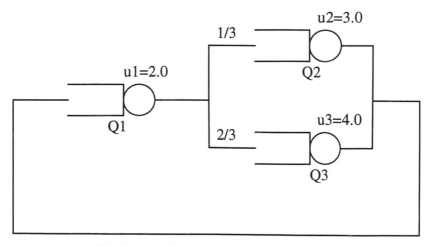

Fig. 4.1: Random Routing Example Network

We are now in a position to determine a number of performance measures from the normalization constants in table 4.4. The mean throughput of the reference queue, Q1, as a function of the number of customers will first be calculated. The general expression for the throughput of the ith queue is

$$\overline{Y}_i(N) = \theta_i \frac{G(N-1)}{G(N)},$$

so

$$\overline{Y}_1(1) = \theta_1 \frac{G(0)}{G(1)} = 1 \times \frac{1.0}{0.7777} = 1.286,$$

$$\overline{Y}_1(2) = \theta_1 \frac{G(1)}{G(2)} = 1 \times \frac{0.7777}{0.4475} = 1.738,$$

$$\overline{Y}_1(3) = \theta_1 \frac{G(2)}{G(3)} = 1 \times \frac{0.4475}{0.2349} = 1.905,$$

$$\overline{Y}_1(4) = \theta_1 \frac{G(3)}{G(4)} = 1 \times \frac{0.2349}{0.1195} = 1.966,$$

As the number of customers increases, the throughput levels off at just under a value of two as this is the service rate of Q1.

The average number of customers in each queue can be calculated using

$$\bar{n}_i(N) = \sum_{n=1}^{N} \left[\frac{\theta_i}{\mu_i}\right]^n \frac{G(N-n)}{G(N)}.$$

We will calculate this quantity for the third queue (and with four customers in the network):

$$\bar{n}_3(4) =$$

$$= \left[\frac{\theta_3}{\mu_3}\right]^1 \frac{G(3)}{G(4)} + \left[\frac{\theta_3}{\mu_3}\right]^2 \frac{G(2)}{G(4)} + \left[\frac{\theta_3}{\mu_3}\right]^3 \frac{G(1)}{G(4)} + \left[\frac{\theta_3}{\mu_3}\right]^4 \frac{G(0)}{G(4)}$$

$$= \frac{0.6666}{4} \frac{0.2349}{0.1195} + \left[\frac{0.6666}{4}\right]^2 \frac{0.4475}{0.1195} + \left[\frac{0.6666}{4}\right]^3 \frac{0.7777}{0.1195}$$

$$+ \left[\frac{0.6666}{4}\right]^4 \frac{1.0}{0.1195}$$

$$= 0.468$$

Finally, the mean waiting time spent in the third queue can be calculated from Little's Law as

$$\bar{\tau}_3(4) = \frac{\bar{n}_3(4)}{\bar{Y}_1(4) \times \theta_3} = \frac{0.468}{0.6666 \times 1.966} = 0.357$$

The mean delay in the third queue consists of the mean service time of 0.25 seconds and the mean queueing time of 0.107 seconds for a total mean waiting time of 0.357 seconds.

Example 4:

A Convolution Algorithm Numerical Example: State Dependent Servers

In this example we will consider a cyclic queueing system of three queues. Naturally, the mean throughput of each queue is equal and $\theta_1 = \theta_2 = \theta_3 = 1.0$. Table 4.5a,b presents the service rates for queue 1 and 2, respectively.

Table 4.5a: Queue 1	
# Customers	Service Rate
1	5
2	6
3	7
4	8
5	9

Table 4.5b: Queue 2	
# Customers	Service Rate
1	4
2	8
3	12
4	16
5	20

Queue 3 has a state-independent server of rate six.

The normalization constants table can be filled in as shown below

Table 4.6			
Stations			
Loads	1	2	3
0	1.0	1.0	1.0
1	$f_1(1)= \dfrac{1}{5} =$ 0.2	$0.2\times1 + 1\times\dfrac{1}{4} =$ 0.45	$0.45\times1 + 1\times\dfrac{1}{6} =$ 0.616
2	$f_1(2)= \dfrac{1}{5}\dfrac{1}{6} =$ 0.0333	$0.0333\times1 +$ $0.2\times\dfrac{1}{4} +$ $1\times\dfrac{1}{4}\times\dfrac{1}{8} =$ 0.1146	$0.1146\times1 +$ $0.45\times\dfrac{1}{6} +$ $1\times\dfrac{1}{6}\times\dfrac{1}{6} =$ 0.217
3	$f_1(3)= \dfrac{1}{5}\dfrac{1}{6}\dfrac{1}{7} =$ 0.00476	$0.00476\times1 +$ $0.0333\times\dfrac{1}{4} +$ $0.2\times\dfrac{1}{8}\times\dfrac{1}{4} +$ $1\times\dfrac{1}{12}\times\dfrac{1}{8}\times\dfrac{1}{4} =$ 0.0219	$0.0219\times1 +$ $0.1146\times\dfrac{1}{6} +$ $0.45\times\dfrac{1}{6}\times\dfrac{1}{6} +$ $1\times\dfrac{1}{6}\times\dfrac{1}{6}\times\dfrac{1}{6} =$ 0.0581
4	$f_1(4) =$ $\dfrac{1}{5}\dfrac{1}{6}\dfrac{1}{7}\dfrac{1}{8} =$ 0.000595	0.00351	0.0132
5	$f_1(5) =$ $\dfrac{1}{5}\dfrac{1}{6}\dfrac{1}{7}\dfrac{1}{8}\dfrac{1}{9} =$ 0.0000661	0.000491	0.00269

From the rightmost column in the table we can read

$$G(1) = 0.616,$$

$$G(2) = 0.217,$$

$$G(3) = 0.0581,$$

$$G(4) = 0.0132,$$

$$G(5) = 0.00269.$$

A number of performance measures can now be determined from the normalization constants in table 4.6. First we will calculate the mean throughput as a function of the number of customers. The expression for this, for both the state-independent and state-dependent cases, is

$$\overline{Y}_i(N) = \theta_i \frac{G(N-1)}{G(N)},$$

so:

$$\overline{Y}_i(1) = \theta_i \frac{G(0)}{G(1)} = 1 \times \frac{1.0}{0.616} = 1.62,$$

$$\overline{Y}_i(2) = \theta_i \frac{G(1)}{G(2)} = 1 \times \frac{0.616}{0.217} = 2.84,$$

$$\overline{Y}_i(3) = \theta_i \frac{G(2)}{G(3)} = 1 \times \frac{0.217}{0.0581} = 3.73,$$

$$\overline{Y}_i(4) = \theta_i \frac{G(3)}{G(4)} = 1 \times \frac{0.0581}{0.0132} = 4.40,$$

$$\overline{Y}_i(5) = \theta_i \frac{G(4)}{G(5)} = 1 \times \frac{0.0132}{0.00269} = 4.91.$$

Next we will calculate the marginal queue length distribution when there are four customers in the cyclic queueing network. We first need the auxiliary function

$$g(N,M-\{i\}) = G(N) - \sum_{n=1}^{N} f_i(n)g(N-n,M-\{i\}).$$

We will assume that the $i=3$rd queue is removed from the network. The specific values, when there are four customers, are

$$g(0,M-\{3\}) = 1.0,$$

$$g(1,M-\{3\}) = G(1) - f_3(1)g(0,M-\{3\})$$

$$= 0.616 - \frac{1}{6} \times 1.0 = 0.45,$$

$$g(2,M-\{3\}) = G(2) - f_3(1)g(1,M-\{3\}) - f_3(2)g(0,M-\{3\})$$

$$= 0.217 - \frac{1}{6}\, 0.45 - \frac{1}{36}\, 1.0 = 0.114,$$

$$g(3,M-\{3\}) =$$

$$= G(3) - f_3(1)g(2,M-\{3\}) - f_3(2)g(1,M-\{3\}) - f_3(3)g(0,M-\{3\})$$

$$= 0.0581 - \frac{1}{6}\, 0.114 - \frac{1}{36}\, 0.45 - \frac{1}{216}\, 1.0 = 0.0219,$$

$$g(4,M-\{3\}) = G(4) - f_3(1)g(3,M-\{3\})$$

$$- f_3(2)g(2,M-\{3\}) - f_3(3)g(1,M-\{3\}) - f_3(4)g(0,M-\{3\})$$

$$= 0.0132 - \frac{1}{6}\, 0.0219 - \frac{1}{36}\, 0.114 - \frac{1}{216}\, .45 - \frac{1}{1296}\, 1 = 0.0035.$$

Again, these values can be seen to match the entries in the middle column of the normalization constants Table 4.6. This is because the results from the columns for stations 1 and 2 are equivalent to results for the network of three queues without the third queue. That is $g(n, M-\{M\}) = g(n, M-1)$. With these values of the normalization constants calculated we can now use the expression

$$p(n_i = n) = \frac{f_i(n)}{G(N)} g(N-n, M-\{i\})$$

to calculate the marginal state probabilities for the third queue. That is:

$$p(n_3 = 0) = \frac{f_3(0)}{G(4)} g(4, M-\{3\}) = \frac{1.0}{0.0132} 0.0035 = 0.265,$$

$$p(n_3 = 1) = \frac{f_3(1)}{G(4)} g(3, M-\{3\}) = \frac{1/6}{0.0132} 0.0219 = 0.277,$$

$$p(n_3 = 2) = \frac{f_3(2)}{G(4)} g(2, M-\{3\}) = \frac{1/36}{0.0132} 0.114 = 0.240,$$

$$p(n_3 = 3) = \frac{f_3(3)}{G(4)} g(1, M-\{3\}) = \frac{1/216}{0.0132} 0.45 = 0.158,$$

$$p(n_3 = 4) = \frac{f_3(4)}{G(4)} g(0, M-\{3\}) = \frac{1/1296}{0.0132} 1.0 = 0.0585.$$

A quick check will show that these marginal state probabilities do indeed sum to one.

The utilization of each queue may be calculated using

$$U_i(N) = 1 - \frac{g(N, M-\{i\})}{G(N)}.$$

For the third queue and four customers this is

$$U_3(4) = 1 - \frac{g(4, M-\{3\})}{G(4)} = 1 - \frac{0.0035}{0.0132} = 0.735.$$

To calculate the utilization for the 1st or 2nd queue would require us to reorder and recompute the normalization constants table 4.6 so that the desired queue would be associated with the last column.

For the expected number of customers for the Mth queue

$$\bar{n}_M(N) = \frac{1}{G(N)} \sum_{n=1}^{N} n f_M(n) g(N-n, M-1),$$

so for four customers and the third (last) queue:

$$\bar{n}_3(4) = \frac{1}{G(4)} \sum_{n=1}^{4} n f_3(n) g(4-n, 2)$$

$$= \frac{1}{G(4)} \left[1 f_3(1) g(3,2) + 2 f_3(2) g(2,2) + 3 f_3(3) g(1,2) + 4 f_3(4) g(0,2) \right]$$

$$= \frac{1}{0.0132} \left[1 \frac{1}{6} 0.0219 + 2 \frac{1}{36} 0.114 + 3 \frac{1}{216} 0.45 + 4 \frac{1}{1296} 1.0 \right]$$

$$= 1.46.$$

As a check, the average number of customers in queue three can be calculated using the marginal probability distribution that we have already calculated for the third queue:

$$\bar{n}_3(4) = 1 \times 0.277 + 2 \times 0.240 + 3 \times 0.158 + 4 \times 0.0585 = 1.46.$$

The results can be seen to match.

Finally, the mean waiting time spent in the third queue can be calculated from Little's Law as

$$\bar{\tau}_3(4) = \frac{\bar{n}_3(4)}{\bar{Y}_3(4)} = \frac{1.46}{4.40} = 0.332.$$

The mean delay in the third queue consists of the mean service time of 0.166 seconds and the mean queueing time of 0.166 seconds for a total mean waiting time of 0.332 seconds.

4.3 Mean Value Analysis

Mean Value Analysis uses a number of fundamental queueing relationships to determine the mean values of throughput, delay, and population size for closed product form queueing networks. It does not make use of the normalization constant, as does the convolution algorithm. MVA was developed by M. Reiser and S. Lavenberg (see [REIS 80,81,82]).

4.3.1 State-(Load) Independent Servers

For simplicity let us consider a closed cyclic network consisting of M queues with N customers. The ith queue will have a service rate of μ_i. Suppose that we write an expression for $\bar{\tau}_i$, the average delay that a customer experiences at the ith queue. This has two components, the service time of that customer and the service time for all the customers before it. This is

$$\bar{\tau}_i = \qquad\qquad (4.54)$$

$$\frac{1}{\mu_i} + \frac{1}{\mu_i} \times (average \ number \ of \ customers \ present \ upon \ arrival).$$

Note that we do not worry about the residual service time of the customer in service because we are dealing with a Markovian system. What do we know about the average number of customers present upon arrival? According to the *Arrival Theorem* ([LAVE 80]) for closed exponential networks, the number of customers present upon arrival has the same distribution as the equilibrium distribution for the network with one customer less. Intuitively, this follows from the Markovian nature of the network. Thus the previous equation becomes

$$\bar{\tau}_i(N) = \frac{1}{\mu_i} + \frac{1}{\mu_i} \times \bar{n}_i(N-1) \qquad\qquad (4.55)$$

where $\bar{n}_i(N)$ is the average number of customers in the ith queue when there are N customers in the network.

The second half of the MVA algorithm depends on Little's Law being applied first to the entire network of queues and then to the ith queue:

$$\bar{Y}(N) \sum_{i=1}^{M} \bar{\tau}_i(N) = N, \qquad\qquad (4.56)$$

$$\bar{Y}(N)\bar{\tau}_i(N) = \bar{n}_i(N). \qquad\qquad (4.57)$$

Here $\overline{Y}(N)$ is the average throughput in the network with N customers. With some simple manipulation these three equations can now be listed in an algorithmic form:

$$
\begin{array}{|ll|}
\hline
\multicolumn{2}{|c|}{\text{MVA Algorithm (State-Independent Servers)}} \\
\hline
\overline{n}_i(0) = 0 & i = 1,2,...M \\[2ex]
\overline{\tau}_i(N) = \dfrac{1}{\mu_i} + \dfrac{1}{\mu_i} \times \overline{n}_i(N-1) & i = 1,2,...M \\[2ex]
\overline{Y}(N) = \dfrac{N}{\displaystyle\sum_{i=1}^{M} \overline{\tau}_i(N)} & \\[2ex]
\overline{n}_i(N) = \overline{Y}(N)\overline{\tau}_i(N) & i = 1,2,...M \\
\hline
\end{array}
$$

These algorithm equations are repeatedly solved for increasing population size.

Examples of the Use of the MVA Algorithm

Example 1: M Cyclic Queues

We will examine a very natural example that appears in [SCHW 87] which consists of an M queue cyclic network with $\mu_1 = \mu_2 = \mu_3 = ... = \mu_M = \mu$. This is the same example that we did before using the convolution algorithm for a three queue cyclic network. The MVA algorithm will allow us to solve for the M queue case.

The assumption that the queues are arranged in a cycle assures that the throughput of each queue is identical. The assumption that all the service rates are the same will also simplify matters since the mean number in each queue and the mean time delay through each queue will be identical.

For one customer over the queues $i=1,2,....M$:

$$\overline{n}_i(0) = 0,$$

$$\overline{\tau}_i(1) = \frac{1}{\mu},$$

$$\overline{Y}(1) = \frac{1}{M\dfrac{1}{\mu}} = \frac{\mu}{M},$$

$$\overline{n}_i(1) = \frac{\mu}{M} \times \frac{1}{\mu} = \frac{1}{M}.$$

For two customers over the queues $i=1,2,....M$:

$$\overline{\tau}_i(2) = \frac{1}{\mu} + \left[\frac{1}{\mu} \times \frac{1}{M}\right] = \frac{1}{\mu}\left[\frac{M+1}{M}\right],$$

$$\overline{Y}(2) = \frac{2}{M \times \dfrac{1}{\mu} \times \left[\dfrac{M+1}{M}\right]} = \frac{2\mu}{M+1},$$

$$\overline{n}_i(2) = \frac{2\mu}{M+1} \times \frac{1}{\mu} \times \frac{M+1}{M} = \frac{2}{M}.$$

For three customers over the queues $i=1,2,....M$:

$$\overline{\tau}_i(3) = \frac{1}{\mu} + \left[\frac{1}{\mu} \times \frac{2}{M}\right] = \frac{1}{\mu}\left[\frac{M+2}{M}\right],$$

$$\overline{Y}(3) = \frac{3}{M \times \dfrac{1}{\mu} \times \left[\dfrac{M+2}{M}\right]} = \frac{3\mu}{M+2},$$

$$\overline{n}_i(3) = \frac{3\mu}{M+2} \times \frac{1}{\mu} \times \left[\frac{M+2}{M}\right] = \frac{3}{M}.$$

Finally, one can deduce a pattern in the answers so that for N customers over the queues $i=1,2,....M$:

$$\overline{\tau}_i(N) = \frac{1}{\mu}\left[\frac{M+N-1}{M}\right],$$

$$\overline{Y}(N) = \frac{N\mu}{M+N-1},$$

$$\overline{n}_i(N) = \frac{N}{M}.$$

These results can be seen to match those obtained previously with the convolution algorithm when $M=3$.

Example 2: Cyclic Queueing Network Numerical Example

We will now re-solve the numerical problem for state-independent servers that we worked out using the convolution algorithm. Again, the closed cyclic queueing network consists of three queues with service rates of $\mu_1 = 2.0$, $\mu_2 = 3.0$, and $\mu_3 = 6.0$. The algorithm is initialized with:

$$\overline{n}_i(0) = 0, \qquad i=1,2,3.$$

For one customer:

$$\overline{\tau}_1(1) = 0.500,$$
$$\overline{\tau}_2(1) = 0.333,$$
$$\overline{\tau}_3(1) = 0.166;$$

$$\overline{Y}(1) = \frac{1}{0.5 + 0.333 + 0.166} = 1.0;$$

$$\overline{n}_1(1) = 1.0 \times 0.500 = 0.500,$$
$$\overline{n}_2(1) = 1.0 \times 0.333 = 0.333,$$
$$\overline{n}_3(1) = 1.0 \times 0.166 = 0.166.$$

For two customers:

$$\overline{\tau}_1(2) = 0.500 + 0.500 \times 0.500 = 0.75,$$
$$\overline{\tau}_2(2) = 0.333 + 0.333 \times 0.333 = 0.444,$$
$$\overline{\tau}_3(2) = 0.166 + 0.166 \times 0.166 = 0.194;$$

$$\overline{Y}(2) = \frac{2}{0.75 + 0.444 + 0.194} = 1.44;$$

$$\overline{n}_1(2) = 1.44 \times 0.75 = 1.08,$$
$$\overline{n}_2(2) = 1.44 \times 0.444 = 0.639,$$
$$\overline{n}_3(2) = 1.44 \times 0.194 = 0.279.$$

For three customers:

$$\overline{\tau}_1(3) = 0.500 + 0.500 \times 1.08 = 1.04,$$
$$\overline{\tau}_2(3) = 0.333 + 0.333 \times 0.639 = 0.546,$$
$$\overline{\tau}_3(3) = 0.166 + 0.166 \times 0.279 = 0.212;$$

$$\overline{Y}(3) = \frac{3}{1.04 + 0.546 + 0.212} = 1.67;$$

$$\overline{n}_1(3) = 1.67 \times 1.04 = 1.74,$$
$$\overline{n}_2(3) = 1.67 \times 0.546 = 0.912,$$
$$\overline{n}_3(3) = 1.67 \times 0.212 = 0.354.$$

For four customers:

$$\overline{\tau}_1(4) = 0.500 + 0.500 \times 1.74 = 1.37,$$
$$\overline{\tau}_2(4) = 0.333 + 0.333 \times 0.912 = 0.637$$
$$\overline{\tau}_3(4) = 0.166 + 0.166 \times 0.354 = 0.225;$$

$$\overline{Y}(4) = \frac{4}{1.37 + 0.637 + 0.225} = 1.79;$$

$$\overline{n}_1(4) = 1.79 \times 1.37 = 2.45,$$
$$\overline{n}_2(4) = 1.79 \times 0.637 = 1.14,$$
$$\overline{n}_3(4) = 1.79 \times 0.225 = 0.40.$$

The values of mean throughput can be seen to match those generated previously by the convolution algorithm. Moreover, the value $\overline{n}_3(4)$ also matches.

4.3.2 A Closer Look at the Arrival Theorem

To verify the arrival theorem a proof due to M. Haviv [HAVI] will be used. Recall from the previous section that the arrival theorem for closed Markovian networks holds that the number of customers present upon arrival has the same equilibrium distribution as for the network with one less customer. The proof here will show this for the mean value only.

First we will need a special expression for the average number of customers in the ith queue when there are N customers in the network. Let

$$\rho_i = \frac{\theta_i}{\mu_i}. \tag{4.58}$$

Then, starting from (4.26),

$$\bar{n}_i(N) = \sum_{n=1}^{N} \rho_i^n \frac{G(N-n)}{G(N)}. \tag{4.59}$$

The $n=1$ term can be separated out:

$$\bar{n}_i(N) = \rho_i \frac{G(N-1)}{G(N)} + \sum_{n=2}^{N} \rho_i^n \frac{G(N-n)}{G(N)}. \tag{4.60}$$

Next, make the change of variables $n \rightarrow n+1$ and factor out $\rho_i/G(N)$:

$$\bar{n}_i(N) = \rho_i \frac{G(N-1)}{G(N)} + \frac{\rho_i}{G(N)} \sum_{n=1}^{N-1} \rho_i^n G(N-n-1). \tag{4.61}$$

But from (4.53) one can recognize utilization, so

$$\bar{n}_i(N) = U_i(N) + U_i(N) \sum_{n=1}^{N-1} \rho_i^n \frac{G(N-n-1)}{G(N-1)}. \tag{4.62}$$

Then using (4.26)

$$\bar{n}_i(N) = U_i(N) + U_i(N)\bar{n}_i(N-1) \tag{4.63}$$

or

$$\bar{n}_i(N) = U_i(N)\left[1 + \bar{n}_i(N-1)\right]. \qquad (4.64)$$

Now we can prove the arrival theorem. The mean waiting time for the ith queue when there are N customers in the network, from Little's law, is

$$\tau_i(N) = \frac{\bar{n}_i(N)}{\bar{Y}_i(N)}. \qquad (4.65)$$

Naturally the mean throughput equals the arrival rate above. From (4.64) and (4.47)

$$\tau_i(N) = \frac{U_i(N)(1 + \bar{n}_i(N-1))}{\theta_i \dfrac{G(N-1)}{G(N)}}. \qquad (4.66)$$

Using (4.53)

$$\tau_i(N) = \frac{1}{\mu_i}\left[1 + \bar{n}_i(N-1)\right]. \qquad (4.67)$$

This is the same as (4.55), and the arrival theorem is proved (though for mean delay only).

4.3.3 State-(Load) Independent Servers (Random Routing)

Suppose now that one allows random routing between the queues where the probability of a queue i departure entering queue j is r_{ij}. Then the Mean Value Analysis algorithm [REIS 80] is:

MVA Algorithm (Random Routing)	
$\bar{n}_i(0) = 0$	$i = 1,2,...M$
$\bar{\tau}_i(N) = \dfrac{1}{\mu_i} + \dfrac{1}{\mu_i} \times \bar{n}_i(N-1)$	$i = 1,2,...M$
$\bar{Y}(N) = \dfrac{N}{\displaystyle\sum_{i=1}^{M} \theta_i \bar{\tau}_i(N)}$ *Reference Queue*	
$\bar{n}_i(N) = \bar{Y}(N)\theta_i\bar{\tau}_i(N)$	$i = 1,2,...M$

These equations differ from those for simple cyclic networks in the inclusion of the θ_i's, the solution of the network traffic equations. Normally the θ_i's are chosen so that one queue's θ is equal to 1. This queue can be referred to as the reference queue. The expression for $\bar{Y}(N)$ then gives the mean throughput for this queue. The mean throughputs of the other queues are naturally $\theta_i\bar{Y}(N)$ for the ith queue. Naturally in the case of a simple cyclic network $\theta_i=1$ for all i and these equations reduce to the earlier set.

The Mean Value Analysis algorithm, as stated, applies to any closed product form network with FIFO, LIFOPR, or PS service disciplines. Note that the MVA algorithm only keeps track of the total number of customers in each queue, not individual customer histories. For the case where the ith service center has an unlimited number of servers of rate μ_i the mean delay time is equal to $1/\mu_i$, so one can modify the MVA algorithm with

$$\bar{\tau}_i(N) = \frac{1}{\mu_i} \qquad (4.55b)$$

for each such service center.

Example: Cyclic Queueing Network Example: Random Routing

Consider now the cyclic queueing network examined previously with the convolution algorithm where Q1 is followed by two queues in parallel, Q2 and Q3. Departures from Q2 and Q3 return to Q1 (see Figure 4.1). The service rates are $\mu_1 = 2.0$, $\mu_2 = 3.0$, and $\mu_3 = 4.0$. The probability of a Q1 departure entering Q2 is 1/3 (and of entering Q3 is 2/3). Naturally $\theta_1 = \theta_{reference} = 1.0$ and $\theta_2 = 1/3$ and $\theta_3 = 2/3$.

$$\bar{n}_i(0) = 0, \qquad i = 1,2,3.$$

For one customer:

$$\bar{\tau}_1(1) = 0.500,$$
$$\bar{\tau}_2(1) = 0.3333,$$
$$\bar{\tau}_3(1) = 0.25;$$

$$\bar{Y}(1) = \frac{1}{1.0 \times 0.5 + 0.3333 \times 0.3333 + 0.6666 \times 0.25} = 1.2857;$$

$$\bar{n}_1(1) = 1.2857 \times 1.0 \times 0.500 = 0.64286,$$
$$\bar{n}_2(1) = 1.2857 \times 0.3333 \times 0.3333 = 0.14286,$$
$$\bar{n}_3(1) = 1.2857 \times 0.6666 \times 0.25 = 0.21429.$$

For two customers:

$$\bar{\tau}_1(2) = 0.500 + 0.500 \times 0.64286 = 0.82143,$$
$$\bar{\tau}_2(2) = 0.3333 + 0.3333 \times 0.14286 = 0.38095,$$
$$\bar{\tau}_3(2) = 0.25 + 0.25 \times 0.21429 = 0.30357;$$

$$\bar{Y}(2) = \frac{2}{1.0 \times 0.82143 + 0.3333 \times 0.38095 + 0.6666 \times 0.30357} = 1.7379;$$

$$\bar{n}_1(2) = 1.7379 \times 1.0 \times 0.82143 = 1.4276,$$
$$\bar{n}_2(2) = 1.7379 \times 0.3333 \times 0.38095 = 0.22069,$$
$$\bar{n}_3(2) = 1.7379 \times 0.6666 \times 0.30357 = .35172.$$

For three customers:

$$\bar{\tau}_1(3) = 0.500 + 0.500 \times 1.4276 = 1.2138,$$
$$\bar{\tau}_2(3) = 0.333 + 0.333 \times 0.22069 = 0.40690,$$
$$\bar{\tau}_3(3) = 0.25 + 0.25 \times 0.35172 = 0.33793;$$

$$\bar{Y}(3) = \frac{3}{1.0 \times 1.2138 + 0.3333 \times 0.40690 + 0.6666 \times 0.33793} = 1.9051;$$

$$\bar{n}_1(3) = 1.9051 \times 1.0 \times 1.2138 = 2.3124,$$
$$\bar{n}_2(3) = 1.9051 \times 0.3333 \times 0.40690 = 0.25839,$$
$$\bar{n}_3(3) = 1.9051 \times 0.6666 \times 0.33793 = 0.42920.$$

For four customers:

$$\bar{\tau}_1(4) = 0.500 + 0.500 \times 2.3124 = 1.6562,$$
$$\bar{\tau}_2(4) = 0.333 + 0.333 \times 0.25839 = 0.41946,$$
$$\bar{\tau}_3(4) = 0.25 + 0.25 \times 0.42920 = 0.35730;$$

$$\bar{Y}(4) = \frac{4}{1.0 \times 1.6562 + 0.3333 \times 0.41946 + 0.6666 \times 0.35730} = 1.9664;$$

$$\bar{n}_1(4) = 1.9664 \times 1.0 \times 1.6562 = 3.2567,$$
$$\bar{n}_2(4) = 1.9664 \times 0.3333 \times 0.41946 = 0.27494,$$
$$\bar{n}_3(4) = 1.9664 \times 0.6666 \times 0.35730 = 0.46838.$$

4.4 PANACEA: Approach for Large Markovian Queueing Networks

4.4.1 Introduction

PANACEA is a software package and associated mathematical technique developed by J. McKenna, D. Mitra, and K. G. Ramakrishnan that is able to solve Markovian queueing networks that are significantly larger than those that can be handled by other computational techniques. The package can solve multiclass, closed, open, and mixed queueing networks. The basic idea used is to express the normalization constant as an integral, which is in turn approximated by an asymptotic power series. Thus PANACEA produces approximations - though very accurate ones - along with bounds (see section 4.4.9). Like the convolution algorithm the key is to calculate the normalization constant efficiently, though by radically different means.

The asymptotic power series used are generally power series in $1/N$, where N is a generic large parameter. The number of terms required to produce a desired accuracy decreases with increasing N. Moreover, multiple classes can be handled with only incremental increases in computing time.

In what follows we will very closely follow the exposition in [McKEN 82]. It deals with the specific case of closed networks of state-independent servers that are not heavily loaded. This material is among the most advanced in this book. However, it is well worth reading as it represents a unique approach to the problem of queueing network calculation.

4.4.2 The Product Form Solution

There are p classes of customers, and j is implicitly used for indexing class. That is, j $(1 \le j \le p)$ will not be explicitly written under summations or products.

First, we will consider some nomenclature. There are s service centers of the BCMP types 1,2,3,4 (see section 3.2.5). It turns out that the PANACEA approach requires a type 3 (infinite number of servers) service center in each route. This is not that overly restrictive as terminals, which are well modeled by type 3 service centers, appear throughout computer systems models. It thus becomes convenient to say that service centers 1 through q will be type 1,2, and 4 centers while service centers $q+1$ through s will be type 3 centers. When class and center indices appear together, the first symbol refers to class.

Next, the product form solution will be considered. The equilibrium probability of finding n_{ji} customers of class j at center i, $1 \le j \le p$, $1 \le i \le s$, is $\pi(\underline{y}_1, \underline{y}_2, \underline{y}_3, \cdots \underline{y}_s)$ where

$$\underline{y}_i = (n_{1i}, n_{2i}, n_{3i}, \cdots n_{pi}), \qquad 1 \le i \le s. \tag{4.68}$$

The product form solution is then

$$\pi(\underline{y}_1, \underline{y}_2, \underline{y}_3, \cdots \underline{y}_s) = \frac{1}{G} \prod_{i=1}^{s} \pi_i(\underline{y}_i) \tag{4.69}$$

where

$$\pi(\underline{y}_i) = (\textstyle\sum n_{ji})! \prod \left[\frac{\rho_{ji}^{n_{ji}}}{n_{ji}!} \right], \qquad 1 \le i \le q. \tag{4.70}$$

$$\pi(\underline{y}_i) = \prod \left[\frac{\rho_{ji}^{n_{ji}}}{n_{ji}!} \right], \qquad q+1 \le i \le s \tag{4.71}$$

where

$$\rho_{ji} = \tag{4.72}$$
$$\frac{expected \ \# \ of \ visits \ of \ class \ j \ customers \ in \ center \ i}{service \ rate \ of \ class \ j \ customers \ in \ center \ i}.$$

The numerator here is equivalent to the solutions of the traffic equations, the θ_i's, discussed in chapter three. Rather than phrasing the physical meaning of the θ_i's as throughput, it is phrased here, equivalently, as the number of visits.

It follows from the normalization of probability that the normalization constant, or partition function as a physicist would call it, is

$$G(\underline{K}) = \sum_{\underline{1'n}_1 = K_1} \cdots \sum_{\underline{1'n}_p = K_p} \prod_{i=1}^{s} \pi_i(y_i) \qquad (4.73)$$

where K_j is the constant customer population of the jth class and $\underline{K}$ is the vector of these populations. Also $\underline{1'n}_1$ is a vector representation of $\sum_{i=1}^{s} n_{ji}$. Expanding the above equation yields

$$G(\underline{K}) = \sum \cdots \sum \left[\prod_{i=1}^{q} \left\{ (\sum n_{ji})! \prod \frac{\rho_{ji}^{n_{ji}}}{n_{ji}!} \right\} \right] \left[\prod_{i=q+1}^{s} \left\{ \prod \frac{\rho_{ji}^{n_{ji}}}{n_{ji}!} \right\} \right]. \qquad (4.74)$$

4.4.3 Conversion to Integral Representation

The first step in the PANACEA technique is to represent the normalization constant as an expression involving integrals. Beginning with Euler's integral:

$$n! = \int_0^\infty e^{-u} u^n du. \qquad (4.75)$$

This representation is used to write

$$(\sum n_{ji})! = \int_0^\infty e^{-u_i} \prod u_i^{n_{ji}} du_i, \quad i = 1,2,3,...q. \qquad (4.76)$$

Substituting this into equation (4.74) and bringing the integrals out results in

$$G = \int_0^\infty \cdots \int_0^\infty \exp(-\sum_{i=1}^{q} u_i) \sum_{\underline{1'n}_1 = K_1} \cdots \sum_{\underline{1'n}_p = K_p} \qquad (4.77)$$

$$\times \left[\prod_{i=1}^{q} \left\{ \prod \frac{(\rho_{ji} u_i)^{n_{ji}}}{n_{ji}!} \right\} \right] \left[\prod_{i=q+1}^{s} \left\{ \prod \frac{\rho_{ji}^{n_{ji}}}{n_{ji}!} \right\} \right] du_1 \cdots du_q.$$

We will now express G in several alternate forms. First, using the multino-
mial theorem:

$$G = (\prod K_j!)^{-1} \int_0^\infty \cdots \int_0^\infty \exp(-\sum_{i=1}^q u_i)$$ (4.78)

$$\times \prod \{ \sum_{i=1}^q \rho_{ji} u_i + \sum_{i=q+1}^s \rho_{ji} \}^{K_j} du_1 \cdots du_q.$$

The notation may be simplified by

$$\rho_{j0} = \sum_{i=q+1}^s \rho_{ji}, \qquad j=1,2,3,...,p.$$ (4.79)

Now the service center index i ranges only over the centers $1 \leq i \leq q$. The
quantity ρ_{j0} is a weighted combination of the mean "think times" of the infinite
service centers in routing the jth class. That is, the reciprocal of the service rate is
the mean time a customer spends "thinking" at a terminal between the time a
prompt appears and the time he or she types return.

If the routing of the jth class contains at least one infinite service (type 3)
center then $\rho_{j0} > 0$, otherwise $\rho_{j0} = 0$. Then let the collection of indices of classes
of the former case be I, and let the collection of indices for the latter case be I^*.
That is:

$$j \varepsilon I \longleftrightarrow \rho_{j0} > 0, \qquad j \varepsilon I^* \longleftrightarrow \rho_{j0} = 0.$$ (4.80)

In this notation G becomes

$$G = [\prod_{j \varepsilon I} \rho_{j0}^{K_j} / \prod_j K_j!] \int_0^\infty \cdots \int_0^\infty \exp(-\sum u_i)$$ (4.81)

$$\times \prod_{j \varepsilon I} \left\{ 1 + \sum_i \frac{\rho_{ji}}{\rho_{j0}} u_i \right\}^{K_j} \prod_{j \varepsilon I^*} \left\{ \sum_i \rho_{ji} u_i \right\}^{K_j} du_1 \cdots du_q.$$

In vector notation, which will be used below,

$$G = [\prod_{j \varepsilon I} \rho_{j0}^{K_j} / \prod_j K_j!] \int_{Q+} e^{-1'u} \prod_j (\delta_{jI} + \underline{r}'_j \underline{u})^{K_j} d\underline{u}$$ (4.82)

where

$$u = (u_1, u_2, u_3, \cdots u_q)\prime$$
$$1 = (1, 1, 1, \cdots 1)\prime$$
$$r_j = (r_{j1}, r_{j2}, \cdots r_{jq})\prime, \quad 1 \leq j \leq p,$$
$$r_{ji} = \rho_{ji}/\rho_{j0} \text{ if } j \varepsilon I,$$
$$r_{ji} = \rho_{ji} \text{ if } j \varepsilon I^*,$$
$$\delta_{jI} = 1 \text{ if } j \varepsilon I,$$
$$\delta_{jI} = 0 \text{ if } j \varepsilon I^*,$$
$$Q^+ = \{u \mid u_i > 0 \text{ for } all \ i\}.$$

At this point in their exposition McKenna and Mitra introduce a large parameter N:

$$\beta_j \equiv K_j/N, \quad 1 \leq j \leq p, \tag{4.83}$$

$$\Gamma_j \equiv N r_j, \quad 1 \leq j \leq p. \tag{4.84}$$

Generally β_j is close to zero and the Γ_{ji} are close to one. While many choices of N are possible, McKenna and Mitra suggest

$$N = \max_{ij} \{\frac{1}{r_{ji}}\}. \tag{4.85}$$

Substituting these expressions for β_i and Γ_j into the last expression for G and changing variables $z = u/N$ results in

$$G(\underline{K}) = [\prod_{j \varepsilon I} \rho_{j0}^{K_j}/\prod_j K_j!] \int_{Q^+} e^{-1\prime u} \prod_j (\delta_{jI} + \underline{r}_j \prime \underline{u})^{K_j} d\underline{u}, \tag{4.86}$$

$$\boxed{G(\underline{K}) = [N^q \prod_{j \varepsilon I} \rho_{j0}^{K_j}/\prod_j K_j!] \int_{Q^+} e^{-Nf(\underline{z})} d\underline{z} \tag{4.87}}$$

where

$$f(\underline{z}) = 1 \prime \underline{z} - \sum_{j-1}^{p} \beta_j \log(\delta_{jl} + \Gamma_j \prime \underline{z}). \qquad (4.88)$$

In all of what follows, it will be necessary to assume that the route for each class contains an infinite server center. That is:

$$\rho_{j0} > 0, \qquad j=1,2,...p. \qquad (4.89)$$

Thus the set I^* is empty and I contains the whole set.

4.4.4 Performance Measures

Such important performance measures as average throughput and average delay are simply related to utilization [BRUE 80]. Therefore, we will concentrate on utilization. Let $u_{\sigma i}(\underline{K})$ be the utilization of the ith processor by customers of the σth class for a population distribution, by class, in the network described by $\underline{K} = (K_1, K_2, K_3, \cdots K_p)$, as before. Then, from [BRUE 80]

$$\mu_{\sigma i}(\underline{K}) = \rho_{\sigma i} \frac{G(\underline{K} - 1_\sigma)}{G(\underline{K})}, \qquad (4.90)$$

$$1 \leq \sigma \leq p, \qquad 1 \leq i \leq q.$$

Here 1_σ is a vector with all components zero except for the σth component, which is one, as before. With some manipulation, see [McKEN 82], this can be rewritten as

$$u_{\sigma i}(\underline{K} + 1_\sigma)^{-1} = \qquad (4.91)$$

$$\left\{ \frac{1}{r_{\sigma i}(K_\sigma + 1)} \right\} \left[\delta_{\sigma l} + \frac{\int_{Q^+} (\Gamma_\sigma \prime \underline{z}) e^{-Nf(\underline{z})} d\underline{z}}{\int_{Q^+} e^{-Nf(\underline{z})} d\underline{z}} \right].$$

A technical note is warranted. In a typical large network under normal (see below) operating conditions $r_{\sigma i}$ can be expected to be on the order of $1/N$, because of the normalization used. Also, K_σ is on the order of N and the term in braces is either close to 0 or 1.

More generally, what has been accomplished is to represent a basic performance measure as a ratio of integrals. This alone would not lead to a computational savings. What helps further is that these integrals can be approximated quite accurately by simple asymptotic series.

4.4.5 "Normal Usage"

The asymptotic expansions that will be developed are for a network under "normal usage". Intuitively this is a network that is not too heavily loaded, say with loads under 80%. More precisely, define

$$\underline{\alpha} \equiv \underline{1} - \sum \beta_j \underline{\Gamma}_j. \tag{4.92}$$

In terms of the original network parameters:

$$\alpha_i = 1 - \sum K_j \frac{\rho_{ji}}{\rho_{j0}}, \qquad i=1,2,3,\dots q. \tag{4.93}$$

Because of the multiplication of β_j and Γ_j, α is independent of N. Physically, α_i is a measure of the unutilized processing capability of the ith service center. Positive α_i corresponds to less than "normal" processor utilizations and negative values to very high utilizations. In what follows it is assumed that $\alpha_i > 0$, $i=1,2,3,\dots q$. This is normal usage. Qualitatively different expansions are needed when some of the α_i are negative (very high utilization).

We close this section by noting that it is shown in [McKEN 82] that, asymptotic with network size, and for all $\alpha_i > 0$

$$u_i = \text{utilization of ith processor} \approx 1-\alpha_i. \tag{4.94}$$

This expression directly relates the parameters α_i's to the actual utilization.

4.4.6 Some Transformations

We will now look at a transformation of the previous integral, one of whose purposes is to bring the expression for utilization into a form that is amenable to an asymptotic expression. We start by adding and subtracting terms in the exponent:

$$\int_{Q^+} e^{-Nf(\underline{z})} d\underline{z} = \int_{Q^+} e^{-Nf(\underline{z})+N\sum \beta_j(\underline{\Gamma}_j'\underline{z})-N\sum \beta_j(\underline{\Gamma}_j'\underline{z})} d\underline{z}, \tag{4.95}$$

$$\int_{Q^+} e^{-Nf\,(\underline{z})} d\underline{z} = \tag{4.96}$$

$$\int_{Q^+} e^{-N\alpha'\underline{z}} \exp{-[N\sum_j \beta_j\{\Gamma_{j'}\underline{z} - \log(1 + \Gamma_{j'}\underline{z})\}]} d\underline{z},$$

$$\int_{Q^+} e^{-Nf\,(\underline{z})} d\underline{z} = \tag{4.97}$$

$$N^{-q} \int_{Q^+} e^{-\alpha'\underline{u}} \exp{-[\sum_j \beta_j\{\Gamma_{j'}\underline{u} - N\log(1 + \frac{1}{N}\Gamma_{j'}\underline{u})\}]} d\underline{u}$$

Here $\underline{u} = N\underline{z}$. Next let us make a change of variables:

$$v_i \equiv \alpha_i \mu_i, \qquad 1 \le i \le q. \tag{4.98}$$

Let us also normalize Γ_{ji}:

$$\tilde{\Gamma}_{ji} \equiv \Gamma_{ji}/\alpha_i. \tag{4.99}$$

Note that

$$\Gamma_{j'}\underline{u} \equiv \tilde{\Gamma}_{j'}\underline{v}. \tag{4.100}$$

From the immediately preceding integral

$$\boxed{\int_{Q^+} e^{-Nf\,(\underline{z})} d\underline{z} = \frac{N^{-q}}{\prod \alpha_i} \int_{Q^+} e^{-1'\underline{v}} H(N^{-1},\underline{v}) d\underline{v} \qquad (4.101)}$$

where

$$H(N^{-1},\underline{v}) \equiv e^{s(N^{-1},\underline{v})}, \tag{4.102}$$

$$s(N^{-1},\underline{v}) \equiv -\sum_{j=1}^{p} \beta_j\{\tilde{\Gamma}_{j'}\underline{v} - N\log(1 + \frac{1}{N}\tilde{\Gamma}_{j'}\underline{v})\}. \tag{4.103}$$

The other reason for making this transformation is that $H(N^{-1},\underline{v})$ will make it possible, later, to determine error bounds for the asymptotic series.

Repeating the transformation for the integral $\int (\Gamma_{\sigma}'\underline{z}) e^{-Nf(\underline{z})} d\underline{z}$ results in a final expression for the utilization

$$u_{\sigma i}(\underline{K+1}_{\sigma})^{-1} = \left\{ \frac{\rho_{\sigma 0}}{\rho_{\sigma i}(K_{\sigma}+1)} \right\}$$ (4.104)

$$\times \left[1 + \frac{1}{N} \frac{\int_{Q^+} e^{-\underline{1}'\underline{v}} (\tilde{\Gamma}_{\sigma}'\underline{v}) H(N^{-1}, \underline{v}) d\underline{v}}{\int_{Q^+} e^{-\underline{1}'\underline{v}} H(N^{-1}, \underline{v}) d\underline{v}} \right]$$

or more compactly

$$u_{\sigma i}(\underline{K+1}_{\sigma})^{-1} = \left\{ \frac{\rho_{\sigma 0}}{\rho_{\sigma i}(K_{\sigma}+1)} \right\} \left[1 + \frac{1}{N} \frac{I_{\sigma}^{(1)}(N)}{I(N)} \right]$$ (4.105)

where $I_{\sigma}^{(1)}(N)$ and $I(N)$ are the integrals of the numerator and denominator, respectively.

4.4.7 Asymptotic Expansions

The PANACEA procedure is to obtain first the power series

$$H(N^{-1}, \underline{v}) = \sum_{k=0}^{\infty} \frac{h_k(\underline{v})}{N^k}$$ (4.106)

and integrate to obtain

$$A_k \equiv \int_{Q^+} e^{-\underline{1}'\underline{v}} h_k(\underline{v}) d\underline{v}.$$ (4.107)

These A_k can be used so that the previous integral is

$$I(N) \approx \sum_{k=0}^{\infty} \frac{A_k}{N^k}. \qquad (4.108)$$

Now

$$h_k(\underline{v}) = \frac{1}{k!} \frac{\partial^k}{\partial(1/N)^k} H(0,\underline{v}), \qquad k=0,1,2,..., \qquad (4.109)$$

and recall that

$$H(N^{-1},\underline{v}) = e^{s(N^{-1},\underline{v})}. \qquad (4.110)$$

From the earlier definition of s it can be noted that for fixed $\underline{v} \varepsilon Q^+$, $s(N^{-1},\underline{v})$ and thus $H(N^{-1},\underline{v})$ are functions of N^{-1} and are analytic in $\mathrm{Re}(\overline{N^{-1}}) > \varepsilon(\underline{v})$ where $\varepsilon(\underline{v}) < 0$.

We can write

$$s^{(k)}(0,\underline{v}) = -k! f_{k+1}(\underline{v}), \qquad k=1,2,3,..., \qquad (4.111)$$

where

$$f_k(\underline{v}) = \frac{(-1)^k}{k} \sum_j \beta_j (\tilde{\Gamma}_j \prime \underline{v})^k, \qquad k=1,2,3,.... \qquad (4.112)$$

The derivatives of H can be expressed as

$$H^{(k+1)}(N^{-1},\underline{v}) = \qquad (4.113)$$

$$\sum_{m=0}^{k} \binom{k}{m} s^{(k+1-m)}(N^{-1},\underline{v}) H^{(m)}(N^{-1},\underline{v}), \qquad k=0,1,2,....$$

From the above the $\{h_k(\underline{v})\}$ can be generated recursively:

$$h_0(\underline{v}) = 1, \qquad (4.114)$$

$$h_{k+1}(\underline{v}) = -\frac{1}{k+1}\sum_{m=0}^{k}(k+1-m)f_{k+2-m}(\underline{v})h_m(\underline{v}), \quad k=0,1,2,\dots .$$

Specifically:

$$h_0(\underline{v}) \equiv 1, \tag{4.115}$$

$$h_1(\underline{v}) = -f_2(\underline{v}),$$

$$h_2(\underline{v}) = -f_3(\underline{v}) + \frac{1}{2}f_2^2(\underline{v}),$$

$$h_3(\underline{v}) = -f_4(\underline{v}) + f_2(\underline{v})f_3(\underline{v}) - \frac{1}{6}f_2^3(\underline{v}).$$

To summarize

$$I(N) \approx \sum_{k=0}^{\infty}\frac{A_k}{N^k} \tag{4.116}$$

where the $\int e^{-1'\underline{v}}h_k(\underline{v})d\underline{v}$ and the $\{h_k(\underline{v})\}$ are obtained recursively from the above equation. The calculation of $I_\sigma^{(1)}(N)$ is quite similar [McKEN 82]. But the calculation of the A_k's can be put into a quite interesting context:

4.4.8 The Pseudonetworks

It turns out that the compositions of the coefficients $\{A_k\}$ and $\{A_{\sigma,k}^{(1)}\}$ are related to the normalization constant of a certain network called the pseudonetwork. To compute the first few elements of $\{A_k\}$ and $\{A_{\sigma,k}^{(1)}\}$ it is only necessary to consider the pseudonetwork with small populations. Thus one can use an efficient algorithm for small populations, such as the convolution algorithm, in the process of calculating the coefficients of the asymptotic series.

For example

$$A_3 = \int_{Q^+} e^{-1'\underline{v}}h_3(\underline{v})d\underline{v}, \tag{4.117}$$

$$A_3 = \int_{Q^+} e^{-1'\underline{v}}\{-f_4(\underline{v}) + f_2(\underline{v})f_3(\underline{v}) - \frac{1}{6}f_2^3(\underline{v})\}d\underline{v}. \tag{4.118}$$

Let us call the third term A_{33}. Using the earlier expansion for $f_2(\underline{v})$:

$$A_{33} = -\frac{1}{48}\sum_j \beta_j^3 \int e^{-1\prime\underline{v}}(\tilde{\Gamma}_j\prime\underline{v})^6 d\underline{v} \qquad (4.119)$$

$$= -\frac{1}{16}\sum_{j\neq k}\beta_j^2\beta_k \int e^{-1\prime\underline{v}}(\tilde{\Gamma}_j\prime\underline{v})^4(\tilde{\Gamma}_k\prime\underline{v})^2 d\underline{v}$$

$$-\frac{1}{48}\sum_{j\neq k\neq l}\beta_j\beta_k\beta_l \int e^{-1\prime\underline{v}}(\tilde{\Gamma}_j\prime\underline{v})^2(\tilde{\Gamma}_k\prime\underline{v})^2(\tilde{\Gamma}_l\prime\underline{v})^2 d\underline{v}$$

The subscripts range over $[1,p]$. The typical integral in the composition of the asymptotic expansion coefficients is, within a multiplicative constant

$$g\,(\underline{m}) = g(m_1,m_2,m_3,\cdots m_p) \equiv \qquad (4.120)$$

$$\frac{1}{(\prod_j m_j!)} \int_{Q^+} e^{-1\prime\underline{u}}\prod_j(\tilde{\Gamma}_j\prime\underline{u})^{m_j} d\underline{u}.$$

This should be familiar since if we take equation (4.86) for the partition function, assume no infinite service centers (I is empty), it becomes the similar

$$G(\underline{K}) = \frac{1}{(\prod_j K_j!)} \int_{Q^+} e^{-1\prime\underline{u}}\prod_j(r_j\prime\underline{u})^{K_j} d\underline{u}. \qquad (4.121)$$

Because of this similarity, $g\,(\underline{m})$ is the normalization constant of a certain network, the "pseudonetwork". The pseudonetwork is closed and has no infinite service centers. As in the original network there are exactly q processing centers and p classes of customers. The service rate of jth class customers in the ith service center is $\tilde{\Gamma}_{ji}$. The population distribution by class in vector form is $(m_1,m_2,m_3,\cdots m_p)$.

Thus we can generate the $\{A_k\}$ as functions of the normalization constants of the pseudonetwork:

$$A_0 = 1, \qquad (4.122)$$

$$A_1 = -\sum_j\beta_j g(2\cdot\underline{1}_j),$$

$$A_2 = 2\sum_j\beta_j g(3\cdot\underline{1}_j) + 3\sum_j\beta_j^2 g(4\cdot\underline{1}_j) + \frac{1}{2}\sum_{j\neq k}\beta_j\beta_k g(2\cdot\underline{1}_j + 2\cdot\underline{1}_k),$$

$$A_3 = -6\sum_j \beta_j g(4 \cdot \underline{1}_j) - 20\sum_j \beta_j^2 g(5 \cdot \underline{1}_j) - 15\sum_j \beta_j^3 g(6 \cdot \underline{1}_j)$$

$$-2\sum_{j\neq k} \beta_j \beta_k g(2 \cdot \underline{1}_j + 3 \cdot \underline{1}_k) - 3\sum_{j\neq k} \beta_j^2 \beta_k g(4 \cdot \underline{1}_j + 2 \cdot \underline{1}_k)$$

$$-\frac{1}{6}\sum_{j\neq k\neq l} \beta_j \beta_k \beta_l g(2 \cdot \underline{1}_j + 2 \cdot \underline{1}_k + 2 \cdot \underline{1}_l).$$

Here j,k,l are class indices in the range $[1,p]$.

It has been found that the above four terms yield a reasonable approximation to $I(N)$. In the above at most three classes have a nonzero number of customers and the total population is at most 6 (for the similar $\{A_{o,k}^{(1)}\}$ it is 7, see [McKEN 82]). The small population sizes involved make the use of the convolution algorithm efficient.

4.4.9 Error Analysis

In a practical sense, it is important to know how accurate the PANACEA approximation is. It turns out that the errors in estimating the integral from the use of m terms is of the same order as the $(m+1)$th term as $N \to \infty$.

That is [McKEN 82],

$$\frac{A_m}{N^m} < I(N) - \sum_{k=0}^{m-1} \frac{A_k}{N^k} < 0, \qquad m=1,3,5,..., \qquad (4.123)$$

$$0 < I(N) - \sum_{k=0}^{m-1} \frac{A_k}{N^k} < \frac{A_m}{N^m}, \qquad m=2,4,6,....$$

This is a consequence of H being completely monotonic with derivatives alternating in sign.

We end here by noting that the convolution algorithm actually is used in the PANACEA package to calculate the normalization constants of the pseudonetworks. Under normal usage it is found that only four leading coefficients give high accuracy estimates.

4.5 Norton's Equivalent for Queueing Networks

4.5.1 Introduction

A basic concept in electric circuit theory is that of the equivalent network (or Norton's or Thevenin's theorem). That is, any collection of impedances and current and voltage sources can, through a series of transformations, be replaced by a simpler equivalent network consisting of a single impedance and either a single current source or a single voltage source. "Equivalence" here means that the equivalent circuit presents electrical characteristics to any external circuitry that are identical and indistinguishable from those of the original circuit.

The advantage of an equivalent circuit is twofold. First of all, an equivalent circuit can greatly simplify analysis and design. Second, it can also greatly simplify numerical circuit calculation in that an equivalent circuit consists of far fewer components than the circuit it replaces. This is particularly true if a subnetwork of the circuit is important for its gross behavior and not for its internal and detailed characteristics. The only restriction on the use of equivalent networks in circuit theory is that the results are only exact for networks comprised of linear components.

Can one take a queueing subnetwork and replace it with an equivalent queue that presents identical queueing characteristics to the remainder of the network? In a paper published in 1975 [CHAN 75] K. M. Chandy, U. Herzog and L. Woo showed that this can indeed be done with certain restrictions. The main restriction is that the network should be a product form network for the results to be exact. If this is so, any two port (one input and one output) subnetwork can be replaced by a single state-dependent Markovian queue presenting identical characteristics to the rest of the network.

The procedure for computing an equivalent queue for a queueing subnetwork in a closed queueing network is as follows. Take the two ports (input and output) and "short" the two ports. That is, establish a zero delay path from the output to the input that effectively bypasses the network outside of the subnetwork. Then compute the throughput through the short supplied by the subnetwork when there are $i=1,2,...N$ customers in the subnetwork. This can be done using the mean value analysis or convolution algorithm. Finally replace the subnetwork with a state-dependent queue whose service rate for i customers equals the previously mentioned shorted throughput for i customers.

In what follows the equivalence produced by this procedure will first be established, following [CHAN 75]. An example of the use of the technique will then be presented.

4.5.2 Equivalence

Let us start with a closed network of M queues. All but the Mth queue will be replaced by an equivalent queue with state-dependent service rate $\mu_{eq}(N)$. Suppose we let $P_M(n)$ be the marginal queue length distribution of queue M when connected to the original sub-network. Also let $P'_M(n)$ be the marginal queue length distribution of queue M when it is connected to the equivalent state dependent queue.

Following Chandy, Herzog, and Woo:

Theorem: The marginal queue length distribution for queue M when connected to the equivalent queue is the same as when connected to the original subnetwork:

$$P'_M(n) = P_M(n), \qquad n=1,2,...N. \tag{4.124}$$

What this practically means is that from the point of view of the external queue M, the substitution of the original subnetwork by the state dependent queue results in identical statistics for queue M. In fact, as far as queue M is concerned, there is no way to tell that a substitution has been made. Let us see why this theorem must be true. From the definition of the product form solution for closed networks (equation (3.36)) one can show that $P'_M(n)$ is proportional to

$$\prod_{j=1}^{n} \frac{1}{\mu_M(j)} \prod_{k=1}^{N-n} \frac{1}{\mu_{eq}(k)}. \tag{4.125}$$

Here the presence of n customers in the Mth queue implies that there are $N-n$ customers in the state-dependent equivalent queue as there are a total of N customers in the entire network.

When queue M is shorted out (while the input/output of the remaining M queues is shorted) its effective service time is zero. Then $f_M(0)=1$ and $f_M(n)=0$ for $n \neq 0$ as the service rate in the denominator of (4.3) is infinite. In this case the service rate of the state dependent equivalent queue is equal to

$$\mu_{eq}(n) = \theta_M \frac{G_{M-1}(n-1)}{G_{M-1}(n)}, \qquad n=1,2,...N. \tag{4.126}$$

This can be recognized as the throughput equation (4.47) set equal to the throughput of the shorted queue with n customers in the entire network. Here $G_{M-1}(n)$ is the normalization constant for the subnetwork of M-1 queues and θ_M is the traffic equation solution for the Mth queue.

If one substitutes (4.126) into (4.125) one can see that $P'_M(n)$ is proportional to

$$\frac{f_M(n)}{(\theta_M)^N} \frac{G_{M-1}(N-n)}{G_{M-1}(0)}, \quad n=0,1,2,...N. \tag{4.127}$$

To arrive at this result use is also made of the definition of $f_M(n)$, equation (4.3). More simply the constants can be removed so that

$$P'_M(n) \quad \alpha \quad f_M(n)G_{M-1}(N-n), \quad n=0,1,2,...N. \tag{4.128}$$

However, it is also true that

$$P_M(n) \quad \alpha \quad f_M(n)G_{M-1}(N-n), \quad n=0,1,2,...N, \tag{4.129}$$

as can be showed from (4.37). Thus $P'_M(n)=P_M(n)$ for $n=0,1,2,...N$, and the proof is complete.

In their original paper Chandy, Herzog, and Woo also discussed creating equivalent networks in the context of open networks. Consider an open network with a two port (input and output) subnetwork. The remainder of the network (external to the subnetwork) can be replaced by an equivalent Poisson source feeding into the subnetwork input. The subnetwork's statistics are preserved under this substitution. Of course, the original network, subnetwork and all, must be a product form network. They also extended the idea of equivalence to multiclass queueing networks satisfying local balance.

A probabilistic interpretation of equivalence for closed queueing networks appears in [HSIA]. Here the service rate $\mu_{eq}(n)$ of the state dependent equivalent queue is expressed as the conditional expectation of the output service rate of the subnetwork replaced by the equivalent queue given that there are n customers in the subnetwork. This expectation must be computed over all the ways that n customers may be arranged in the subnetwork. Note that some arrangements lead to zero output. One advantage of the conditional expectation formulation is that conditional expectations are estimates that can be obtained through time delay measurements during the normal operation of a network.

Example:

Consider the queueing network in Figure 4.2. It is desired to replace the circled subnetwork with a single state-dependent equivalent queue. Shorting the subnetwork's output and input results in the reduced network of Figure 4.3. What has to be computed is the throughput in the lowermost path. However, to do this it is simpler to combine the upper and lower return paths and replace the network of Figure 4.3 by a simple cyclic two queue network. The MVA algorithm can be used to compute the throughput of this network:

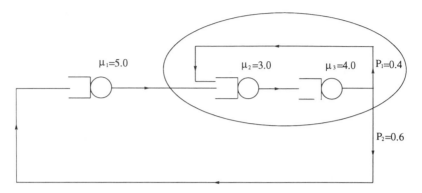

Fig. 4.2: Original Network

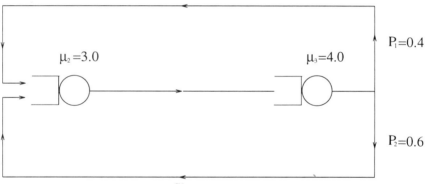

Short

Fig. 4.3: Reduced (shorted) Network

$$Y(1) = 1.71,$$
$$Y(2) = 2.27,$$
$$Y(3) = 2.54.$$

The desired throughput is just 0.6 times these values or just the throughput of the lower path in Figure 4.3:

$$Y(1) = 1.03,$$
$$Y(2) = 1.36,$$
$$Y(3) = 1.52.$$

Thus the service rate of the state-dependent equivalent queue is set to these values:

$$\mu_{eq}(1) = 1.03,$$
$$\mu_{eq}(2) = 1.36,$$
$$\mu_{eq}(3) = 1.52.$$

The following section on the simulation of communication networks is authored by J. F. Kurose and by H. T. Mouftah. It originally appeared as part of [KURO], copyright 1988 IEEE, and is reprinted here with the permission of the authors and the IEEE.

4.6 Simulation of Communication Networks

By J. F. Kurose and H. T. Mouftah

4.6.1 Introduction

As seen in the previous sections, a tractable analytic model often restricts the range of system characteristics that can be *explicitly* considered in a performance model. For example, systems with complex job-scheduling policies, finite buffers (which cause blocking), simultaneous resource possession by a job, complex timing constraints, or tightly coupled interactions between various parts of the model cannot be explicitly modeled using the general-purpose queueing paradigm of the previous sections. In such cases, approximate analytic models may sometimes be adopted [LAVE 83], [SAUE 81] or *simulation* may be used to solve a performance model.

The tradeoffs between an analytic and a simulation approach toward modeling are in the relative amounts of time spent on model formulation and model solution. Since analytic models require a higher degree of abstraction, considerable effort and skill may be required on the part of the network modeler to develop a performance model that accurately reflects the system under study. The analytic model itself, however, can generally be solved rather quickly. In a simulation approach, the system may be modeled to any arbitrary degree of detail; the process model formulation thus becomes a more straightforward task, although it remains advantageous from a solution standpoint to abstract out as many secondary system details as possible. The solution of a simulation model requires significantly more computer time. In some cases, however, simulation is the only viable approach.

The simulation technique typically used to solve a model of a communication network or protocol is *stochastic discrete event simulation*. In discrete event simulation, various components of the actual network under study (e.g., the communication links, buffers, access strategies, network control structures) are represented within a computer program. The *events* that would occur during the actual operation of the network (e.g., the arrival, transmission, routing, and departure of messages, error conditions such as message loss or corruption, link/node failures and recovery) are then mimicked during the execution of the program. The function of the simulation program is thus simply to generate events and then simulate the network's response. (The simulation program typically also performs other ancillary tasks such as recording and later analyzing performance data as well.)

Our goal in this section is not to provide a tutorial introduction to the topic of discrete event simulation; several excellent texts are readily available on both the general topic [FISH 78a], [LAVE 83], [LAW 82], as well as on the use of discrete event simulation for performance modeling, analysis, and design of computer networks [SAUE 83], [SCHO]. Moreover, there are many software tools available that free the practicing modeler from the lower level details of discrete event simulation. Instead, our goal is to provide an overview of several current research areas that are currently advancing the state-of-the-art in using simulation in communication network design and performance analysis.

4.6.2 The Statistical Nature of a Simulation

The generation of events by the simulation program is driven by a stream of *pseudorandom numbers*. These random numbers might be used, for example, to generate the lengths of messages, interarrival times of messages at a given node, time between failures of a link, or probability of a transmission error. Since the event-generation process depends on the pseudorandom number stream, the performance measures output by the model (e.g., queueing delays, device utilizations, buffer occupancies, message loss, etc.) are themselves random in nature.

A simulation thus represents a statistical experiment [WELC], and the performance results should thus be subjected to careful statistical analysis. For a given sequence of random numbers, a particular set of values is obtained for the performance measures of interest (e.g., message delay, buffer occupancies, device utilization, throughput, etc.). If the simulation had been run for either more or less time, or if a different stream of random numbers had been used, different values would have been obtained for these performance measures. How then does one determine whether the performance values obtained from a particular run are in some sense the true or correct values?

1) Transient Versus Steady-State Performance Values: In some cases, the network modeler is interested in performance measures such as the average delay of the first 100 messages through a node (e.g., following a simulated link failure) or the buffer occupancy statistics over some relatively short period of time. In such cases, the performance results may depend quite strongly on the initial conditions of the simulation, e.g., the number of messages initially in the node or buffer when the simulation begins. Performance measures that depend on the initial state of the simulation are referred to as *transient* measures. In general, the amount of simulated time needed to obtain these transient statistics is well defined (as above) by the nature of transient measures of interest.

In other cases, a modeler may be interested in the long-term or steady-state results of a simulation. In this case, the simulation must typically be run long enough so that the effects of the initial state of the simulation on the performance measures of interest are negligible. Alternatively, the transient portion of a simulation run may be discarded and performance statistics collected only after the simulation has reached steady state. The problem of determining the end of the transient phase is a difficult one, since there is no well-specified point in time at which the transient phase ends. Rather, the effects of the initial configuration become less important as the length of the simulation increases. The problem of identifying the transient phase of a simulation is addressed in [SCHR] and [WELC].

2) Confidence Intervals: Since the performance results predicted by one run of the simulation depend on the particular stream of pseudorandom numbers used to drive the simulation, the results of a performance model will typically vary from one run to another. If we were to run the simulation ten times and obtain ten different values, all within one percent of each other, our confidence in that value would be high. On the other hand, if the ten values obtained varied greatly, our confidence in any one value, or even in the average of the ten values, would be small. In a simulation experiment, *confidence interval techniques* are used to quantify such confidence in a performance estimate. Roughly speaking, if some interval (a,b) is an χ percent confidence interval for the performance measure μ, then if the simulation were to be independently repeated some number of times, the estimated value for μ obtained from the simulation would fall in the interval (a,b) in approximately χ percent of these runs.

Several techniques have been devised for generating confidence intervals. In the method of independent replications, the simulation is run n independent times and n estimates are thus obtained for each performance measure of interest. Each set of n values thus represents n independent samples of the quantity to be estimated and standard statistical techniques can be used to construct a confidence interval [LAW 83], [WELC]. Note that in the case of steady-state performance measures, the transient portion of *each* of the *n* simulation runs must be discarded, thus making the method of independent replications a potentially expensive confidence interval technique. In the method of batch means, a *single* run of the simulation is divided into *N* equal-length periods of time (after discarding the initial transient phase). The values of the performance measure during each of these *N* periods are then taken as *N* (approximately) independent samples and a confidence interval can again be constructed for this performance measure using standard statistical techniques. Note that this technique has the advantage of discarding just a single transient phase. However, the problem of determining an appropriate value of *N* is crucial yet difficult [FISH 78b], [WELC]. If *N* is too small, the samples will be correlated; if *N* is too large, an excessive amount of simulation time will be used.

A third technique for generating confidence intervals is known as the regenerative method. This technique is also based on partitioning a single simulation run into independent subruns; it thus also enjoys the advantage of requiring that only a single transient phase be discarded. In the regenerative approach, the simulation run is partitioned on the basis of a modeler-defined *regeneration state,* a state such that the future evolution of the simulation is statistically identical following each entry into this state. For example, in an open queueing model of a network with Poisson arrivals, the regeneration state might be the state with no messages in the system. The regeneration state thus serves to divide the simulation into independent and identically distributed partitions of time of random length. The technique for generating confidence intervals using the regenerative method is described in [CRAN]. One drawback of this method is that a regeneration state may be hard to identify. Also, large models may require an excessive amount of simulation time to pass through enough regeneration points to produce valid confidence intervals.

The three confidence interval techniques discussed above are based on partitioning the overall simulation and obtaining (at least approximately) independent and identically distributed values of performance measures from the partitions. A fourth confidence interval technique, known as the spectral method [HEID 81], is a single run method and explicitly takes into account the correlation between data gathered by the simulation (e.g., between successive queueing times at a queue).

4.6.3 Sensitivity Analysis of Simulation Results

An important part of any modeling study is that of sensitivity analysis-- determining how changes in model parameters affect system performance. For example, a network modeler might be interested in the effect of an increased arrival rate or a decreased channel capacity on the average message delay. Sensitivity analysis also helps identify both critical model parameters, as well as those that have relatively little influence on performance. It thus also provides an indication of the quality and general validity of the model. If performance is extremely sensitive to a certain parameter value, the model may not be applicable over a wide range of parameter value. On the other hand, if performance is insensitive to certain parameters, this would suggest that the model may be needlessly complex and that these parameters might possibly be abstracted out of the model with no loss of applicability.

The most straightforward (and common) approach toward sensitivity analysis is to first run the simulation to determine the baseline performance for the given parameter values. The simulation is then run again several times, each time with a slightly perturbed value of a single model parameter. Clearly, if there are a large number of parameters, this can be a very costly process--if there are N parameters, the simulation must be run N times to determine model sensitivity for the baseline parameter values.

Recently, two approaches have emerged for estimating such sensitivity or gradient information from a *single* run of a simulation. Although their applicability is often restricted to certain classes of models or simulation techniques, they offer a promising methodology for efficiently obtaining sensitivity information. The first approach is known as perturbation analysis (PA) [CASS], [HO 83a], [SURI 87]. One version of this approach, known as infinitesimal perturbation analysis (IPA), is based on the assumption that if an extremely small change had been made in a parameter value before a simulation run, the timing, but not the relative ordering of the simulation events, would have changed. IPA routines can be incorporated into a simulation to track these relative timing changes as the simulation progresses and then produce a sensitivity estimate at the end of the single simulation run. PA has been studied both in the context of single queue systems [SURI 84], [ZAZA] and networks of queues [HO 83b], [HO 84]. Some potential limitations of perturbation analysis are discussed in [HEID 86].

A second approach for obtaining sensitivity estimates from a single simulation run is based on the use of likelihood ratios [REIM]. This technique, which requires only that the occurrence of certain events be counted during the simulation, utilizes the natural variation in the random processes underlying the parameter of interest to generate the sensitivity estimate. In [REIM], it is noted that the likelihood ratio method has an advantage over PA in that sensitivities can be computed with respect to any parameter in a simple and uniform manner. It has the disadvantage, however, of being applicable primarily to regenerative simulation and generally providing sensitivity estimates with a higher variance than those produced using PA.

As discussed above, sensitivity estimates provide important information about the robustness of the simulation results. An additional use of these estimates is in the automated Monte Carlo optimization of the network designs. Typically, network performance results from a complex interaction among many parameters, or design variables, and an explicit closed-form expression seldom exists for the performance measures of interest, much less their derivatives. In such cases, one possible approach is to use simulation to optimize system performance with respect to the design variables. In this case, the system is simulated, sensitivities with respect to the design variables are determined, and the model parameters are then changed in the direction for which the performance increase is greatest. The system is then simulated again and this iterative process repeated until the design has been optimized. Note that the optimization process is complicated by the fact that the gradient values computed via simulation represent noisy estimates of the true gradient values. Hence, a stochastic approximation (optimization) method must be used [GLYN 86b]. A probably convergent simulation-based optimization algorithm is given in [GLYN 86a] which obtains sensitivities using a likelihood ratio approach. The use in optimization of sensitivity estimates via perturbation is studied in [HO 83b].

4.6.4 Speeding Up a Simulation

The major disadvantage of a simulation (as opposed to analytic) approach toward network performance modeling is the amount of time needed to simulate a model. Generally, as network protocols become more sophisticated and complex, so too do their performance models. Moreover, rapidly increasing computation, communication, and switching speeds of communication networks are resulting in ever-increasing message traffic rates. For example, the designs of some fast packet switches (e.g., [TURN]) permit up to 1.2 million packets to be switched *per second.* Clearly, simulating the operation of even one such switch for any nontrivial amount of time is a formidable computational task.

    1) Distributed and Parallel Simulations: One approach toward decreasing the amount of time needed to perform a simulation study is to distribute the computational burden of simulation among several processors in a distributed or parallel multiprocessor system [CHAN 79], [CHAN 81], [COMF], [DAVI], [DECE], [JEFF], [MISR], [PEAC]. In this approach, the simulation is partitioned into numerous *logical processes,* each of which is responsible for simulating one or more of the physical processes in the system being modeled. For example, the operation of each network node might be simulated by a different logical process; in a high-speed packet switch, a logical process might be responsible for simulating the packet processor for a specific input line or an element within the switching fabric.

    In a multiprocessor environment, different logical processes may be executed in parallel on different processors, with the goal of decreasing the time required until the simulation is completed. Of course, processes may not proceed completely asynchronously, since causal interactions between processes in the physical system must also be maintained between the logical processes of the simulation. The manner in which this synchronization is maintained in the simulated system is of fundamental importance, since it is this which prevents a K-fold speedup from being obtained when a simulation is distributed over K processors. Synchronization is typically implemented via message passing between the logical processes. In [DECE], [JEFF], an optimistic approach toward synchronization is taken. Logical processes are permitted to continue a simulation (assuming the absence of a synchronization message) even though a later-arriving synchronization message may require a process to resimulate a previous part of the simulation. This resimulation is accomplished using a simple method for rolling-back the state of the logical process's simulation and undoing its effects on other processes. The experimental results reported in [DECE] suggest that this method yields a linear increase in effective computing power as the number of nodes increases.

In [CHAN 79], [CHAN 81], [MISR], and [PEAC], less asynchrony is permitted but rollback is not required. This approach has been reported to produce significant speedups in the simulation of acyclic queueing networks [MISR]. However, recent studies of actual implementations on a multiprocessor system have shown that for queueing networks in which customer feedback is important (e.g., as in a central server model [SAUE 81]), the synchronization overhead of certain distributed simulation techniques may be so high that simulation distributed over five machines runs several times *slower* than if the simulation had been run on but one of these machines [REED]. A third approach towards distributing a simulation has been investigated in [LUBA]. In this work, a time-driven (as opposed to event-driven) approach was taken, i.e., the simulation clock was incremented by fixed-length time intervals rather than on the basis of event times. For synchronous networks, this was shown to result in a significant speedup of the simulation.

An alternative to distributing a single simulation over K processors is to run K independent copies of the simulation in parallel, one on each of the K processors. In this case, no synchronization is required among the K processors. At the end of the simulation, the results of the K simulations are averaged and the statistical significance of these results determined. As shown in [HEID 85], the simulation termination criteria must be carefully chosen in order to avoid introducing sampling bias into the estimates of the performance measures. A model is also developed in [HEID 85] for comparing distributing a single simulation over K processors with running K independent copies of simulation on K processors. The purpose of this comparison is to determine which approach is statistically more efficient, i.e., produces performance estimates with a smaller mean squared error for the same amount of computing resources. It is shown that if the run length is long or the initial transient period is short, replicating a simulation is statistically more efficient than distributing a simulation. The question of combining the two methods of simulation is also investigated.

2) Hierarchical Decomposition and Hybrid Techniques: Hierarchical decomposition is a *modeling* technique for speeding up a simulation. In this approach, portions of the model are grouped into submodels and each submodel is replaced by a *single* composite queue with a queue-length dependent service rate. If the submodel has a product form solution (see chapter 3), this reduced model is statistically identical to the original model [CHAN 75]. In the case of nearly decomposable systems [COUR 77], the decomposition is approximate but often quite accurate (see, e.g., the model of SNA virtual route pacing in [SCHW 82]).

The potential speedup results from the fact that the submodel need only be solved once (for each possible customer population) in order to determine the service rates of the composite queue. Then, in the simulation of the entire model, the potentially large number of events that would have occurred when a customer passed through the submodel are replaced by only two events: the arrival and eventual departure of the customer from the composite queue. Experimental results of decomposing extended queueing network models are reported in [BLUM] and several rules of thumb are given for cases in which decomposition would be advantageous.

Note that the submodel itself may be solved either through simulation or analysis, even if the overall model requires a simulation solution. The possibility of such a *hybrid* approach toward modeling [SHAN], i.e., combining both analysis and simulation in a single model, may result in even further reduction in model solution time. Studies examining the use of hybrid models in network modeling can be found in [BHAT], [FROS 86], [SAUE 76], and [VANS]. A hybrid model for studying file transfer protocols in token ring networks was studied in [WONG]. The hybrid model was found to produce performance results typically within 5 percent of those found via a detailed simulation, and did so on the average 50 times faster than the detailed simulation. In the hybrid model of CSMA/CD in [OREI], the operation of a few of the network stations was simulated in detail, while the effects of the remaining stations were modeled by an analytic algorithm that produced sequences of channel busy/idle times. This approach was shown to be more efficient than a detailed simulation of all stations when the overall number of stations in the network was large.

3) Variance Reduction: A third method for decreasing the run time of a simulation is the use of variance reduction techniques. These techniques exploit known statistical properties of the system being modeled in order to reduce the amount of time needed to obtain performance estimates of a given accuracy. While these techniques have been extensively studied in the literature, they have not often been used in real applications [HEID 84]. An overview of the application of variance reduction techniques in computer network modeling is given in [FROS 88]; a general discussion of variance techniques can be found in [LAW 82]. A promising recent variance reduction technique that has been applied to the simulation of queueing networks is based on the theory of large deviations and is described in [WALR].

To Look Further

The book by Bruell and Balbo [BRUE 80] is an excellent introduction to the convolution algorithm for closed queueing networks. It includes a variety of specialized expressions not covered here as well as performance measures for multiclass networks.

The paper by Reiser [REIS 81] describes the Mean Value Analysis algorithm for state-dependent servers. It also relates it to the LBANC algorithm of Chandy and Sauer [CHAN 80] and to the convolution algorithm. The occurrence of overflows in the algorithms is discussed.

Conway and Georganas [CONW 86, 89b] have developed an exact algorithm for multiple chain closed queueing networks, which is known as RECAL. Its basis is a new recursive expression that relates the normalization constant of a network with r closed routing chains to those of a set of networks with r-1 routing chains. This algorithm's advantage is that the time and space requirements of the algorithm are polynomial in the number of chains. The algorithm is efficient when there are a large number of chains. By way of contrast, MVA and the convolution algorithm have time and space requirements that increase exponentially with the number of chains. However, RECAL's requirements are combinatorial in the number of service centers. The implementation of RECAL on a shared memory multiprocessor is discussed in [GREE].

Conway, De Souza E Silva, and Lavenberg [CONW 89] have developed an algorithm called Mean Value Analysis by Chain (MVAC). The recursion used in this algorithm is different from that used in Mean Value Analysis but similar in structure to the recursion used in RECAL. MVAC does not compute the normalization constant and so it is felt that it avoids the underflow/overflow problems associated with implementing RECAL (or the convolution algorithm for that matter). The computation and storage costs of MVAC are similar to those for RECAL but different compared to MVA.

Another recent algorithm is Distribution Analysis by Chain (DAC) by De Souza E Silva and Lavenberg [DESO]. It computes joint queue length distributions for product form queueing networks with single server fixed rate, infinite server, and queue-dependent service centers. It also computes mean queue lengths and throughputs somewhat more efficiently than RECAL and MVAC.

The usefulness of the PANACEA approach has been extended in a number of papers. State-dependent service rates are covered in [McKEN 86]. The calculation of mean queue lengths is discussed in [McKEN 84]. Finally, the use of the PANACEA technique in calculating sojourn time (i.e. time delay experienced by a customer) distribution function is developed in [McKEN 87]. Other work on asymptotic expansions includes [BIRM] and [KOGA].

A recent paper of interest on distributed simulation is [WAGN].

Problems

4.1 Consider the convolution algorithm for a single class of customers. Show, as is done in [BRUE 80], that only $N+1$ memory locations are needed for the operation of the algorithm for both the cases of state dependent and state independent service rates.

4.2 Derive the boxed equation for $g(n,m)$ for state independent servers from the boxed equation before it for state dependent servers.

4.3 Prove that

$$E[n_i] = \sum_{n=1}^{N} np\,(n_i{=}n) = \sum_{n=1}^{N} p(n_i{\geq}n).$$

You can prove this graphically.

4.4 Prove that $g\,(n,M - \{M\}) = g\,(n,M{-}1)$. It should only take two or three lines to do this.

The following problems are designed so that the results can be cross checked with one another.

4.5 Use the convolution algorithm to calculate performance measures for a cyclic network of three Markovian queues where $\mu_1 = \mu$, $\mu_2 = 2\mu$ and $\mu_3 = 3\mu$. Specifically:

a) Calculate the normalization constants table and calculate $G(N)$ for $N=1,2,...5$.

b) Calculate the mean throughput, $\overline{Y}_i(N)$ for $N=1,2,...5$.

c) Calculate the mean number of customers, $\overline{n}_i(N)$ for $i=1,2,3$ and $N=1,2,3$.

d) Calculate the mean time delay through each queue, $\overline{\tau}_i(N)$, for $i=1,2,3$ and $N=1,2,3$.

4.6 Use the MVA algorithm to calculate performance measures for the network of the previous problem. Specifically, calculate for $i=1,2,...5$:

a) The mean number in each queue.

b) The mean time delay through each queue.

c) The mean throughput in each queue.

4.7 Use the convolution algorithm to calculate performance measures for a cyclic queueing network of three Markovian queues where $\mu_1 = 1.0$, $\mu_2 = 2.0$ and $\mu_3 = 3.0$. Specifically, calculate:

a) The normalization constants table and the normalization constants G(N) for $N=1,2,...5$.

b) The mean throughput $\overline{Y}_i(N)$ for $N=1,2,...5$.

c) The marginal queue length distribution when there are four customers in the network using the equation for state-independent servers.

d) The marginal queue length distribution for (c) using the equation for state-dependent servers. You will need to first calculate $g(N,M-\{i\})$ for $N=0,1,...4$.

e) The utilization, $U_i(4)$, for each queue in the network with four customers.

f) The mean number in the third queue, $\bar{n}_3(4)$, when there are four customers.

g) Check (f) using the marginal probability density.

h) The mean waiting time for the third queue with four customers.

4.8 Use the MVA algorithm to compute performance measures for the network of the previous example. Specifically, calculate:

a) The mean number in each queue.

b) The mean time delay through each queue.

c) The mean throughput in each queue.

4.9 Use the convolution algorithm to calculate performance measures for a Markovian queueing network with random routing where $\mu_1=\mu_2=\mu_3=2.0$. Let the routing probabilities be $r_{12}=1.0$, $r_{23}=1.0$, $r_{31}=0.4$ and $r_{32}=0.6$. Specifically:

a) Draw and label the queueing schematic.

b) The normalization constants table and the normalization constants $G(N)$ for $N=1,2,...5$.

c) The mean throughput $\overline{Y}_3(N)$ for $N=1,2,...5$.

d) The marginal queue length distribution for the the third queue when there are four customers in the network using the equation for state-independent servers.

e) The utilization, $U_i(4)$, for each queue in the network with four customers.

f) The mean number in the third queue, $\bar{n}_3(4)$, when there are four customers.

g) Check (f) using the marginal probability density.

h) The mean waiting time for the third queue with four customers.

4.10 Use the MVA algorithm to compute performance measures for the network of the previous problem. Specifically calculate:

a) The mean number in each queue.

b) The mean time delay through each queue.

c) The mean throughput in each queue.

4.11 Use the convolution algorithm for state dependent servers to calculate various performance measures for a three Markovian queue cyclic network where the service rates for queue 1 and 2 are:

Queue 1	
# Customers	Service Rate
1	1
2	5
3	8
4	10
5	11

Queue 2	
# Customers	Service Rate
1	3
2	5
3	7
4	9
5	11

Queue 3 has a state-independent server of rate eleven.

a) Calculate the entries of the normalization constants table. Then calculate $G(N)$ for $N=1,2,...5$.

b) Calculate the mean throughput, $\overline{Y}_i(N)$ for $N=1,2,...5$.

c) Calculate the marginal queue length distribution for the third queue when there are four customers.

d) Calculate, $U_3(4)$, the utilization of the third queue when there are four customers in the network.

e) Calculate, $\overline{n}_3(4)$, the mean number in the third queue when there are four customers in the network using the state-dependent formula.

f) Confirm (e) using the marginal state probabilities.

g) Calculate $\overline{\tau}_3(4)$, the mean time delay in the third queue when there are four customers in the network.

4.12 Consider a cyclic queueing network of $M+1$ Markovian queues. Each queue has service rate μ. Find the state dependent service rate of a Norton equivalent queue for M of the $M+1$ queues. Hint: Refer to one of the examples in the MVA section of this chapter.

4.13 Consider a closed Markovian queueing network of $M+1$ queues, all with service rate μ. A customer departing queue 0 enters queue i ($i=1,2,...M$) with probability p_i. Departures from queue i enter queue 0. Compute the Norton equivalent state-dependent service rate for the bank of parallel queues ($i=1,2,...M$) for 1 through 3 customers in the network.

4.14 **Computer Project**: Write and execute a computer program that executes the convolution algorithm for state-independent servers. You should be able to input the traffic equations solutions and numerical values for the service rates. The program should print out all the performance measures discussed in this chapter. Hand in a report consisting of a program flow chart, the code listing, and an example of the program's use.

4.15 **Computer Project**: Write and execute a computer program that executes the convolution algorithm for state-dependent servers. You should be able to input the traffic equations solutions and numerical values for the service rates. The program should print out all the performance measures discussed in this chapter. Hand in a report consisting of a program flow chart, the code listing, and an example of the program's use.

4.16 **Computer Project**: Write and execute a computer program that executes the mean value analysis algorithm for state-independent servers. You should be able to input the traffic equations solutions and numerical values for the service rates. The program should print out all the performance measures discussed in this chapter. Hand in a report consisting of a program flow chart, the code listing, and an example of the program's use.

Chapter 5: Stochastic Petri Nets

5.1 Introduction

Petri Nets provide a means for modeling and graphically representing the possible behavior of systems in which concurrency, serializability, synchronization and resource sharing are important considerations [PETE], [FILM]. They can be used for the understanding and prediction of the behavior of computer systems, communication protocols, biological, economic and other complex systems.

Petri Net models originated from the doctoral dissertation of C. A. Petri [PETR], written in 1962 at the University of Bonn, West Germany and have been extended over the years [BRAU], [DENN], [HOLT]. Such aspects of Petri Nets as determining the set of reachable states have received considerable attention. More recently attention has been given to including appropriate timing mechanisms in Petri Net models.

Stochastic Petri Nets (SPN's) [FLOR], [MOLL 82] are based on the assumption that the time between successive events can be modeled as a random variable. Most work to date has assumed a Markovian statistical framework. We will refer to such Stochastic Petri Nets as Markovian Petri Nets (MPN's). We do not consider the immediate firing transitions of Generalized Stochastic Petri Networks [MARS 84].

The major approaches for the numerical solution of arbitrarily configured Stochastic Petri Nets are the use of a Markov chain solver or simulation. In [DUGA] the features of three existing Stochastic Petri Net solution packages are described and combined into the design of a unified solution package. This package can solve ergodic or transient problems. This includes phase type timing [MARS 85]. The use of regeneration in simulation is discussed in [HAAS].

Work has also been conducted on state aggregation [AMMA], calculating cycle time [WONG CY], and throughput bounds [BRUE 86], [MOLL 86]. Stochastic Petri Net models of data link protocols [MOLL 82], local area networks [GRES], and bus-oriented multiprocessor systems [MARS 83] [MARS] have appeared in the literature.

A major difficulty encountered in modeling systems using the Petri Net representation is that the network schematic often becomes unwieldy as the complexity of the system increases. This gives rise to dauntingly large state transition diagrams [DIAZ]. Analytical or even numerical calculations of equilibrium probabilities become difficult. The actual difficulty, however, is not the large state transition diagram itself but rather dealing with state transition diagrams with an *arbitrary structure*. For instance, as we have seen in chapter 3, the state transition diagrams of certain classes of queueing networks [BASK 75], [GORD], [JACK 57], [KELL] exhibit enough structure so that simple product form solutions exist for the equilibrium state probabilities. Yet, the state transition diagram can be arbitrarily large or even infinite in extent.

In this chapter a class of Markovian Petri Net models with relevance to practical applications is analyzed. For this class, the state transition diagram of the associated Markov chain exhibits a rich algebraic topological structure that can be readily characterized. Geometrically, the state transition diagram can be embedded in a toroidal manifold. This embedding alleviates the visual clutter of the usual Petri Net schematic diagram. In addition, the equilibrium probabilities satisfy local balance equations and, therefore, exhibit the classical product form. This will serve to reinforce some of the concepts of chapter 3. These results first appeared in, and this chapter follows closely, [LAZA 86], [LAZA 87b] (copyright 1987 IEEE)].

The chapter is organized as follows. In section 5.2 MPN's are motivated by means of an example of a multiprocessor bus-oriented interconnection network. In section 5.3 the general algebraic topological structure of this class of resource-sharing protocols is presented and the product form of the equilibrium probabilities derived. In section 5.4 a steady-state version of the well-known dining philosophers problem is investigated. An aggregate CSMA protocol model appears in section 5.5. The alternating bit protocol is modeled as a Petri Net in section 5.6. Although shorter in length, sections 5.4, 5.5, and 5.6 discuss for their respective settings the theoretical issues raised in the previous sections. In many general MPN's product form results do not exist. The reason for this is examined in section 5.7.

5.2 A Bus-oriented Multiprocessor Model

Consider a dual processor computer system as shown in Figure 5.1. It consists of two processors, P_1 and P_2, and a common shared memory, CM. P_1 or P_2, or both, may attempt transfers to CM through the bus. Conflicts occur as only one processor may utilize the bus and, hence, access memory CM, at one time.

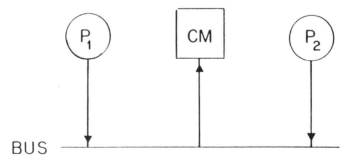

Fig. 5.1: Dual Processor, Single Memory, Multiprocessor

We note that this and the other bus-oriented multiprocessor example in this chapter, are based on a modified version of the TOMP architecture [MARS 83]. TOMP is the TOrino MultiProcessor developed at the Politecnico of Torino in Torino, Italy.

A Markovian Petri Net model of this system appears in Figure 5.2. Formally, the MPN is defined as the sixtuple:

$$\mathbf{P} = (P, T, I, O, M, Q). \tag{5.1}$$

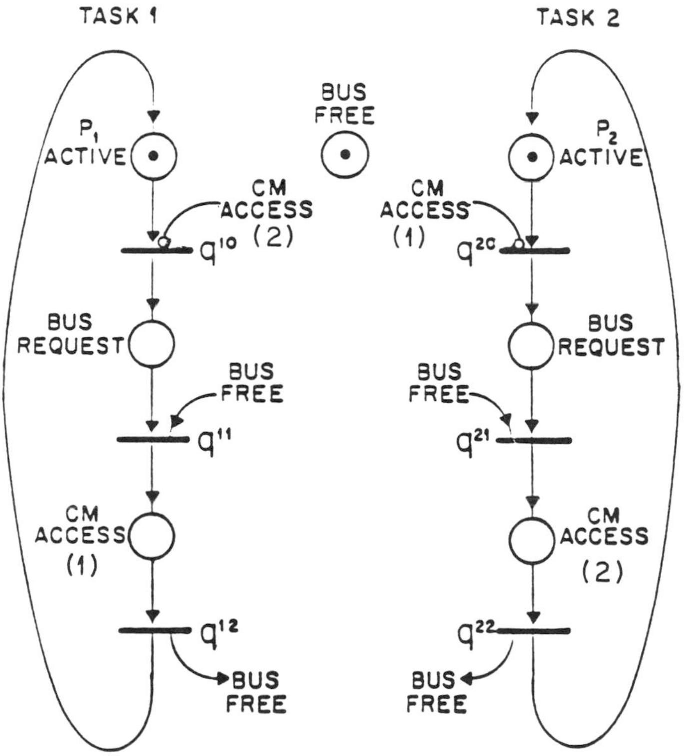

Fig. 5.2: Petri Net of Figure 5.1

Here P is a set of places (drawn as circles), and T is a set of transitions (drawn as horizontal bars). I is an input function that maps each transition to one or more places (drawn as directed arcs) and O is an output function that maps each place to one or more transitions (drawn as continuous or dotted directed arcs). A marking, M, is an assignment of a nonnegative number of tokens (drawn as dots) to individual places. For a given Petri Net, the marking defines the state of the system. Q is the set of rates associated with the transitions.

The rules of Petri Net operation are simple. If there is at least one token in each place incident (through solid or dotted arcs) to a transition, that transition is enabled for firing. This involves removing one token from each place incident to the transition (except from places connected to the transition with dotted arcs) and adding one token to each place connected to the transition by an outgoing arc. The use of these nonstandard dotted arcs will become clear as we proceed. For a MPN the time between an enabled transition and its firing is an exponentially distributed random variable. In a deterministically timed Petri Net this time would be a deterministic quantity (zero for an "immediate" transition [MARS 84]).

Note that the presence of a small circle where an arc meets a transition indicates a complemented dependency. For instance, in Figure 5.2 one can move to a BUS REQUEST place only if P1 is ACTIVE and there is *no* ongoing CM ACCESS.

The class of MPN's considered in sections 5.2 through 5.6 of this chapter is safe, that is, a place may hold at most one token at any given time. Nonsafe Petri Nets are discussed in section 5.7. As we are interested in steady-state solutions, all transitions also exhibit liveness, i.e., deadlock cannot occur.

Finally, while they are not considered in this chapter, it is possible in a non-safe Petri Net to have multiple arcs from a transition to a place or from a place to a transition. If there are multiple arcs leading from a transition to a place when the transition fires it delivers as many tokens to the place as there are arcs between the transition and place. If there are multiple arcs from a place to a transition, a necessary condition for the transition to be enabled for firing is that the place hold at least as many tokens are there are arcs. Moreover when the transition does fire it removes as many tokens from the place as there are arcs between the place and transition.

In the dual processor MPN model of Figure 5.2 one can see a total of two linear sequences of events originating from the places indicating that the processors are internally active (externally idle). Each such sequence will be referred to as a *task sequence* and the events that comprise it as *subtasks*. The first state is often an idle event (i.e., task waiting to be initiated). The two tasks of Figure 5.2 thus correspond to P_1 or P_2 attempting to perform a transfer with CM.

Generally, one can distinguish between places that represent subtasks and those which represent resources (e.g., GB FREE). *The essence of the Petri Net paradigm is that a task sequence may be blocked (temporarily or permanently) if resources are not available.* The resource places and associated transitions could be replaced by a set of control rules making enablement dependent on the state of selected task sequences. In [GHAN] a similar distinction is made between "job tokens" and "control tokens".

In Figure 5.2, q^{kl} is the exponential rate associated with the lth sub-task ($l = 0,1,2$) of the kth task sequence ($k=1,2$). The Petri Net transitions should not be confused with the transitions of the state transition diagram of the Markov process associated with the Petri Net. The corresponding state transition diagram appears in Figure 5.3. The horizontal axis corresponds to P_1 and task sequence 1 while the vertical axis corresponds to P_2 and task sequence 2. The equilibrium state probabilities are given below for all states (i,j) of the indicated task sequences:

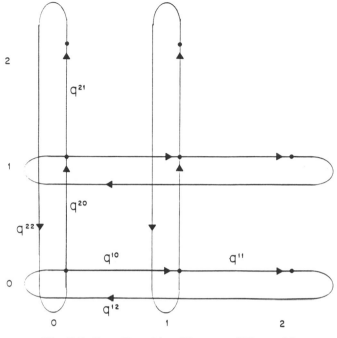

Fig. 5.3: State Transition Diagram of Figure 5.2

$$p(i,j) = \frac{q^{10}}{q^{1i}} \frac{q^{20}}{q^{2j}} p(0,0). \tag{5.2}$$

Note that these probabilities have the characteristic product form, which is known to exist for certain classes of queueing networks.

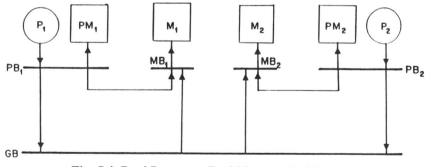

Fig. 5.4: Dual Processor, Dual Memory, Multiprocessor
© 1987 IEEE

Consider now the more sophisticated multiprocessor system shown in Figure 5.4. It consists of two processors, P_1 and P_2, and the corresponding memories, M_1 and M_2. P_1 (or alternately, P_2) may attempt transfers to $M_1(M_2)$ through the local memory bus $MB_1(MB_2)$ connected to the private processor bus $PB_1(PB_2)$. Alternatively, $P_1(P_2)$ may attempt transfers to $M_2(M_1)$ via the global bus (GB) and the local memory bus $MB_2(MB_1)$. Conflicts occur as only one processor may utilize the local bus or the global bus at a time. Note that the processor memories, PM_1 and PM_2, are only accessible from their respective processors on the private buses PB_1 and PB_2, respectively. Unlike in [MARS 83], we assume that processor $P_1(P_2)$ attempts an access only when the local bus and if necessary the global bus are free. As will be seen, this is necessary to guarantee the existence of the product form solution.

In the multiprocessor MPN model of Figure 5.5 one can see a total of four task sequences of events originating from the places indicating that the processors are (internally) active or idle with respect to the rest of the system. The four task sequences of Figure 5.5 correspond to P_1 or P_2 attempting to perform a transfer with M_1 or M_2.

In Figure 5.5, q^{kl} is the exponential rate associated with the lth subtask ($l = 0,1,2,...$) of the kth task sequence ($k=1,2,3,4$). The corresponding state transition diagram appears in Figure 5.6. The horizontal axis corresponds to P_1 and task sequences 1 and 3 while the vertical axis corresponds to P_2 and task sequence 2 and 4. The equilibrium state probabilities, for each quadrant of the diagram, are given below for all states of nonzero probability. For all states (i, j) of the indicated task sequences we have

$$p(i,j) = \frac{q^{10}}{q^{1i}} \frac{q^{20}}{q^{2j}} \, p(0,0) \tag{5.3}$$

for tasks 1 and 2,

$$p(i,j) = \frac{q^{30}}{q^{3i}} \frac{q^{40}}{q^{4j}} \, p(0,0)$$

for tasks 3 and 4,

$$p(i,j) = \frac{q^{10}}{q^{1i}} \frac{q^{40}}{q^{4j}} \, p(0,0)$$

for tasks 1 and 4, and

$$p(i,j) = \frac{q^{20}}{q^{2i}} \frac{q^{30}}{q^{3j}} \, p(0,0)$$

for tasks 2 and 3.

Note that these probabilities have the characteristic product form that is known to exist for certain classes of queueing networks. The basis of the surprising result of (5.2) and (5.3) will be explained in the next section in terms of the algebraic topological structure of the state transition diagram of the Markov chain.

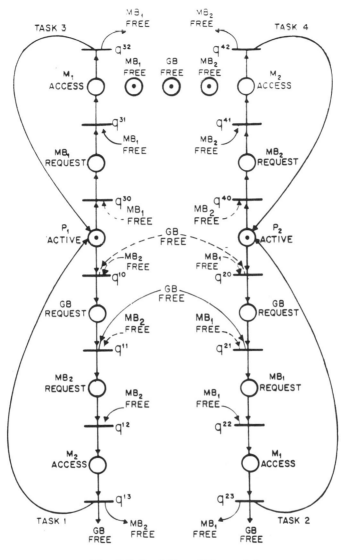

Fig. 5.5: Petri Net of Figure 5.4
© 1987 IEEE

5.3 Toroidal MPN Lattices

To understand why the MPN of Figure 5.2 and Figure 5.5 has a product form solution, consider two *independent* task sequences. That is, sufficient resources exist so that the state of each sequence does not affect the state of the other sequence. This leads to a state transition diagram like that of Figure 5.7. Let task 1 correspond to the horizontal axis and task 2 correspond to the vertical axis.

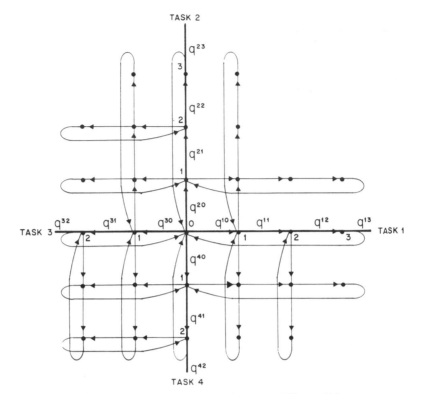

Fig. 5.6: State Transition Diagram of Figure 5.5
© 1987 IEEE

While the position of the states appears to conform to a rectangular geometry, when one considers the "wrap around" character of the boundary transitions it becomes apparent that the natural topological space [MASS] for embedding the state transition diagram is the surface of a torus. This toroidal embedding is natural in that it is symmetrical and planar. The class of MPN's that can be naturally embedded on a torus will be referred to as *Toroidal Markovian Petri Networks.*

The torus, a two-dimensional manifold [MASS], can be thought of as a smooth surface without a boundary. The state transition diagram of the dual processor system of Figure 5.1 and Figure 5.2 embedded in a torus in R^3 is shown in Figure 5.8. The embedding of the four quadrant state transition diagram of Figure 5.6 into torii is shown in Figure 5.9. Only the cycles used for pasting torii together are shown here. In this pasting the top and bottom torii rest upon each other and the left and right torii pass partially through each other.

It is apparent in Figure 5.7 that the global balance equations at each point in the state transition diagram are structurally the same. Boundary states, which often appear in queueing based state transition diagrams, need not be considered separately.

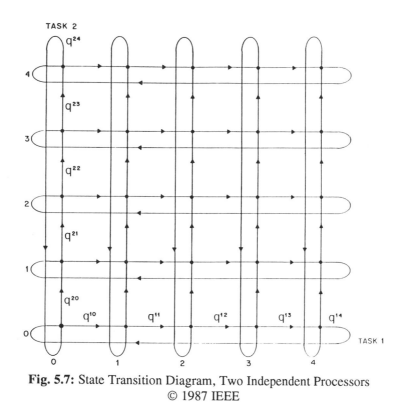

Fig. 5.7: State Transition Diagram, Two Independent Processors
© 1987 IEEE

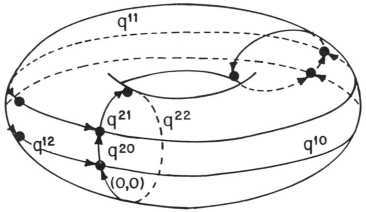

Fig. 5.8: Transition Diagram of Figure 5.3 Embedded in Torus

While any state may be chosen as a reference state, in applied terms the natural choice is the state that corresponds to both independent tasks being active (idle). With this state being given by the coordinates (0,0), we seek an expression for equilibrium probabilities $p(i,j)$. The global balance equation for the conservation of probability flux at vertex (i,j) is

$$(q^{1i} + q^{2j}) \, p(i,j) = q^{1,i-1} p(i-1,j) + q^{2,j-1} p(i,j-1). \qquad (5.4)$$

As has been mentioned in chapter 3, previous work [LAZA 84a,b,c], [WANG 86] on the algebraic topological structure of queueing network state transition diagrams has shown that the classical Jackson-type product form solution of the equilibrium probabilities can be traced back to a decomposition of the state transition diagram into elementary building blocks. For a class of state transition diagrams [WANG 86], cyclic flows can be associated with these building blocks. A cyclic flow is associated with the transitions of each building block and has equal magnitude of probability flux on each transition. Note that the state transition diagram structure in Figure 5.7 suggests the presence of cyclic flows about the torus (e.g., (0,0), (1,0), (2,0), (3,0), (4,0) to (0,0)). It will be shown that these cycles are the fundamental building blocks [LAZA 84a] of the state transition diagram.

This observation is equivalent (Duality Principle [WANG 86]) to the hypothesis that the following local balance equations are satisfied :

$$q^{1,i-1} p(i-1,j) = q^{1i} p(i,j), \qquad (5.5)$$

$$q^{2,j-1} p(i,j-1) = q^{2j} p(i,j).$$

These local balance equations are suggestive of a natural recursion for $p(i,j)$:

$$p(i,j) = \frac{q^{10}}{q^{1i}} \frac{q^{20}}{q^{2j}} \, p(0,0). \qquad (5.6)$$

This does indeed satisfy the global balance equations (5.4). For N independent tasks the expression can be generalized in a straightforward manner to

$$P(k_1, k_2, \cdots, k_N) = \prod_{l=1}^{N} \frac{q^{l0}}{q^{lk_l}} \, p(0,0,...,0) \qquad (5.7)$$

where k_l is the k_lth subtask of the lth task sequence. That we have achieved a product form solution is not surprising at this point since task independence was assumed.

This is indeed the case in the lower left quadrant of Figure 5.6, which corresponds to concurrent access by each processor to its corresponding memory. What of the other quadrants? Each can be seen to have the same structure as the state transition diagram of Figure 5.7 *with specific cyclic building blocks removed*. The removal of a building block(s) does not affect the form of the solution, except to cause a renormalization of p(0,0,...,0). Consistency of the local balance equations is preserved since a form of geometric replication applies [LAZA 84a,c]. Therefore, as has been discussed in chapter 3, the flow on a cycle is isolated [WANG 86], i.e., it neither contributes nor removes probability flux from the other parts of the state transition diagram.

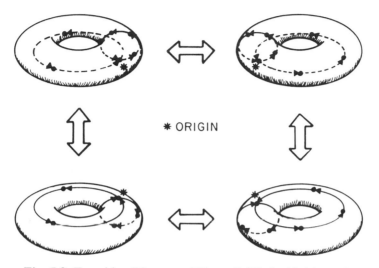

Fig. 5.9: Transition Diagram of Figure 5.6 Embedded in Torii
© 1987 IEEE

The remaining question to be answered is why the *pasting* (an algebraic topological concept) of the four-quadrant state transition diagrams of Figure 5.6 preserves consistency and the product form solution. Note that the boundary between any two quadrant state transition diagrams consists of a cycle (see Figure 5.9). For each of the quadrants taken in isolation, the equilibrium probabilities of the states of this cycle have the same product form when expressed as a function of an arbitrary state of the cycle. This property preserves, therefore, the *continuity* of the form of the state probabilities. Thus, when two-quadrant state transition diagrams are pasted together (see Figure 5.9), the form of the solution of the equilibrium probabilities for each quadrant remains the same, except for the renormalization factor.

We can summarize the properties of the MPN investigated above as:

Theorem: A safe MPN consisting of a number of task sequences that are comprised of a series of sequential subtasks has a product form solution for the equilibrium state probabilities if the state transition lattice can be naturally associated with a Cartesian coordinate system (alternatively: naturally embedded into an aggregation of toroidal manifolds) and if the state transition lattice is comprised of integral building blocks and corresponding consistent set of local balance equations.

Proof: As explained above, the assumed solution form is verified to be a solution of the system's local balance equations.

The properties of the MPN investigated above have been characterized in terms of the algebraic topological structure of the state transition diagram of the associated Markov chain. This characterization can be translated into the Petri Net schematic representation as follows:

Theorem: Consider a safe Markovian Petri Net consisting of a number of task sequences that are comprised of a series of sequential subtasks. The product form solution for the equilibrium state probabilities exists, if and only if, a task sequence is only allowed to proceed if there is a nonzero probability that it can return to its current state without the need for a state change in other task sequences.

Proof: This characterization is equivalent to the state transition lattice being comprised of a consistent set of integral building blocks.

Essentially, the characterization precludes the possibility of a form of blocking. It is well known that the product form solution does not exist in blocking environments. This is a fundamental limitation that the algebraic topology of the state transition lattice imposes. In short, this characterization is useful for determining which models do and do not have a product form solution.

As an example, note that the MPN of Figure 5.5 differs from the more practically oriented one in [MARS 83] in the presence of the dotted arcs in the Petri Net schematic. These provide additional restrictions on when the task sequence is allowed to proceed. Without these restrictions the state transition diagram of Figure 5.6 gains six transitions:

$$(3,0) \rightarrow (3,1), \qquad (0,3) \rightarrow (1,3),$$
$$(2,0) \rightarrow (2,1), \qquad (0,2) \rightarrow (1,2),$$
$$(3,0) \rightarrow (3,-1), \qquad (0,3) \rightarrow (-1,3),$$

which do not belong to a cycle and do not fit into the above characterization. These transitions modify the circulatory pattern of the state transition diagram and the classical Jackson-type product form solutions do not hold for this case.

The discussion so far pointed out a general methodology for determining whether the equilibrium probabilities have a product form that can be obtained by inspection. Consequently, the set of equations (5.7) holds for a multiprocessor system with an arbitrary number of processors, as long as its state transition diagram can be constructed using the previous simple rule.

In short, this characterization is useful for determining which models do and do not have a product form solution. It is somewhat disappointing that the product form solution does not exist in the presence of blocking, but this is also the same situation that occurs in queueing networks. This is a fundamental limitation that the algebraic topology of the state transition diagrams imposes upon us.

5.4 The Dining Philosophers Problem

This well-known model, due to Dijkstra [DIJK], is a good example of a distributed protocol for shared resources and has often appeared in Petri Net form [PETE], [ZENI]. A number of philosopher's, say five, are seated around a circular table and are dining with chopsticks. A *single* chopstick is placed on the table between each philosopher. Naturally, if a philosopher picks up the chopsticks on either side of him (her) to dine with, the philosopher's neighbors may not dine.

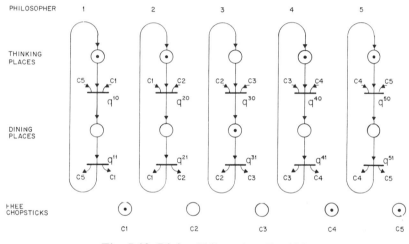

Fig. 5.10: Dining Philosophers Petri Net
© 1987 IEEE

A MPN model of the dining philosophers problem, with five philosophers, is depicted in Figure 5.10. This is a steady-state version and is thus deadlock free. The task sequence for each philosopher can be seen to consist of simply an idle and a dining state.

In the state transition diagram of the associated Markov chain (see Figure 5.11) there are five states that correspond to a single philosopher dining and five states that correspond to the maximum of two philosophers dining. The reference state of the state transition diagram corresponds to all philosophers being in the thinking state. This is a sufficient state description since the system is memoryless.

In states which correspond to only one philosopher dining, the only task sequences that may be initiated are those that correspond to the other two *allowable* philosophers making a transition into the dining state.

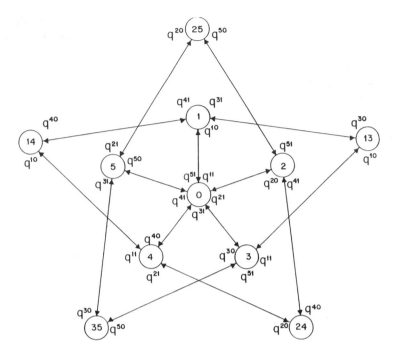

Fig. 5.11: State Transition Diagram of Figure 5.10
© 1987 IEEE

Let q^{k0} be the rate at which the kth philosopher makes a transition to the dining state and q^{k1} be the rate at which the same philosopher leaves it. Then, over the set of allowable states:

$$p(k,k\pm2) = \frac{q^{k\pm2,0}}{q^{k\pm2,1}}\, p(k) = \frac{q^{k\pm2,0}}{q^{k\pm2,1}}\, \frac{q^{k0}}{q^{k1}}\, p(0) \qquad (5.8)$$

for all k, $1 \le k \le 5$, where $\pm$ is modulo 5 addition and subtraction.

This result is the direct consequence of the algebraic topological structure of the state transition diagram. The "wrap-around" character of the state transition diagram depicted in Figure 5.11 leads to its embedding into a five-dimensional toroidal manifold.

(An N-dimensional manifold exhibits locally the properties of the N-dimensional Euclidean space. A standard technique to visualize a multidimensional torus is by topological identification. For example, a 2-dimensional torus can be obtained by identifying opposing sides of a square as being equivalent. A 3-dimensional torus is equivalent with a cube with the opposite planes identified as being equivalent [MASS], [FIRB].)

Generalization of the dining philosophers problem to an arbitrary number of philosophers is straightforward. The natural space to embed the state transition diagram is the N-dimensional toroidal manifold, where N is the number of philosophers. Since geometric replication applies, a simple product form solution of the type given by equation (5.8) holds.

A closed form expression for the normalization constant of the dining philosopher system has been developed by B. Ycart [YCAR]. For this closed form solution let the thinking service rate be $\lambda(=q^{\cdot 0})$, and the dining service rate is 1.0 $(=q^{\cdot 1})$. Then

$$G(N) = \frac{(1+(4\lambda+1)^{1/2})^L+(1-(4\lambda+1)^{1/2})^L}{2^L}. \tag{5.9}$$

In order to produce this expression Ycart models the system as an interacting particle system. These have been studied in physics [LIGG] for problems where one has collections of particles where the state of a particle depends on the state of its neighbors.

5.5 A Station-oriented CSMA/CD Protocol Model

Since bus-oriented multiprocessor architectures can be modeled by Petri Nets, the modeling of bus type local area networks is another potential application. Petri Net models of bus type local area networks have appeared in [GRES] and [MOLL 84].

Bus type local area network models may either attempt to model the various stations and their interaction in detail or to model the behavior of a single station in the context of the aggregate statistical behavior of the remainder of the network. A Carrier Sense Multiple Access Protocol with Collision Detection (CSMA/CD) model of the latter type will now be introduced. It should be noted that the difficulty in taking the former approach for local area networks, as opposed to multiprocessor bus interconnections, lies in the nonnegligible propagation delay along the bus.

A MPN of the LAN station appears in Figure 5.12 and the corresponding state transition diagram in Figure 5.13. Messages originate at the station with rate q^{10}. They are transmitted after an average time of $1/q^{11}$ spent in channel scanning. Once transmission has been initiated, collisions occur with probability ϵ [SHAP] after an average time of $1/q^{12}$. The average duration of a successful transmission is $1/q^{12} + 1/q^{13}$. The average collision resolution time is $1/\tau$.

Note that the q^{11} transition in the Petri Net does not depend on the bus availability. While bus availability could be modeled stochastically, more correctly it depends on the states of other stations and their connection through the bus with significant propagation delay. The simplified model of Figure 5.13 will be used to illustrate some relevant ideas concerning the toroidal class rather than as a

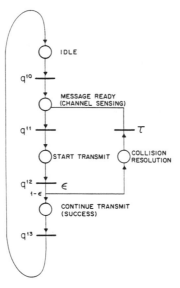

Fig. 5.12: CSMA/CD Petri Net

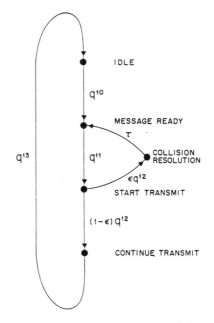

Fig. 5.13: State Transition Diagram of Figure 5.12

detailed representation of the system operating under the CSMA/CD protocol.

The equilibrium probabilities of the states along the state transition diagram's main loop are

$$p(k) = \frac{q^{10}}{\alpha q^{1k}} p(0) \tag{5.10}$$

where α is $(1-\varepsilon)$ when k represents the MESSAGE READY and START TRANSMIT states and 1 otherwise. One way of viewing the state transition diagram of Figure 5.13 is as a cycle embedded in a one dimensional toroidal manifold with a "feedback" loop attached that corresponds to the possibility of collisions. The product form is maintained since the split of probability flux between the two loops is proportional to $1-\varepsilon$ and ε.

However, when searching for a natural topological space for embedding the entire state transition diagram must be considered. While this state transition diagram can be embedded on a 2-dimensional torus, this could also be accomplished using a 2-dimensional spherical manifold. A circle is a one-dimensional toroidal manifold as well as a degenerate one-dimensional spherical manifold.

The key point in the analysis above is the random bifurcation of flows. Thus it would seem most appropriate to view the state transition diagram as being embedded on a manifold obtained by the "pasting together" (an algebraic topological concept) of two 1-dimensional toroidal manifolds. The pasting occurs along shared transitions and has been previously noted for queueing networks with random routing [LAZA 86]. A detailed example of pasting along nodes appears in [LAZA 84a,b,c], [WANG 86].

5.6 The Alternating Bit Protocol

The alternating bit protocol has been previously modeled in state machine [KRIT] and Petri Net form [MOLL 82]. A bit is alternatively set to 0 or 1 in successive packets in an acknowledgment and timeout based link. The alternating bit allows the receiver to disregard packets that are erroneously retransmitted due to acknowledgment loss.

A MPN of the protocol appears in Figure 5.14 and the corresponding state transition diagram in Figure 5.15. Note that the state transition diagram consists largely of a single cycle with some minor feedback loops. Token movements off the main cycle occur with probability ε_P and ε_A [SHAP]. Here the probability of packet loss is ε_P and that of acknowledgment packet loss is ε_A. The nominal task sequence consists of starting from the idle state (with the acknowledgment bit at 0 or 1) and then sending the packet, checking the correctness of the received packet, sending an acknowledgement packet, and checking the correctness of the acknowledgment packet. These tasks are then repeated with the bit complemented.

The loops are due to either the loss in transit through the packet network of either the packet or the acknowledgment packet. In the alternating bit protocol a timer at the transmitter, started after packet transmission, indicates when an acknowledgment is overdue ("time out"). After time out a retransmission occurs. The states off the main cycle help to model this timeout mechanism.

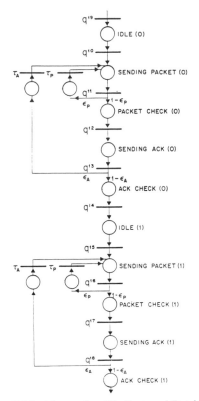

Fig. 5.14: Alternating Bit Protocol Petri Net
© 1987 IEEE

Let τ_P and τ_A be the rate modeling the average timeout duration for message and acknowledgment packet loss, respectively (see Figure 5.14). In Figure 5.15, τ_P originates from a state corresponding to packet loss; there is also a state with outgoing rate τ_A modeling timeouts due to acknowledgement packet loss. The transmitter timer can not distinguish between message packet and acknowledgment packet loss so that the average timeout setting must be the same in either case. Thus $1/\tau_P = 1/\tau_A + 1/q^{12} + 1/q^{13}$ for the upper part of the diagram and a similar equation holds for the lower part.

The equilibrium probabilities along the main cycle have the form

$$p(k) = \frac{q^{10}}{\alpha\, q^{1k}} p(0) \qquad\qquad (5.11)$$

where α is equal to $(1-\varepsilon_P)\times(1-\varepsilon_A)$ or $(1-\varepsilon_A)$, depending on whether the state k represents the SENDING PACKET or the PACKET CHECK and SENDING ACK states, respectively. For all other states belonging to the main cycle $\alpha=1$.

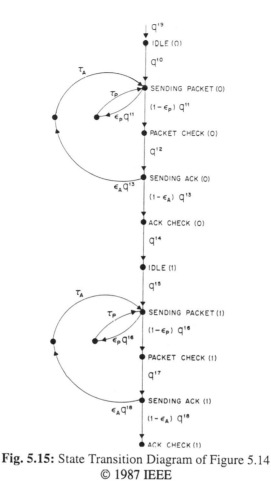

Fig. 5.15: State Transition Diagram of Figure 5.14
© 1987 IEEE

The equilibrium probabilities of the states in the feedback loops in Figure 5.15 can also easily be found. The product form of a single cycle is maintained, in spite of the feedback loops, because the splits of flow are proportional to ε_P and $(1-\varepsilon_P)$ and to ε_A and to $(1-\varepsilon_A)$. Again, a number of one-dimensional toroidal manifolds are pasted together to embed the state transition diagram. Although the two outer feedback loops in Figure 5.15 change the character of the embedding manifold, the product form of the equilibrium probabilities still applies.

5.7 SPN's without Product Form Solutions

5.7.1 Introduction

The existence of the tractable product form solution for certain classes of queueing networks has been known since the work of Jackson in 1957 [JACK 57] and has been further generalized over the years [BASK 75][GORD]. Stochastic Petri networks are analogous to queueing networks in that they consist of a high-level schematic representation and an underlying Markov chain foundation. They differ from queueing networks in that the movement of tokens from places [PETE] is more heavily dependent on the states of other places than the degree to which movement of customers from queues usually depends on the customer status in other queues.

In spite of the presence of such strong dependencies in stochastic Petri networks and the fact that the product form solution is usually thought of as an independence result, several researchers have attempted to define classes of stochastic Petri networks possessing a product form solution. The first such work was by Marsan, Balbo, Chiola, Ponateli, and Conte in 1984 [MARS], [MARS 86], [MARS 84b] and was in the context of multiprocessor systems. Work by the author and co-authors [LAZA 87], [WANG 89] since 1986 has characterized conditions sufficient for the existence of a product form solution for *safe* stochastic Petri networks consisting of a concurrent number of linear sequences of places and transitions known as "task sequences". The characterization holds that a product form solution exists if, whenever a token is allowed to proceed in a task sequence, there is a nonzero probability that the task sequence can return to its original state without a need for state changes in other task sequences to allow this progress. Finally Henderson and Taylor in 1989 [HEND], [HEND 89], Florin and Natkin in 1991 [FLOR 91] and Frosch and Natarajan in 1992 and 1993 [FROS 92], [FROS 93] describe stochastic Petri networks with a product form solution.

It is the purpose of this section to show that basic Petri network models of concurrent resource sharing and synchronization among concurrent processes generally do not admit a (closed form) product form solution, except in special cases.

This section is organized as follows. In section 5.7.2, the case of nonsafe stochastic Petri nets modeling resource sharing is examined. In section 5.7.3 the case of nonsafe Petri nets modeling synchronization is considered. The interested reader can consult [PETE] for the details of stochastic Petri network operation. In this section all transitions are assumed to fire a negative exponential distributed amount of time after being enabled.

5.7.2 Nonsafe Resource Sharing Models

First consider the canonical nonsafe resource sharing model of Figure 5.16. A nonsafe Petri network may have more than one token in each place. Figure 5.16 consists of a linear task sequence of three places indicating the basic functions IDLE, REQUESTING RESOURCE, and ACCESSING RESOURCE. A "task sequence" is a series of sequential tasks that a user repetitively cycles through [LAZA 87][WANG 89]. There is also a RESOURCE place. After being enabled transitions fire at rates of, respectively, q^0, q^1, and q^2. Such models, developed for multiprocessor modeling, appear in [MARS].

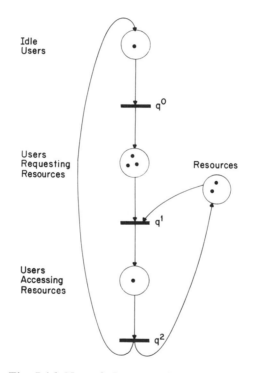

Fig. 5.16: Nonsafe Resource Sharing Petri Net

Figure 5.17 shows the state transition diagram of this model where the number of users making resource accesses is the vertical coordinate and the number of resource requests is the horizontal coordinate. Transitions have the rates indicated in the legend. The state transition diagram has an upper boundary due to the constraint that the number of concurrent resource accesses must be less than the number of resources. There is also a diagonal boundary at the right due to the constraint that the sum of the number of concurrent resource requests and the number of concurrent resource accesses must be less than the number of users.

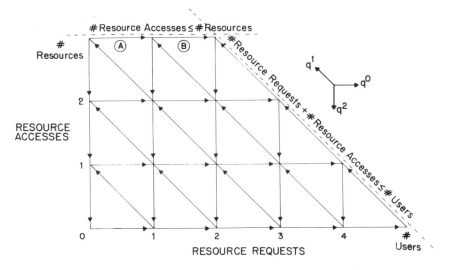

Fig. 5.17: State Transition Diagram of Figure 5.16

In chapter 3 it is pointed out that the probability flux of stochastic networks with a product form solution has a distinctive circulatory structure. Specifically, the overall circulation is actually an aggregation of smaller, "isolated" circulations. An isolated circulation consists of the probability flux along a subset of state transition diagram transitions and states for which there is conservation of the net flow at each adjacent state that remains unchanged, except for a renormalization, when the subset is embedded in the overall state transition diagram. Subsets of transitions and states with an isolated circulation are referred to as "building blocks".

The state transition diagram of Figure 5.17, almost, but not completely, has the requisite structure to produce a product form solution. The flaw is the presence of transitions labeled "A" and "B". These transitions do not belong to consistent integral building blocks so their presence [LAZA 87], [WANG 89] disrupts the flow of probability flux so that isolated circulations and a product form solution do not result.

In fact, if these transitions were removed from the state transition diagram a product form solution would exist. The corresponding Petri network schematic is shown in Figure 5.18. It is like that of Figure 5.16 except for the presence of the dotted arc. The presence of this nonstandard arc means that a necessary condition for the q^0 transition to fire is that there must be at least one resource in the resource place. Note that the dotted arc also indicates that resources are not removed from the resource place when the q^0 transition fires.

Essentially this means that a user can only move from being idle to requesting a resource if a resource is presently free. In the model of Figure 5.16 a user can go from being idle to requesting a resource and then be "stuck" there if no resources are available. This can be thought of as a form of blocking. It is well

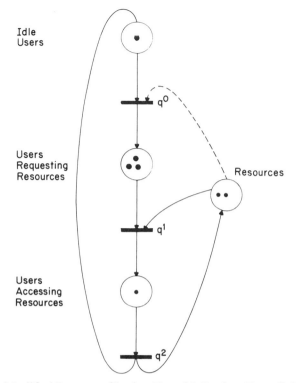

Fig. 5.18: Modified Resource Sharing Net with Product Form Solution

known that blocking queueing networks generally do not have a product form solution. Another way of looking at the model of Figure 5.18 is to note that requests are immediately cleared from the system if the necessary resources are not immediately available. By way of contrast, in the model of Figure 5.16 users requesting resources are allowed to wait in the USERS REQUESTING RESOURCES place until they become available.

The product form solution for the model of Figure 5.18 is

$$p(n_r, n_a) = \left[\frac{q^0}{q^1} \right]^{n_r} \left[\frac{q^0}{q^2} \right]^{n_a} p(0,0) \tag{5.12}$$

where n_r is the number of users requesting resources and n_a is the number of users accessing resources.

An alternate way of modeling resource sharing appears in Figure 5.19. Here each user has a separate Petri subnet to represent the states it can assume. These subnets are safe. There is, again, a single place for holding multiple idle resources.

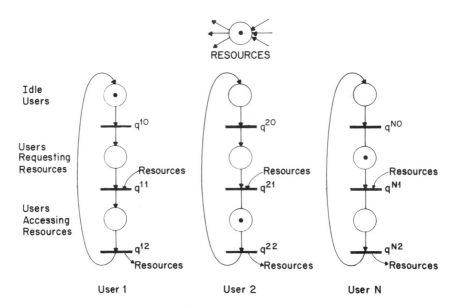

Fig. 5.19: Alternative Representation of Resource Sharing

The state transition diagram for this model when there are two users and a single resource is shown in Figure 5.20. Here the coordinates for a user indicates if it is idle ("0"), requesting a resource ("1") or accessing a resource ("2"). Work on safe Petri nets [LAZA 87], [WANG 89] with product form solutions indicates that this model would have a product form solution if it was not for the presence of transitions "A" and "B". Once again, the problem is that a user can move from being idle to requesting a resource and then be "stuck" there if the resource is being accessed by the other user. This form of blocking precludes the existence of the product form solution.

Increasing the dimensionality, Figure 5.21 shows the state transition diagram for three users and two resources. Again, the presence of transitions "A", "B" and "C" precludes the existence of the product form solution. As the dimensionality is increased further, the same problem of blocking remains as long as there are fewer resources than users.

To show that a state transition diagram like those of Figure 5.20 and Figure 5.21, without the A,B,C... transitions has a product form solution one can consider the drawing of Figure 5.22. It shows one of the building blocks (see chapter 3) that should have an isolated circulation if the product form solution exists. Specifically it is a building block of the kth task sequence (i.e., of user k), with the other user states assumed to be held fixed. We can assume a product form solution of the form:

$$p(0) = \frac{q^{10}}{q^{1i}} \frac{q^{20}}{q^{2j}} \cdots \frac{q^{N0}}{q^{Nz}},$$

(5.13)

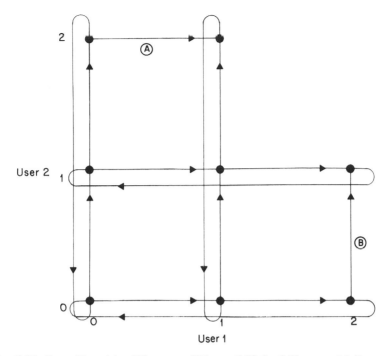

Fig. 5.20: State Transition Diagram of Figure 5.19 for 2 Users and 1 Resource

$$p(1) = \frac{q^{k0}}{q^{k1}} p(0), \qquad (5.14)$$

$$p(2) = \frac{q^{k0}}{q^{k2}} p(0) \qquad (5.15)$$

and show that it satisfies the local balance equations of this building block. Here $(i,j,k,...z)$ indicates the states of the task sequences. Similar equations satisfy each of the possible building blocks of the state transition diagram, so the product form solution exists.

The overall solution is

$$P(k_1,k_2, \cdots k_N) = \prod_{l=1}^{N} \frac{q^{l0}}{q^{lk_l}} p(0,0, \cdots 0) \qquad (5.16)$$

where k_l is the k_lth sub-task (place) of the lth linear task sequence of places.

The previous discussion can be summarized in the following theorem:

Theorem: Consider a nonsafe linear task sequence of places (or a number of safe concurrent task sequences of places) modeling the state of users. Assume also that for some place (or concurrent and like places) that forward progress in the task sequence(s) can only be made by acquiring the use of a limited resource that

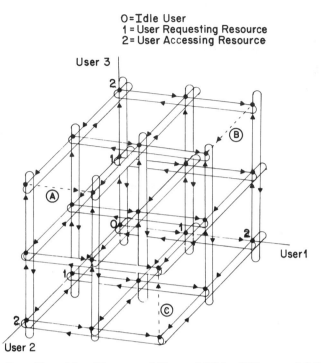

Fig. 5.21: State Transition Diagram of Figure 5.19 for 3 Users and 2 Resources

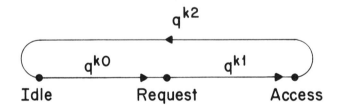

Fig. 5.22: Building Block of Product Form Versions of Figures 5.20 and 5.21.

is released at a successive place. Then a product form solution for the equilibrium state probabilities exists if and only if a user is allowed to change state only if there is a nonzero probability that the user can return to this original state without a need for other users to change their state.

Proof: The condition above is equivalent to having a consistent integral building block structure (or equivalently a consistent set of local balance equations) for the state transition diagram. This is a necessary and sufficient condition for the existence of a product form solution (see chapter 3).

∇

The theorem is somewhat more general than the three place examples of Figures 5.16 and 5.19. For instance, one could have a two nonsafe place net with IDLE USERS and USERS ACCESSING RESOURCES places, along the lines of Figure 5.16. That is, there is no USERS REQUESTING RESOURCES place. The movement of tokens from the IDLE USERS place to the USERS ACCESSING RESOURCES place is conditioned on the availability of resources. In this case it can be seen that the state transition diagram of the corresponding Markov chain is topologically equivalent to that of a finite buffered M/M/1 queue and a (trivial) product form solution exists. The reason that a product form solution does exist is that a token can only leave the IDLE USERS place if resources are presently available, so the condition of the theorem is satisfied. Similarly, a product form solution exists for two place concurrent task sequences along the lines of Figure 5.19.

Conversely, if one takes the Petri network of Figure 5.16 and adds additional sequential places to the task sequence, then generally a product form solution does not exist unless the condition of the theorem is met through the use of dotted transitions to condition task sequence token movement as in Figure 5.18.

In summary, we have delineated two classes of resource sharing models. For one, users may proceed from being idle to requesting a resource and then be blocked from proceeding if resources are not available. This class is a non-product form network. For the other class, requests for resources that are not immediately available are immediately cleared from the system and a product form solution exists. It seems that the former class would be more prevalent in applications. It should be noted that the preceding information is of interest for situations where the canonical models of this paper are subnets of larger, more complex nets.

5.7.3 Synchronization Models

Besides resource sharing, another common paradigm that can be modeled by Petri nets is that of synchronization. In this paper the view is taken that synchronization can be modeled as the need for tokens to arrive into each of a number of places that are incident to the same transition before the transition can fire. This may represent, for instance, parallel fragments of a process that must be completed before the next, serial fragment can be executed. In this section a number of increasingly elaborate synchronization models will be examined.

The first synchronization model appears in Figure 5.23. It, and all the models of this section, are not safe. There must be at least one token in each of the upper places before the λ transition can fire. For this simple minded model the synchronization problem is trivial since the μ transition releases tokens into each of the upper places simultaneously. Letting the number of tokens in the lower place be the state variable, the state description of Figure 5.24 results. It can be seen to be equivalent to the state transition diagram of a single queue with finite buffer size and has the characteristic one dimensional product form solution

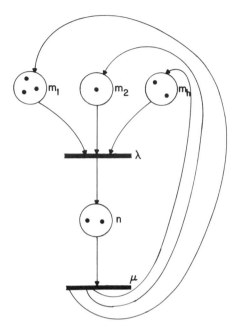

Fig. 5.23: One-dimensional Synchronization Model

$$p(n) = \left[\frac{\lambda}{\mu}\right]^n p(0), \qquad n = 1, 2, \dots M. \tag{5.17}$$

$$M = \min\ (\max\ m_1, \max\ m_2 \cdots \max\ m_h)$$

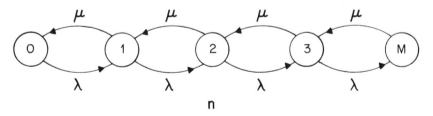

Fig. 5.24: State Transition Diagram of Figure 5.23

A model with a more substantial stochastic component appears in Figure 5.25. Here there must be tokens in each of the l_i places before the λ_1 transition can fire, in each of the m_i places before the λ_2 transition can fire, and in each of the n_i places before the μ transition can fire. The stochastic synchronization part of this model has to do with the n_1 and n_2 places and μ transition. Tokens in one of these places may have to wait if the other place is empty. The waiting time is distributed exponentially with the rate parameter of the appropriate λ transition.

The state transition diagram of the model of Figure 5.25 appears in Figure 5.26. The state variables are the number of tokens in the n_1 and n_2 places. The

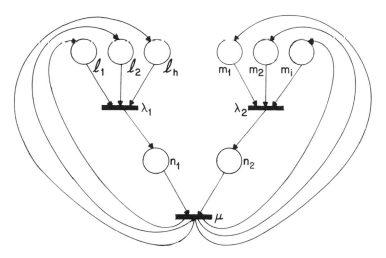

Fig. 5.25: Two-dimensional Synchronization Model

transition rates are shown in the legend. Expressions for the extent of the right and upper boundaries are also listed. These boundaries are not necessarily equal in size.

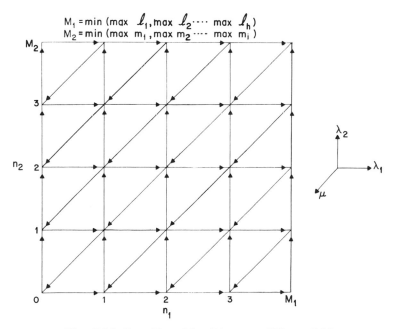

Fig. 5.26: State Transition Diagram of Figure 5.25

If one makes the change of variables

$$n'_1 = n_1, \tag{5.18}$$

$$n'_2 = M_2 - n_2, \tag{5.19}$$

the state transition diagram can be seen to be equivalent to that for two tandem queues with finite buffers. It is well known that such a model has no closed form analytical solution (if one buffer is of size 1 or 2, recursive solutions are possible (see chap. 3)).

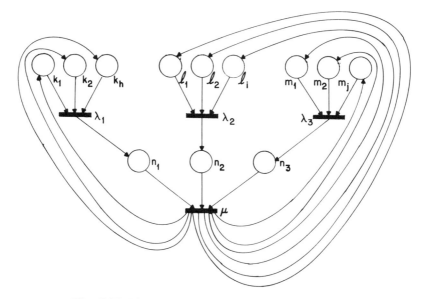

Fig. 5.27: Three-dimensional Synchronization Model

The next model appears in Figure 5.27. The state description is now three dimensional (n_1, n_2, n_3) and the state transition diagram appears in Figure 5.28. The transitions oriented in rectilinear directions correspond to arrivals to the n_1, n_2, n_3 places and the diagonal transitions correspond to the firing of the μ transition. The overall structure is rectangular in 3 dimensions.

The state transition diagram is composed of elements like that of Figure 5.29 with adjacent elements sharing element transitions. These elements are not building blocks in the product form sense.

To see if the state transition diagram of Figure 5.28 can support a product form solution one must be able to find a decomposition of the state transition diagram into consistent integral building blocks with isolated circulations. This does not seem possible. For instance, one might consider a decomposition into tetrahedral-shaped building blocks. There are six such potential decompositions.

$$M_1 = \min(\max k_1, \max k_2 \cdots \max k_h)$$
$$M_2 = \min(\max \ell_1, \max \ell_2 \cdots \max \ell_i)$$
$$M_3 = \min(\max m_1, \max m_2 \cdots \max m_j)$$

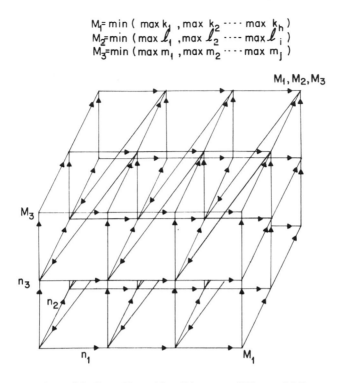

Fig. 5.28: State Transition Diagram of Figure 5.27

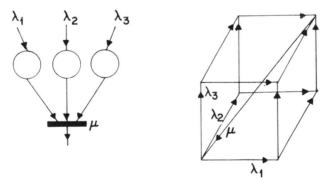

Fig. 5.29: Basic Element of Figure 5.28

However, after attempting to make each such decomposition there are transitions left over that do not belong to integral building blocks. This indicates a lack of product form solution.

After examining the models of Figure 5.23 and Figure 5.25 one might conjecture that the state transition diagram of Figure 5.28 is equivalent to that of three tandem queues with finite buffers. In fact this is not true. There are key topological differences between the state transition diagram of Figure 5.28 and that of the finite buffer three tandem queue network:

→ The state transition diagram origin of the three queue tandem network has one incoming transition and one outgoing transition. None of the corners of Figure 5.28 (or Figure 5.29) has this topology.

→ In the tandem model state transition diagram a "diagonal" transition increments a single state variable (i.e., queue length) and decrements another. In the state transition diagram of Figure 5.28 a diagonal transition decrements three state variables simultaneously.

Experience with such synchronization state transition diagrams for 1,2, and 3 dimensions would appear to support the following conjecture:

Conjecture: Except for the trivial one-dimensional case, synchronization models of higher dimension do not have a product form solution for the equilibrium state probabilities.

The conjecture seems reasonable, at least intuitively, for two reasons. One is that synchronization inherently involves some token(s) waiting for the arrival of some other token(s) and so seems similar to a blocking network. It is well known that the presence of blocking generally precludes the existence of a product form solution. The other is that it is hard to see how one might associate synchronization events with integral building blocks so as to produce a consistent set of local balance equations and a product form solution.

5.8 Conclusion

A class of Markovian Petri Net models with equilibrium probabilities that can be expressed in the classical product form has been described. Examples have been given involving resource sharing (a bus-oriented system and the dining philosophers problem) as well as communication protocol problems (the alternating bit protocol) that arise in practical applications.

The results obtained here have been based on the algebraic, geometric and topological properties of the associated state transition diagrams. The basic principle derived represents a structural result characterizing the existence of Markovian models for resource sharing having simple product form equilibrium probabilities.

The widespread use of Petri Net models has been somewhat inhibited by a perceived unwieldiness of the Petri Net schematic. It has been found by the authors of [LAZA 86,87b] that drawing the task sequences in a linear form and not completely drawing directed arcs to and from resource places tends to enhance the clarity and readability of Petri Net schematics. While such a structured approach has sometimes been implicit in previous work, this work provides an explicit basis for its use.

The consequence of these new results, as well as previous results [FLOR 91], [MARS], is to indicate that *generally* stochastic Petri Nets modeling concurrent resource sharing and synchronization among concurrent processes do not have a product form solution for the equilibrium state probabilities. In the context of resource sharing the situation where a product form solution exists is when requests for resources are cleared when the necessary resources are not immediately available. It would seem that the cases where a product form solution exists would be outnumbered in practice by those where it does not exist. This observation confirms the use of Markov chain solvers and simulation for analyzing such nets in the general case.

To Look Further

The characterization of the product form solution for stochastic Petri Nets discussed in this chapter can be extended to the case where the enabled transitions fire after a time that is a sum of exponential random variables [WANG 89].

Some interesting effort on Petri Nets includes work on generalized stochastic Petri net (GSPN) reward models [CIAR 91], GSPN models of Markovian multi-server multiqueue systems [MARS 90], SPN models of vacation queueing systems [IBE] and on applying the theory of insensitivity (see To Look Further in chapter 3) to stochastic Petri Nets [HEND 93].

An excellent source for material concerning timed Petri Nets are the proceedings of a series of international workshops that have been held on the subject. The first was the International Workshop on Timed Petri Nets, held in Torino, Italy during 1985. The second is the '87 International Workshop on Petri Nets and Performance Analysis held in Madison, Wisconsin USA in 1987. The third is the 3rd International Workshop on Petri Nets and Performance Models held in Kyoto-shi, Japan during December 1989. The last is the 4th International Workshop on Petri Nets and Performance Models held in Melbourne, Australia in December 1991. The proceedings of these workshops are published by IEEE Computer Society Press.

A package for numerically solving stochastic Petri Net problems is SPNP (Stochastic Petri Net Package). It was developed at the Computer Science Dept. of Duke University, Durham N. C., by G. Ciardo, J. Muppala and K. S. Trivedi. Based on C, it runs on a variety of systems. It is capable of solving generalized stochastic Petri Nets in both the transient and steady state case. For details see [CIAR]. This article includes a comparison with two other packages: GreatSPN [CHIO] from the University of Torino, Italy and METASAN [SAND] from the University of Michigan at Ann Arbor and Industrial Technology Institute, Ann Arbor. These latter two packages have simulation capabilities.

Problems

5.1   a) Draw and label a safe Petri Net for the following situation. In a power
plant an alarm sometimes activates. This causes the separate dispathes of a
supervisor and a technician to check the generator. Each may arrive at the
generator at different times. When they are both present they reset the equipment, leading to the alarm deactivating and they immediately return to their
posts.

b) Consider appropriate timing for each of the transitions. Take into account
the different time scales of each event.

5.2   A remote earthquake monitoring station has three redundant transmitters.
Because of the cost of a service call the system is only repaired when two of
the transmitters fail (are "down"). Draw a safe Petri Net to illustrate the
cycle of operation and failure. Would immediate transitions be appropriate
in any part of the net?

5.3   Construct a Petri net consisting of two linear task sequences that models two
users in New York who can access the same satellite link to Houston. However, only one may utilize the link at any one time. There are two resources:
the uplink from New York to the satellite and the downlink from the satellite
to Houston. A user may seize the uplink first and then, after some satellite
processing delay, the user may seize the downlink. The places in each linear
task sequence are IDLE, REQUESTING UPLINK, REQUESTING DOWN-
LINK, and TRANSMIT. Can deadlocks occur?

5.4   Consider the state transition diagram of Figure 5.3. Using equation (5.2)
verify that the *global* balance equations are satisfied at nodes (0,0), (1,0) and
(1,1).

5.5   For the state transition diagram of Figure 5.3 verify that the *local* balance
equations are satisfied for nodes (0,0), (1,0) and (1,1) using equation (5.2).

5.6   **Computer Project:** a)Write software to calculate the equilibrium state probabilities of the system in Figures 5.1-5.3 using equation (5.2).

b) Write software to calculate the equilibrium state probabilities of the state
transition diagram of Figure 5.3 with additional transitions $(2,0) \rightarrow (2,1)$ and
$(0,2) \rightarrow (1,2)$. Assume Markovian statistics. Since this is a non-product
form Markov chain, calculate the equilibrium state probabilities by solving
the global balance equations of this system. Note that one global balance
equation is replaced by the normalization equation. Draw the Petri net
corresponding to the system.

c) Use the software of parts (a) and (b) to calculate the fraction of time that a
CM ACCESS takes place (really the common memory utilization). Choose
values of the transition rates that show a difference in performance between

the systems of (a) and (b).

5.7 Consider the state transition diagram of Figure 5.6. Verify that the *global* balance equations are satisfied at states (1,-1), (1,0) and (2,-1) in the lower right quadrant using equation (5.3).

5.8 For the state transition diagram of Figure 5.6 verify that the *local* balance equations are satisfied for states (1,-1), (1,0), and (2,-1) in the lower right quadrant using equation (5.3).

5.9 In the Toroidal MPN Lattices section of this chapter it is mentioned that if the dotted arcs in Figure 5.5 are removed, this leads to the state transition diagram of Figure 5.6 with six additional transitions. Add these transitions to the state transition diagram and attempt to write a global balance equation for state (3,1) in the upper right quadrant using the product form solution of equation (5.3). You will see that the equation does not balance as the additional transitions create a non-product form network.

5.10 Show that the product form solution of equation (5.6) satisfies the *global* balance equation of (5.4).

5.11 Show that the product form solution of equation (5.6) satisfies the *local* balance equations of (5.5).

5.12 a) Calculate the normalization constant for the stochastic dining philosopher model of Figure 5.11 (5 philosophers) when $\lambda=5.0$ using Ycart's formula.

b) Verify this result by calculating the product form solution of equation (5.8) for each state.

5.13 a) Consider a dining philosopher model as in Figure 5.10 where philosophers **do not** have to wait for resources (chopsticks) before moving from thinking to dining. Let the q^{-0} transition rates be λ and the q^{-1} transition rates be μ. Calculate the equilibrium probability that l of L philosophers are "dining".

b) Do the same as in (a) for the actual model of Figure 5.10 when L=5 philosophers.

c) **Computer Project:** Plot the probabilities of one, two, and zero philosophers dining versus λ for (a) and (b) with L=5 philosophers when $\mu=1.0$ from $\lambda=0.1$ to $\lambda=2.0$ in steps of 0.2.

5.14 Explain why the dining philosopher model of Figure 5.10 has a product form solution without the need for the restrictions appearing in the earlier multiprocessor models (i.e. dotted arcs etc...). Phrase your answer in terms of one of the theorems in the Toroidal MPN lattices section.

5.15 Consider the CSMA/CD Petri Net of Figure 5.12. Develop an expression for the mean number of times that the COLLISION RESOLUTION place is visited as a token moves from IDLE to CONTINUE TRANSMIT. Hint: Make use of a useful summation from the appendix. Plot the result vs. ε.

5.16 a) For the CSMA/CD model of Figure 5.12 develop an expression for the approximate utilization of the channel (fraction of time spent in the SUCCESS state) using the product form expression of equation (5.10).

b) **Computer Project:** Plot utilization for various values of ε and the transition rates.

5.17 For the alternating bit protocol of Figure 5.14 calculate the mean time for the protocol to move through a complete cycle. Hint: Make use of the useful summations in the appendix.

5.18 Using balance equations derive expressions for the equilibrium-state probabilities of *all* the states in the alternating bit protocol state transition diagram of Figure 5.15. Hint: Set Prob(IDLE(0)) = 1.0, and work from the bottom of the diagram upward.

5.19 Create a two place (nonsafe) task sequence with IDLE USERS and USERS ACCESSING RESOURCES places, along the lines of Fig. 5.16. That is, there is no USERS REQUESTING RESOURCE place. The movement of tokens from the IDLE USERS place to the USERS ACCESSING RESOURCES place is conditioned on the availability of a resource in a separate RESOURCES place. Once a user token leaves the USERS ACCESSING RESOURCES place its associated resource is released. Let q^0 be the rate at which enabled users leave the IDLE USERS place and let q^1 be the rate at which users leave the USERS ACCESSING RESOURCE place. The net statistics are Markovian.

a) Draw and label the associated Markov chain. Let the state variable be the number of users in the USERS ACCESSING RESOURCES place.

b) Let M be the largest state. To what is it equal?

c) Develop an expression for the equilibrium-state probabilities.

5.20 Consider a two place (nonsafe) task sequence in a Petri net, along the lines of the previous problem. Let the two places in the task sequence be IDLE USERS and USERS ACCESSING RESOURCES. This time there are two resource places, RESOURCE A and RESOURCE B. A user token in the IDLE USERS place can only move to the USERS ACCESSING RESOURCES place if there is at least one resource in each of the resource places. That is, a user will access two resources simultaneously. Both resources are released when the user leaves the USERS ACCESSING RESOURCES place. Again, let q^0 be the rate at which enabled users leave the IDLE USERS place, and let q^1 be the rate at which users leave the

USERS ACCESSING RESOURCES place.

a) Draw and label the associated Markov chain. Let the state variable be the number of users in the USERS ACCESSING RESOURCES place.

b) Let M be the largest state. What is it equal to?

c) Develop an expression for the equilibrium state probabilities.

5.21 Construct a three-place (nonsafe) task sequence. There are also two resource places (R1 and R2). A user token moves (at rate q^0) from the first, IDLE, place to the next, ACCESSING RESOURCE 1 place only if at least one resource token is present in the R1 place. The user token then can move (at rate q^1) to the ACCESSING BOTH RESOURCES place if at least one resource token is present in the R2 place. A user token leaving the ACCESS-ING BOTH RESOURCES place (at rate q^2) releases both resources and returns to the IDLE place. The net statistics are Markovian.

a) Draw and label the associated Markov chain. Let the number of user tokens in the ACCESSING RESOURCE 1 place be one coordinate (horizontal) and let the number of user tokens in the ACCESSING BOTH RESOURCES place be the other coordinate (vertical).

b) Is this a product form or a non-product form network?

5.22 Consider the Markovian state transition diagram of Figure 5.17 without transitions A and B. The corresponding product form solution is given by equation (5.12).

a) In Figure 5.17 write the *local* balance equations at nodes (1,2) and nodes (1,3).

b) If transitions A and B are added to the diagram then the model is a non-product form one. Attempt to write local balance equations at nodes (1,2) and (1,3) using the solution of (5.12). What happens?

5.23 **Computer Project:** Write software to evaluate the equilibrium state probabilities of Figure 5.17 both with and without transitions like A and B. Set up the software to handle arbitrary numbers of RESOURCE ACCESSES and RESOURCE REQUESTS. Solve for the equilibrium state probabilities of the model *without* transitions like A and B through equation (5.12), and solve for the equilibrium state probabilities of the model *with* transitions like A and B by solving the global balance equations with a linear equation solver. Note that one global balance equation must be replaced by the normalization equation. Try to determine under what parameter values the two systems act in a similar fashion and under what parameter values they act differently.

5.24 a) Show that a building block like those in Figure 5.21 (without the ABC transitions) has its local balance equations satisfied by equations (5.13) through (5.15).

b) Why does the inclusion of transitions A,B, and C prevent the existence of a product form solution?

5.25 a) In Figure 5.24 verify the definition of M.

b) In Figure 5.26 which transitions cause the state transition diagram to be of the nonproduct form type?

c) Verify the two arrowed differences in the text between the state transition diagram of Figure 5.28 and that of a finite buffer three tandem queue network.

Chapter 6: Discrete Time

Queueing Systems

6.1 Introduction

The study of continuous time queueing systems goes back to the early work of Erlang. However a somewhat different model, that of discrete time queueing systems, has recently been the subject of increasing attention. In a discrete time model time is assumed to be "slotted". That is, time consists of a concatenation of fixed length intervals known as slots. Events are constrained to take place during these slots. For instance, a discrete time queue might accept at most one packet during a slot and service at most one packet during a slot. On a network-wide basis, multiple events may occur during each slot.

Interest in discrete time queueing systems has largely been driven by potential applications that make use of slotted time. Sometimes this is for reasons of flexibility. For instance, many packet network architectures, such as ATM [DEPR], call for the use of fixed length packets. The time to transmit a fixed-length packet, the slot, thus becomes a basic parameter of the architecture and associated analysis. Sometimes the reason people are interested in discrete time systems is efficiency. It is well known, for instance, that slotted Aloha multiple access radio systems have twice the maximum throughput of continuous time asynchronous Aloha implementations ([ROM], [ROBE 72], [TANE]).

This look at discrete time queueing systems will start with a discussion of the general solution of such a system's global balance equations. Similarities and differences between discrete and continuous time queueing systems will be examined. This includes a study of a canonical discrete time arrival process. A discrete time queueing system analogous to some of those studied for continuous time will be presented.

This chapter concludes with three case studies involving high-speed networks. Problems of this type represent a new challenge for the application of queueing theory. In section 6.6 we present an interesting case study based on recent work by M. Karol, M. Hluchyj and S. Morgan. It involves the calculation of throughput for a space division packet switch. It is a good case study as it brings together ideas concerning networks of queues, non-Markovian queues and transform techniques in the context of a problem of practical interest.

In section 6.7 a second case study is presented based on work by Y. C. Jenq [JENQ]. It involves a Banyan interconnection network with single buffers at each switching element. It is an interesting case study as it shows one way to approach a distributed queueing system with blocking.

Finally, in section 6.8, a case study based on recent work by M. Garrett and S.-Q. Li is presented. It deals with the placement of erasure nodes in the DQDB metropolitan area network. An ingenious continuous formulation of the problem allows the calculation of optimal erasure station location and system throughput.

6.2 Discrete Time Queueing Systems

In most of the queueing systems that have been discussed so far, time has been assumed to be continuous. That is, arrivals and departures may take place at any instant of time. Certain practical systems operate in a different mode involving discrete time. Here events may only be allowed to occur at periodic, equispaced instants.

As an example, consider two stations (Figure 6.1a) connected to a common communication medium such as a coaxial cable or radio channel. Assume that time is "slotted", being comprised of equiwidth slots (Figure 6.1b). Packet length, slot width, and channel speed are such that one packet can be transmitted by one station on the bus during one slot.

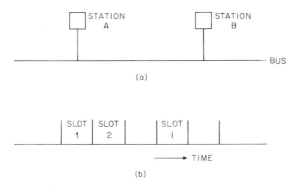

Fig. 6.1: Two-station Slotted Network

A packet will arrive at a station during a slot with probability a. For the purposes of this example we will assume that each station can hold only a single packet. Once a packet is in a station, further arrivals to the station are blocked.

If a station has a packet, it transmits it during each slot with probability s. A packet can only be transmitted successfully during a slot by one of the stations if the other station does not transmit a packet at the same time. If packets are transmitted by both stations during the same slot they overlap ("collide") and become garbled so that neither is successfully transmitted. Both packets must be retransmitted later. Each station can detect the occurrence of such a collision event.

Even though the protocol being described results in occasional collisions, it has the virtue of being decentralized and straightforward to implement.

We will take a numerical approach that involves calculating the state probabilities of the system and then calculating performance measures from these state probabilities. As an example, the state transition diagram of this system is shown in Figure 6.2. There are clearly four states and a number of transitions between them.

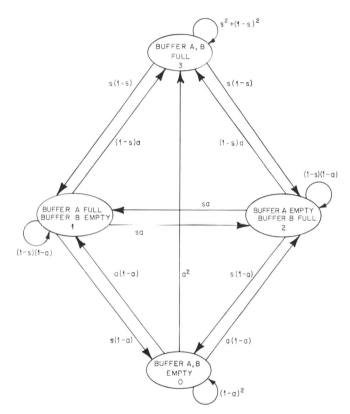

Fig. 6.2: Two-station State Transition Diagram

This state transition diagram is different from the ones that have been seen before, except for the M/G/1 queueing system, as time is now discrete (note that in the M/G/1 queue the spacing between events was not a constant). The values associated with the transitions are not rates but the probabilities of making each transition during each slot. Put another way, each slot finds the system in some state. At the end of the slot the system must transit one of the transitions leaving the state so that it can enter a state (perhaps the same one) for the next slot. Some transition must be chosen, and the sum of the probabilities of all the outgoing transitions must equal one.

Note that whereas in a Markovian continuous time queueing system the time spent in a state is an exponential random variable, for this discrete time system it is a geometrically distributed random variable (see below).

Let us examine some of the transitions in Figure 6.2. If both buffers are empty, each may receive a packet with probability a^2 to push the system into state 3. If during the next slot there is either a collision (with probability s^2) or no attempted transmissions (with probability $(1-s)^2$), the system will remain in the same state. Perhaps then station B will successfully transmit its packet with probability $s(1-s)$ followed by a successful transmission by station A with probability $s(1-a)$.

One can write global balance equations for such a discrete time Markov chain for each state j of N states where p_{ij} is the probability of going from state i to j and π_j is the equilibrium probability of state j:

$$\sum_i \pi_j p_{ji} = \sum_i \pi_i p_{ij}, \tag{6.1}$$

$$\pi_j \sum_i p_{ji} = \sum_i \pi_i p_{ij}, \tag{6.2}$$

$$\pi_j = \sum_i \pi_i p_{ij}, \quad j=1,2,...N. \tag{6.3}$$

Naturally

$$\sum_i p_{ji} = 1, \qquad \sum_i \pi_i = 1. \tag{6.4}$$

In matrix form equation (6.3) can be written as

$$\underline{\pi} = \underline{\pi} \underline{P} \tag{6.5}$$

where $\underline{\pi} = [\pi_0 \ \pi_1 \ \pi_2 \cdots]$ and $\underline{P}$ is the transition matrix. That is, the ijth entry of $\underline{P}$ is p_{ij}.

A different and computationally preferable solution [WIES] solution approach goes as follows. Let the equilibrium probability vector start with a (probably incorrect) guess of the state probabilities at slot 0. Then the state probabilities can be generated at each slot through the recursion:

$$\pi^{(n)} = \pi^{(n-1)} \underline{P}. \qquad (6.6)$$

If the recursion is run long enough, it will converge to the equilibrium state probabilities. A test for convergence such as

$$|\pi_i(n) - \pi_i(n-1)| \, / \, |\pi_i(n-1)| \leq \varepsilon \quad \text{for } each \ i \qquad (6.7)$$

can be used [WIES]. Here ε is a positive number much smaller than one. Work by J. E. Wieselthier and by A. Ephremides [WIES] on discrete time models of multiple access protocols has used this solution method for state spaces of up to 700 states. It was found that a renormalization of the probabilities should be performed each iteration to prevent round-off errors from becoming a problem.

As is often the case, performance measures can be calculated for the simple model of Figure 6.2 as functions of the state probabilities. For instance, a little thought will show that the average throughput of the channel is

$$Mean \ Throughput = \pi_1(s) + \pi_2(s) + \pi_3(2s(1-s)). \qquad (6.8)$$

The average number of packets waiting in the two packet buffers is

$$Mean \ \# \ of \ Packets = \pi_1 + \pi_2 + 2\pi_3. \qquad (6.9)$$

Using Little's Law, the average delay for packets that succeed in entering a buffer is

$$Mean \ Delay = \frac{\pi_1 + \pi_2 + 2\pi_3}{\pi_1(s) + \pi_2(s) + \pi_3(2s(1-s))}. \qquad (6.10)$$

Finally, the blocking probability, the probability that an arriving packet is blocked from entering a buffer, is

$$Prob \ [Blocking] = 0.5\pi_1 + 0.5\pi_2 + \pi_3. \qquad (6.11)$$

Note that in analyzing multiple access systems with N stations the state variable is usually chosen to be the number of stations with packets ready to transmit. This leads to a one-dimensional state description, which simplifies analysis [ROM].

6.3 Discrete Time Arrival Processes

6.3.1 The Bernoulli Process

The most basic of arrival processes in continuous time models is the Poisson process. Its analog in discrete time modeling is the Bernoulli process. For a Bernoulli process modeling arrivals, during each slot one has

$$Prob \, (1 \; arrival) = p_1 = p \qquad\qquad (6.12)$$

$$Prob \, (0 \; arrivals) = p_0 = 1-p \qquad\qquad (6.13)$$

$$Prob \, (\geq 2 \; arrivals) = 0. \qquad\qquad (6.14)$$

Here p is a constant probability of arrival. The motivations for such an arrival process are applications such as synchronous high-speed packet switches (e.g. ATM technology [BAE], [DEPR]) where at most one packet can be transmitted over a link during each slot. However, there are situations where multiple arrivals may occur in a single slot. One occurs in considering a number of Bernoulli arrival processes operating in parallel. For instance, a high-speed packet switch may have N inputs, each of which can be modeled as a Bernoulli process. In a single time slot anywhere from 0 to N packets may arrive at the switch. Another example occurs in slotted Aloha radio system. While the arrival of attempted transmissions may be modeled as a continuous time Poisson process, their success depends on the number of arrivals in each fixed length time slot. Aloha models involving a finite and also an infinite number of attempts can be developed [ROM].

The Bernoulli arrival process shares some of the characteristics of the Poisson process. For instance, it is a memoryless process. That is, a knowledge of the past arrival pattern does not help in predicting current or future arrivals. This is easy to see as the outcome of each slot (arrival/no arrival) is independent of the outcomes of all other slots. This point will be returned to below in discussing the geometric distribution.

Another similarity to the Poisson process has to do with multiplexing and de-multiplexing. If N Bernoulli processes with identical p's and slot widths are multiplexed (interleaved) together then the resulting process is also a Bernoulli process, though it is N times faster (with $1/N$ the original slot width). On the other hand if a Bernoulli process is de-multiplexed into N arrival processes then these are also Bernoulli though each is N times slower (N times the original slot width). This invariance under multiplexing and de-multiplexing extends the usefulness of the Bernoulli process.

Let us now calculate some of the statistics of the Bernoulli process [LEON]. One has

$$p_0 = 1-p, \qquad (6.15)$$

$$p_1 = p. \qquad (6.16)$$

So the mean number of arrivals per slot is

$$\mu = \sum_{i=0}^{1} i p_i, \qquad (6.17)$$

$$\mu = 0 \cdot p_0 + 1 \cdot p_1, \qquad (6.18)$$

$$\boxed{\mu = p. \qquad (6.19)}$$

And the variance of the number of arrivals per slot is

$$\sigma^2 = \sum_{i=0}^{1} (i - \mu)^2 p_i, \qquad (6.20)$$

$$\sigma^2 = (0 - p)^2 (1 - p) + (1 - p)^2 p, \qquad (6.21)$$

$$\boxed{\sigma^2 = p(1-p). \qquad (6.22)}$$

Finally, the transform of the Bernoulli process is

$$P(z) = \sum_{i=0}^{1} p_i z^i, \qquad (6.23)$$

$$P(z) = (1 - p)z^0 + pz^1, \qquad (6.24)$$

$$P(z) = (1-p) + pz. \qquad (6.25)$$

6.3.2 The Geometric Distribution

Let us calculate the probability of the next slot at which an arrival occurs. If the current slot is slot 0 and the next arrival occurs in slot i then

$$Prob\,(Arrival\ In\ Slot\ i) = Prob(i-1\ empty\ slots)Prob(1\ arrival), \quad (6.26)$$

$$Prob\,(Arrival\ In\ Slot\ i) = (1-p)^{i-1}p, \quad i=1,2,3,\dots . \qquad (6.27)$$

This geometric distribution is the discrete time analog to the continuous time exponential distribution. Recall that for a continuous time Poisson process the time until the next arrival is given by the exponential distribution. Like the exponential distribution, the geometric distribution is memoryless (it is the only discrete time distribution to be so). Intuitively this can be seen as each arrival/nonarrival is analogous to a coin flip with probability p for heads and probability $1-p$ for tails. Certainly knowing the past history of coin flips is of no help in predicting the future.

It is easy to see that the mean value of this distribution is

$$\mu = \sum_{i=1}^{\infty} ip(1-p)^{i-1} = \frac{1}{p}. \qquad (6.28)$$

This makes intuitive sense. If p=0.5, on average two slots will pass before an arrival. If p=0.1, on average 10 slots will pass before an arrival.

To calculate the variance of the geometric distribution:

$$\sigma^2 = \sum_{i=1}^{\infty} (i-\mu)^2 p_i, \qquad (6.29)$$

$$\sigma^2 = \sum_{i=1}^{\infty} i^2 p_i - \mu^2. \tag{6.30}$$

Now

$$\sum_{i=1}^{\infty} i^2 p_i = \sum_{i=1}^{\infty} i^2 p(1-p)^{i-1}, \tag{6.31}$$

$$\sum_{i=1}^{\infty} i^2 p_i = p \sum_{i=1}^{\infty} i^2 (1-p)^{i-1}. \tag{6.32}$$

But

$$\sum_{i=1}^{\infty} i^2 (1-p)^{i-1} = \frac{(2-p)}{p^3}. \tag{6.33}$$

So substituting yields

$$\sigma^2 = \frac{(2-p)}{p^2} - \frac{1}{p^2}. \tag{6.34}$$

And the variance of the geometric distribution is

$$\boxed{\sigma^2 = \frac{(1-p)}{p^2}. \qquad (6.35)}$$

To calculate the transform of the geometric distribution one has

$$P(z) = \sum_{i=1}^{\infty} p(1-p)^{i-1} z^i, \tag{6.36}$$

$$P(z) = zp \sum_{i=1}^{\infty} (1-p)^{i-1} z^{i-1}. \tag{6.37}$$

Making the change of variables $i \rightarrow i+1$

$$P(z) = zp \sum_{i=0}^{\infty} \left[(1-p)z \right]^i, \tag{6.38}$$

$$P(z) = \frac{zp}{1 - (1-p)z}. \qquad (6.39)$$

6.3.3 The Binomial Distribution

Consider packets arriving to a discrete time queue according to a Bernoulli arrival process. An interesting and practical question is what is the probability that i of N consecutive slots contain arrivals if the probability of a single arrival in a single slot is p. An equivalent question is if a high-speed packet switch has N input queues, what is the probability of i arrivals in a single slot to the switch. Here we assume that the probability of an arrival to each queue during a slot is p. The binomial distribution allows one to simply answer these questions.

Let us go back to the first example of Bernoulli arrivals to a single discrete time queue. A sample arrival sequence might be

$$A \ N \ A \ A \ N. \qquad (6.40)$$

Here there are three slots with arrivals (A's) and two slots with no arrivals (N's). We know that

$$Prob(A) = p, \qquad (6.41)$$

$$Prob(N) = 1-p. \qquad (6.42)$$

Since each arrival is independent of all other arrivals

$$Prob(ANAAN) = p(1-p)pp(1-p) = p^3(1-p)^2. \qquad (6.43)$$

Note that this is the probability of arrivals occurring *in this particular sequence*, or really any sequence with three arrivals and two idle slots:

$$Prob(NNAAA) = (1-p)(1-p)ppp = p^3(1-p)^2, \qquad (6.44)$$

$$Prob(AANNA) = pp(1-p)(1-p)p = p^3(1-p)^2. \qquad (6.45)$$

It is easy to see that in general if there are N slots with i arrivals in *some* sequence then

$$Prob\,(sequence) = p^i(1-p)^{N-i}. \qquad (6.46)$$

What one is usually interested in is not the probability of a particular sequence but the probability that there are exactly i arrivals in N slots for some value of p. To calculate this quantity one must multiply Prob(sequence) by the number of sequences that produce i arrivals out of N slots. A little thought will show that this number is simply the number of combinations of i arrivals taken over N slots at a time

$$\binom{N}{i} = \frac{N!}{(N-i)!\,i!}. \qquad (6.47)$$

Now let us call the probability of i arrivals in N slots for some p to be $b(i,N,p)$. Then

$$b\,(i,N,p) = \binom{N}{i}p^i(1-p)^{N-i}, \qquad i=0,1,2,...N. \qquad (6.48)$$

This is the binomial distribution. It is plotted in Figure 6.3 for $p=0.375$ and $N=8$. Note that while the binomial distribution is unimodal for this choice of parameters if p is close to 0 or 1, then the distribution is monotonically decreasing or increasing, respectively.

A fact that will prove useful in calculating the mean and transform of this distribution [THOM] is that $b(i,N,p)$ is the ith term in the binomial expansion of $(q+r)^N$ where q and r are arbitrary numbers:

$$(q+r)^N = \sum_{i=0}^{N}\binom{N}{i}q^i r^{N-i}. \qquad (6.49)$$

To calculate the mean of the binomial distribution, from first principles one has

$$\mu = E[i] = \sum_{i=0}^{N}ib\,(i,N,p), \qquad (6.50)$$

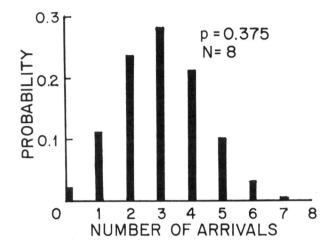

Fig. 6.3: An Example Binomial Distribution

$$\mu = \sum_{i=0}^{N} i \binom{N}{i} p^i (1-p)^{N-i}, \tag{6.51}$$

$$\mu = \sum_{i=1}^{N} i \binom{N}{i} p^i (1-p)^{N-i}, \tag{6.52}$$

$$\mu = \sum_{i=1}^{N} i \frac{N(N-1)(N-2)\cdots(N-i+1)}{i(i-1)(i-2)\cdots 2\cdot 1} p^i (1-p)^{N-i}, \tag{6.53}$$

$$\mu = Np \sum_{i=1}^{N} \binom{N-1}{i-1} p^{i-1} (1-p)^{(N-1)-(i-1)}. \tag{6.54}$$

Making the change of variables $i \rightarrow j+1$:

$$\mu = Np \sum_{j=0}^{N-1} \binom{N-1}{j} p^j (1-p)^{N-1-j}. \tag{6.55}$$

And from the binomial expansion:

$$\mu = Np(p+1-p)^{N-1} \tag{6.56}$$

or

$$\mu = Np. \quad (6.57)$$

Intuitively this makes sense. The number of positive outcomes of N "experiments" should be, on average, Np. Note that in Figure 6.3 that $Np=8\text{x}0.375=3$, which is where, in this case, the distribution peaks.

The transform of the binomial distribution can be found from

$$P(z) = \sum_{i=0}^{\infty} z^i b(i,N,p), \quad (6.58)$$

$$P(z) = \sum_{i=0}^{\infty} \binom{N}{i} (zp)^i (1-p)^{N-i} \quad (6.59)$$

And from the binomial expansion

$$P(z) = (zp + 1 - p)^N. \quad (6.60)$$

To calculate the variance of the binomial distribution one can first calculate the second moment. To do this one can make use of the fact [PAPO 84] that

$$E^{(k)}\left[z^i\right] = E\left[i(i-1)(i-2) \cdots (i-k+1)z^{i-k}\right] \quad (6.61)$$

where the superscript k indicates the kth derivative with respect to z is being taken and E is the expectation operator. The special case of this required is where $k=2$ and $z=1$

$$E^{(2)}[1] = E[i^2] - E[i] \quad (6.62)$$

or

$$E[i^2] = E^{(2)}[1] + E[i]. \quad (6.63)$$

From the previous transform for the binomial distribution

$$E^{(2)}[1] = N(N-1)(zp+1-p)^{N-2}p^2\,|_{z=1},\qquad(6.64)$$

$$E^{(2)}[1] = N(N-1)p^2.\qquad(6.65)$$

Since it has already been determined that $E[i]=\mu=Np$, one has

$$E[i^2] = N(N-1)p^2 + Np.\qquad(6.66)$$

Finally

$$\sigma^2 = E[i^2] - \mu^2,\qquad(6.67)$$

$$\sigma^2 = N(N-1)p^2 + Np - (Np)^2,\qquad(6.68)$$

$$\sigma^2 = Np(1-p).\qquad(6.69)$$

Note that the variance of the binomial distribution is maximized when $p=0.5$ and is minimized when $p=0$ or 1 as one then has deterministic arrivals.

6.3.4 Poisson Approximation to Binomial Distribution

The binomial distribution can be accurately approximated for large N and small p by the Poisson distribution. As an example, consider a large switch fabric with N slotted inputs and with arrival probability per input per slot of p. The total arrival rate to the switch, per slot, is $Np = \lambda$.

The binomial probability of zero arrivals is [THOM]

$$b(0,N,p) = \binom{N}{0}p^0(1-p)^N,\qquad(6.70)$$

$$b(0,N,p) = (1-p)^N.\qquad(6.71)$$

But $p=\lambda/N$, so

$$\lim_{N \to \infty} b(0,N,p) = \lim_{N \to \infty} \left[1 - \frac{\lambda}{N} \right]^N = e^{-\lambda}. \tag{6.72}$$

Continuing, the binomial probability of one arrival is

$$b(1,N,p) = \begin{bmatrix} N \\ 1 \end{bmatrix} p (1-p)^{N-1}, \tag{6.73}$$

$$b(1,N,p) = \frac{Np}{1-p} b(0,N,p). \tag{6.74}$$

And

$$\lim_{N \to \infty} b(1,N,p) = \lim_{N \to \infty} \frac{\lambda}{1 - \frac{\lambda}{N}} e^{-\lambda} = \lambda e^{-\lambda}. \tag{6.75}$$

The binomial probability of two arrivals is:

$$b(2,N,p) = \begin{bmatrix} N \\ 2 \end{bmatrix} p^2 (1-p)^{N-2}, \tag{6.76}$$

$$b(2,N,p) = \frac{N(N-1)}{2} p^2 (1-p)^{N-2}, \tag{6.77}$$

$$b(2,N,p) = \frac{(N-1)p}{2(1-p)} b(1,N,p). \tag{6.78}$$

And

$$\lim_{N \to \infty} b(2,N,p) = \lim_{N \to \infty} \frac{(N-1)p}{2\left[1 - \frac{\lambda}{N}\right]} \lambda e^{-\lambda}, \tag{6.79}$$

$$\lim_{N \to \infty} b(2,N,p) = \frac{\lambda^2}{2} e^{-\lambda}. \tag{6.80}$$

This reasoning can be continued by induction to yield the Poisson distribution as an approximation for the binomial distribution for large N and small p:

$$\lim_{N\to\infty} b\,(i,N,p) = \frac{\lambda^i}{i\,!}e^{-\lambda}, \quad \lambda=Np, \quad i=0,1,2,3,\dots . \tag{6.81}$$

6.4 The Geom/Geom/m/N Queueing System

In chapter 2 a number of state-dependent continuous time queueing systems were presented. In this section a discrete time queueing model whose state transition diagram has the Type A structure of chapter 3 is examined. Thus a simple recursive solution of the equilibrium probabilities is possible. This model is based on the work in [HUAN].

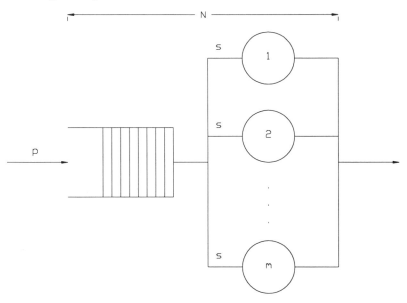

Fig. 6.4: Geom/Geom/m/N Queueing System

　　Consider a discrete time queue as shown in Figure 6.4 where p is the arrival probability of a customer during a time slot and s is the service completion probability of a customer in a server during a slot. Consequently this corresponds to a Bernoulli arrival process and a geometric service time distribution. Also define N as the capacity of the queueing system and m as the number of servers. This is a Geom/Geom/m/N queueing system, which is the discrete time analog of the continuous time M/M/m/N queueing system in discrete time. The system is memoryless so that the state of the queue is defined to be the number of customers in the queue (including the ones in service). We assume that customers can both enter and leave a full queue in the same slot---leading to no net change in state. Figure

6.5 illustrates the state transition diagram of this queueing system. It can be seen to have the Type Λ nonproduct form structure of section 3.4.

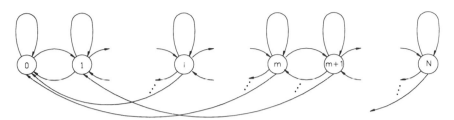

Fig. 6.5: Geom/Geom/m/N State Transition Diagram

Our goal is to find the equilibrium state probabilities given p,s,m and N. Since this system's state transition diagram is Type A, a recursive solution for the equilibrium state probabilities is possible by drawing vertical boundaries between adjacent states and equating the flow of probability flux from left to right to that from right to left across the boundaries. Of course, since this is a discrete time Markov chain, there are transition *probabilities*, not continuous time rates, associated with each transition. If X_k is the state at slot k then the transition probability a_{ij} is defined as:

$$a_{ij} = \lim_{k \to \infty} Prob\,(X_{k+1}{=}j \mid X_k{=}i), \quad i,j \ \varepsilon \ [0,N]. \tag{6.82}$$

In one possible case a customer can arrive at the queue with probability p, and, after entering a server, it can finish the service in i time slots with probability $s\,(1{-}s)^{i-1}$. This means that it is possible for a customer to enter and leave an empty queue in the same slot. To find the transition probabilities, all possible combinations of customers arrivals and departures have to be considered. Two sets of equations can be obtained. First let

$$[n,m]^* = min\,(n,m). \tag{6.83}$$

Also let the combinatorial expressions below equal 0 if the upper term is less than the lower term or if the lower term is less than 0. Then the one-step transition probabilities from state n to n-l (except from N to N) are

$$a_{n,n-l} = p \begin{bmatrix} [n{+}1,m]^* \\ l{+}1 \end{bmatrix} s^{l+1}(1{-}s)^{[n+1,m]^*-l-1} \tag{6.84}$$

$$+ (1{-}p) \begin{bmatrix} [n,m]^* \\ l \end{bmatrix} s^l(1{-}s)^{[n,m]^*-l},$$

$$-1 \leq l \leq m, \qquad n{\neq}N,$$

and the transition probability from state N to N is:

$$a_{N,N} = p\left[\binom{m}{1}s(1-s)^{m-1}+(1-s)^m\right]+(1-p)(1-s)^m. \tag{6.85}$$

Now in a second case if a customer cannot enter and leave an empty queue in the same slot, there will be a different set of state transition probabilities. This delay can be called the synchronization time. One example where this might arise is when time is required to process a header in a packet network node. The one-step transition probabilities from state n to n-l (except from N to N) in this case are

$$a_{n,n-l} = p\left[\begin{matrix}[n,m]^*\\l+1\end{matrix}\right]s^{l+1}(1-s)^{[n,m]^*-l-1} \tag{6.86}$$

$$+ (1-p)\left[\begin{matrix}[n,m]^*\\l\end{matrix}\right]s^l(1-s)^{[n,m]^*-l},$$

$$-1 \le l \le m, \qquad n \ne N.$$

The transition probability from state N to N is

$$a_{N,N} = p\left[\binom{m}{1}s(1-s)^{m-1}+(1-s)^m\right]+(1-p)(1-s)^m. \tag{6.87}$$

To solve this Markov chain one can make use of the fact that in equilibrium the flow of probability flux across a vertical boundary between state i-1 and i balances in both directions. For the state transition diagram of Figure 6.5 it can be expressed as

$$a_{i-1,i}P_{i-1} = \sum_{n=i}^{m+i-1}\sum_{j=n-m}^{i-1} a_{n,j}P_n, \qquad i=N,...\,1. \tag{6.88}$$

Here P_i is the equilibrium state probability of there being i customers in the queueing system. Note that in chapter 2 a lower case p was used to describe the same quantity. The (unnormalized) state probabilities can be evaluated starting with state N-1 and proceeding toward state 0. One can then choose P_N so that

$$\sum_{n=0}^{N}P_n = 1 \tag{6.89}$$

and then compute the normalized equilibrium state probabilities. The complete procedure is

(1)   Let $P_N = 1.0$.

(2)   Initialize $a_{i,j}$'s.

(3) $i = N-1$

(4) $P_i = \dfrac{1}{a_{i,i+1}} \displaystyle\sum_{n=i+1}^{i+m} \sum_{j=n-m}^{i} a_{n,j} P_n$.

(5) $i = i - 1$

(6)   Repeat step (4) and (5) until $i < 0$.

(7)   Find $\sum P_i$.

(8)   Divide all P_i's acquired in steps (1) and (4) by the sum of step (7). This produces the normalized equilibrium state probabilities.

Overflow problems are possible with the use of these equations for large N. However, one can use scaling techniques to scale up or down intermediate results automatically. It should also be noted that a reflected version ($n=N-n$) of the state transition diagram of this section can be used to model a discrete time queue with multiple arrivals and single departures [SZYM].

Of course, once the equilibrium state probabilities are calculated various performance measures, which are functions of these state probabilities, can also be calculated:

$$P_b = Prob(Arriving\ Customer\ is\ Blocked), \qquad\qquad (6.90)$$

$$P_b = Prob(Queue\ Full) \times Prob(No\ Service\ Completion),$$

$$P_b = P_N s_0 \quad (where\ s_0 = (1-s)^m);$$

$$\overline{Y} = Mean\ Throughput, \qquad\qquad (6.91)$$

$$\overline{Y} = Prob\,(Customer\ Arrival) \times Prob\,(No\ Blocking),$$

$$\overline{Y} = p(1 - P_b);$$

$$\overline{n} = Mean\ Queue\ Length, \qquad\qquad (6.92)$$

$$\overline{n} = \sum_{n=1}^{N} n P_n;$$

$$\overline{\tau} = Mean\ Delay, \qquad\qquad (6.93)$$

$$\overline{\tau} = \frac{\overline{n}}{\overline{Y}} \quad (Little's\ Law).$$

6.5 The Geom/Geom/1/N and Geom/Geom/1 Queueing Systems

A special case of the Geom/Geom/m/N queueing system of the previous section occurs when there is a single server. This queueing system is the discrete time analog of the M/M/1/N continuous time queueing system. Actually there are two types of Geom/Geom/1/N queueing systems that can be studied depending on whether or not an arriving customer to an empty queue can be serviced in the same slot. Let us first consider a model where a customer arriving to an empty queue *can* be serviced in the same slot. This is similar to "virtual cut-through" routing described in the literature [BADR], [KERM] where a complete packet need not be received before the received portion of the packet can begin transmission on the next hop of its path.

Using the formula of the previous section for the Geom/Geom/m/N queueing system where $m=1$, or just working from first principles, one can deduce that the transition probabilities of this system are

$$a_{0,0} = ps + (1-p),\tag{6.94}$$

$$a_{n,n} = ps + (1-p)(1-s),$$

$$a_{N,N} = p(s + (1-s)) + (1-p)(1-s),$$

$$a_{n,n+1} = p(1-s), \qquad n=0,1,...N-1,$$

$$a_{n,n-1} = (1-p)s, \qquad n=1,2,...N.$$

The state transition diagram for the Geom/Geom/1/N queueing system is similar in its topology to that of the M/M/1/N queueing system except that there are now transitions from each state to itself. To solve for the equilibrium state probabilities one can draw vertical boundaries between adjacent states and equate the flow of probability flux across the boundary in both directions. Letting P_i be the equilibrium state probability that there are i customers in the queueing system results in

$$p(1-s)P_0 = s(1-p)P_1,\tag{6.95}$$

$$p(1-s)P_1 = s(1-p)P_2,$$

$$p(1-s)P_2 = s(1-p)P_3,$$

$$p(1-s)P_{n-1} = s(1-p)P_n,$$

$$p(1-s)P_{N-1} = s(1-p)P_N.$$

From these equations one can easily write the recursion:

$$P_1 = \frac{p(1-s)}{s(1-p)}P_0, \tag{6.96}$$

$$P_2 = \frac{p(1-s)}{s(1-p)}P_1,$$

$$P_3 = \frac{p(1-s)}{s(1-p)}P_2,$$

$$P_n = \frac{p(1-s)}{s(1-p)}P_{n-1},$$

$$P_N = \frac{p(1-s)}{s(1-p)}P_{N-1}.$$

Substituting one into each other leads to the compact expression

$$P_n = \left[\frac{p(1-s)}{s(1-p)}\right]^n P_0, \quad n=1,2,...N. \tag{6.97}$$

Solving for P_0 through the conservation of probability results in

$$P_0 = \frac{1}{\sum\limits_{i=0}^{N}\left[\frac{p(1-s)}{s(1-p)}\right]^i}. \tag{6.98}$$

And

$$P_0 = \frac{1 - \dfrac{p(1-s)}{s(1-p)}}{1 - \left[\dfrac{p(1-s)}{s(1-p)}\right]^{N+1}}. \quad (6.99)$$

From (6.97) and (6.99) the equilibrium state probabilities can be calculated (see the algorithmic description of the previous section). Note that for an infinite buffer Geom/Geom/1 queueing system $p(1-s)/(s(1-p))$ is less than one when the queue is stable ($p<s$). Equation (6.97) holds though over an infinite number of states. Then

$$P_0 = 1 - \frac{p(1-s)}{s(1-p)}. \quad (6.100)$$

Now if one has a Geom/Geom/1/N queueing system where a customer arriving to an empty queue must wait at least until the next slot for service, one has the same transition probabilities as in (6.94) except that $a_{0,0}=1-p$ and $a_{0,1}=p$. One can show in a way similar to the above (see problem 6.23) that

$$P_n = \frac{p^n(1-s)^{n-1}}{s^n(1-p)^n} P_0, \quad n=1,2,...N. \quad (6.101) \tag{6.101}$$

And

$$P_0 = \frac{1-s}{\dfrac{1 - \left[\dfrac{p(1-s)}{s(1-p)}\right]^{N+1}}{1 - \dfrac{p(1-s)}{s(1-p)}} - s}. \quad (6.102)$$

Naturally for an infinite buffer Geom/Geom/1 queueing system this becomes

$$P_0 = \frac{1-s}{\dfrac{1}{1 - \dfrac{p(1-s)}{s(1-p)}} - s}. \quad (6.103)$$

or

$$P_0 = 1 - \frac{p}{s}. \tag{6.104}$$

Note that in calculating the mean throughput for the queueing system with virtual cut through there is a contribution to mean throughput when the queue is "empty" due to packets that enter and leave during the same slot.

It should be mentioned that the finite queueing systems of this and the preceding section assume that if the queue is full, an arriving customer can enter the system during the same slot that a customer leaves the server --- leading to no net change in state. In terms of the timing this could happen if departures occurred at the beginning of the slot and arrivals occurred at the end of the slot. However, if the timing was reversed and arrivals occurred at the beginning of the slot and departures occurred at the end of the slot then an arriving packet could be blocked (at the beginning of the slot) even if a customer were to leave eventually (at the end of the slot). In this case for the Geom/Geom/1/N systems $a_{N,N}=1-s$ and $a_{N,N-1}=s$.

6.6 Case Study I: Queueing on a Space Division Packet Switch

6.6.1 Introduction

As long as there have been telephones, and more recently with multiprocessor systems, a basic problem has been the interconnection of N inputs and N outputs. That is, how does one design a box, as in Figure 6.6, that simultaneously provides connection paths from the multiple inputs to multiple outputs. In traditional telephony the paths are actual circuits through the interconnection network (or simply the "switch") that are in use for the duration of a call. In the more modern packet-switching technology, packets of bits pass through the interconnection network. A packet is a finite series of digital bits, some of which represent control information such as the destination address and some of which represent the actual information being transmitted.

A simple way to design an interconnection network is with a cross-bar architecture. In an $N{\times}N$ crossbar switch, as in Figure 6.7 inputs are connected to say, horizontal wires (buses) and outputs to vertical buses. There is a switch, which can be independently closed or opened, everywhere a horizontal bus crosses a vertical bus. These are the interconnection crosspoints. An "X" in the figure indicates a closed switch, creating a path from input to output. Such a crossbar switch is said to be a "space division" switch in that the distinct paths through it

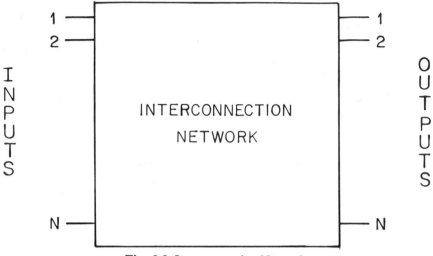

Fig. 6.6: Interconnection Network

are spatially separated.

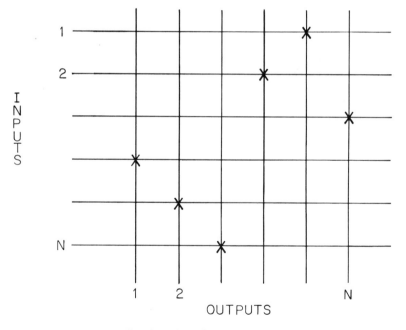

Fig. 6.7: $N{\times}N$ Crossbar Switch

There is a wide variety of interconnection architectures available [WU], [ROBE 93]. Some are blocking networks. These reduce the number of crosspoints at the expense of blocking. That is, not every input can be connected to a different output at the same time. An input that can not find a path to an output is said to be blocked. Blocking networks can be advantageous as they reduce the number of crosspoints, a traditional measure of switch complexity, and because it may be that traffic patterns are such that only a small number of inputs will be active at one time. However, studies have shown that in terms of chip area blocking Banyan networks and nonblocking crossbar networks have similar requirements [FRAN], [SZYM 86].

In what follows though, we will consider the case of a nonblocking packet switch, such as the crossbar, with queues at either the inputs (Figure 6.8b) or outputs (Figure 6.8a). The results and methodology we will use were published by M. J. Karol, M. C. Hluchyj and S. P. Morgan in 1986 [KARO 86,87,88]. An alternative approach to determining the limiting throughput for input queueing, by J.Y. Hui and E. Arthurs was published in 1987 [HUI].

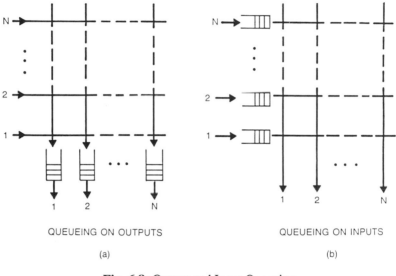

QUEUEING ON OUTPUTS   QUEUEING ON INPUTS

(a)   (b)

Fig. 6.8: Output and Input Queueing
© 1987 IEEE

But why would one consider placing queues at the input or output of a packet crossbar switch? The answer is that queueing arises naturally as packets may arrive at a number of inputs, all destined for the same output. Only one of these packets can use an output trunk (line) at a time, and the others must be queued.

In all that follows it is assumed that all packets are of the same fixed length. Time is slotted, and one packet just fills one time slot. The switch is synchronous. That is, arriving packets all arrive during the same time slot (see Figure 6.9).

The switch can be designed in two ways, with the queues at the inputs or the outputs. If the switch runs at the same speed (switches) as the input and the output, then only one packet can reach a specific output during a time slot and

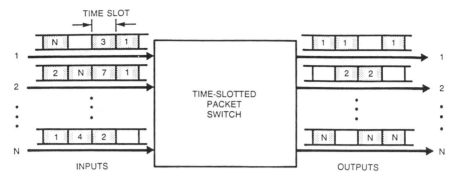

Fig. 6.9: Time-Slotted Packet Switch
© 1987 IEEE

queues must be present on the switch inputs to hold packets that must wait, as in Figure 6.8b. On the other hand, if the switch is run N times as fast as the inputs and outputs, then up to N packets destined for one output can pass through the switch at one time and the queueing will take place at the outputs, as in Figure 6.8a.

Intuitively, one would expect that queueing at inputs will result in longer average queue lengths, and thus longer average waiting times, than queueing at outputs. This is because for input queueing a packet that may be able to pass through the switch in a given time instant may be forced to wait in the input queue behind a packet whose output is busy. In what follows, this superiority of output queueing over input queueing will be quantified.

6.6.2 Output Queueing

It will be assumed that packet arrivals on the N inputs obey independent and identical Bernoulli processes. That is, the probability that a packet arrives at each input during each time slot is p. Naturally, p is the utilization of each input. Each incoming packet has a probability of $1/N$ of being destined for any specific output. Successive packets on the same input are independently addressed.

Let us consider a specific output queue called the "tagged" queue. Let the random variable A be the number of arriving packets at the tagged queue during a specific time slot. A should have the binomial probabilities:

$$a_i \equiv Prob[A=i] = \binom{N}{i}(p/N)^i(1-p/N)^{N-i}, \qquad i=0,1,2,...N. \qquad (6.105)$$

In the binomial distribution the term p/N is present since a packet arriving at the tagged output queue must have arrived at some input (with probability p) and then must have been destined for the tagged output (with independent probability

1/N). The moment-generating function of the binomial distribution is

$$A(z) \equiv \sum_{i=0}^{N} z^i Prob[A=i] = (1-\frac{p}{N}+z\frac{p}{N})^N. \tag{6.106}$$

If $N\to\infty$, the distribution of A approaches a Poisson distribution [THOM]. We have

$$a_i \equiv Prob[A=i] = \frac{e^{-p}p^i}{i!}, \qquad i=0,1,2,.... \tag{6.107}$$

The moment-generating function is

$$A(z) \equiv \sum_{i=0}^{N} z^i Prob[A=i] = e^{-p(1-z)}. \tag{6.108}$$

Now we will write down a discrete time equation governing the tagged output queue. Let Q_m be the number of packets in the tagged queue at the end of the mth time slot. Let A_m be the number of arriving packets for the mth time slot. Then

$$Q_m = \max(0, Q_{m-1} + A_m - 1). \tag{6.109}$$

The -1 in (6.109) is due to the packet that is transmitted by the output queue during each time slot. The number in the tagged queue is drawn in terms of a discrete time state transition diagram in Figure 6.10. A little thought will show that it is identical to the state transition diagram of Figure 2.26 for the M/G/1 queue. Thus, using the same approach as in chapter 2, one winds up with a result for the moment-generating function of the distribution of the number in the queue, which is remarkably similar to equation (2.248):

$$Q(z) = \frac{(1-p)(1-z)}{A(z)-z}. \tag{6.110}$$

Here, compared to (2.248), ρ is replaced by p and $K(z)$ is replaced by $A(z)$. There is no $A(z)$ multiplying this expression, as $K(z)$ does in (2.248), as the queueing system under consideration does not have a customer in a separate server.

Substituting $A(z)$ for the binomial distribution from before one has:

STATE TRANSITION PROBABILITIES

$$a_i \equiv \Pr(A = i) \qquad\qquad i = 0, 1, 2, \ldots$$

$$= \begin{cases} \dfrac{p^i e^{-p}}{i!} & \text{IF } N = \infty \\[3mm] \dbinom{N}{i}\left(\dfrac{p}{N}\right)^i \left(1 - \dfrac{p}{N}\right)^{N-i} & \text{IF } N < \infty \end{cases}$$

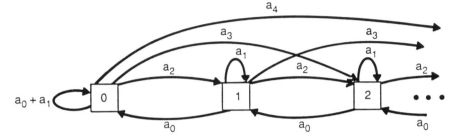

Fig. 6.10: Tagged Queue State Transition Diagram
© 1987 IEEE

$$Q(z) = \frac{(1-p)(1-z)}{(1-\frac{p}{N}+z\frac{p}{N})^N - z}. \tag{6.111}$$

One approach now would be to invert $Q(z)$. An alternative approach, discussed in [KARO 87], is a recursive algorithm, similar to that in section 3.4, to compute the steady-state probabilities of the number in the tagged output queue. We will turn our attention though to simply calculating the average number in the queue. This can be accomplished by differentiating the above equation with respect to z and then letting $z \to 1$ (i.e. $\frac{d}{dz}\sum_{i=0}^{\infty} p_i z^i \big|_{z=1} = \sum_{i=0}^{\infty} i p_i$ for some probability density p_i). The average number in the queue is then given by

$$\overline{Q} = \frac{(N-1)}{N} \times \frac{p^2}{2(1-p)} = \frac{(N-1)}{N}\, \overline{Q}_{M/D/1}. \tag{6.112}$$

Here $\overline{Q}_{M/D/1}$ is the average queue length for an M/D/1 queue modified into the context of this discrete time problem. From the above it is apparent that as $N \to \infty$,

$\overline{Q} \rightarrow \overline{Q}_{M/D/1}$.

Actually, as $N \rightarrow \infty$ the equilibrium probabilities of the number in the queue also converge to those of an M/D/1 queue. To see this, one uses the previous expression for $A(z)$ for the Poisson distribution to obtain

$$\lim_{N \rightarrow \infty} Q(z) = \frac{(1-p)(1-z)}{e^{-p(1-z)} - z}. \tag{6.113}$$

This is the moment-generating function of a comparable discrete time M/D/1 queue.

Finally we will calculate the average waiting time experienced by a packet arriving at the tagged queue during the mth time slot. The service discipline is FIFO with respect to packets arriving during different time slots. However, packets arriving during the same time slot are assumed to be transmitted in random order.

To calculate the average waiting time, use will be made of Little's Law. The arrival rate into the tagged queue is the total switch arrival rate (Np) multiplied by the probability that incoming packets are destined for the tagged queue ($1/N$). Thus

$$\overline{W} = \frac{\overline{Q}}{p} = \frac{(N-1)}{N} \times \frac{p}{2(1-p)} = \frac{(N-1)}{N} \overline{W}_{M/D/1}. \tag{6.114}$$

Here $\overline{W}_{M/D/1}$ is the average waiting time for a discrete time M/D/1 queue.

Figure 6.11 [KARO 86] shows the mean waiting time as a function of p for various values of N. It can be seen that the case of $N \rightarrow \infty$ provides an upper bound on the average waiting time with respect to the case of finite N.

6.6.3 Input Queueing

As for the case of output queueing, arriving packets to the N inputs of the switch follow independent and identical Bernoulli processes. That is, the probability that a packet arrives at each input during each time slot is p. Each packet is directed to one of the N outputs with probability $1/N$.

In what follows, if k packets are destined for the same output, the switch controller picks one at random (with probability $1/k$) and the others wait for a new selection in the next slot. Other policies are possible [KARO 87]. One could, for example, pick a packet from the longest input queue or one can establish priorities.

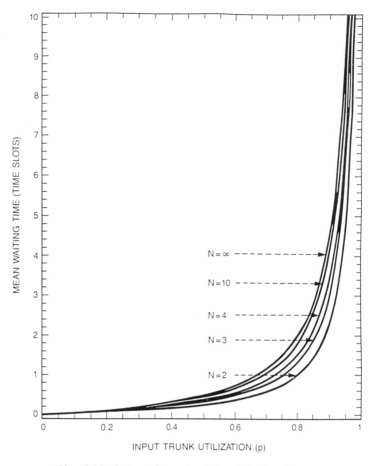

Fig. 6.11: Output Queueing Mean Waiting Time
© 1987 IEEE

In what follows, the case of input saturation is considered. This is the case when the input queue always has a packet waiting.

Let R_m^i be the number of packets at the heads of the input queues that are blocked at the end of the mth time slot and are destined for output i. That is, R_m^i is the number of packets destined for output i that are *not* selected by the switch controller during the mth time slot.

Let A_m^i be the number of packets moving to the head of the free input queues, destined for output i, during the mth time slot. An input queue is said to be *free* during the mth time slot if, and only if, a packet from it was transmitted during the $(m-1)$st time slot.

Again, one can write a discrete time equation:

$$R_m^i = \max{(0, R_{m-1}^i + A_m^i - 1)}. \quad (6.115)$$

This equation is similar to the previous one for output queueing.

The distribution of A_m^i, the number of arriving packets to free input queues, destined for output i, during the mth time slot, has the binomial probabilities

$$Prob[A_m^i = k] = \quad (6.116)$$

$$\begin{bmatrix} F_{m-1} \\ k \end{bmatrix} (1/N)^k (1-1/N)^{F_{m-1}-k}, \quad k=0,1,2,...,F_{m-1},$$

where

$$F_{m-1} \equiv N - \sum_{i=1}^{N} R_{m-1}^i. \quad (6.117)$$

In the above binomial distribution $1/N$ is used instead of p/N as $p=1$ under input saturation. The variable F_{m-1} is the number of free input queues at the end of the $(m-1)$st time slot, or the total number of packets passed through the switch during the $(m-1)$st time slot. Thus F_{m-1} is also the total number of input queues, during the mth time slot, with new packets at their heads or:

$$F_{m-1} = \sum_{i=1}^{N} A_m^i. \quad (6.118)$$

A little thought will show that it is true that $\overline{F}/N = \rho$, where ρ is the utilization of the output lines or the switch throughput. That is, $\rho = \overline{F}/N$ is the fraction of time slots on the outputs that have packets. Moreover, the equilibrium number of packets destined for output i and moving to the head of the input queues each time slot, A^i, becomes Poisson with rate ρ as $N \to \infty$. What these observations and the previous discrete time equation imply is that the output queueing M/D/1 results can be used to get an expression for the average equilibrium value of R^i. That is, as $N \to \infty$

$$\overline{R^i} = \frac{\rho^2}{2(1-\rho)}. \qquad (6.119)$$

While R_m^i does not correspond to a physical queue, all the arguments of the previous section carry through. Now an additional expression for $\overline{R^i}$ can be obtained from

$$F_{m-1} \equiv N - \sum_{i=1}^{N} R_{m-1}^i. \qquad (6.120)$$

Rearranging

$$\frac{\sum_{i=1}^{N} R_{m-1}^i}{N} = 1 - \frac{F_{m-1}}{N}. \qquad (6.121)$$

In equilibrium

$$\overline{R^i} = 1 - \rho. \qquad (6.122)$$

Expressions (6.119) and (6.122) for $\overline{R^i}$ can only be true if as $N \to \infty$ and under saturation:

$$\rho = (2 - \sqrt{2}) = 0.586. \qquad (6.123)$$

This is an upper bound on the throughput of any nonblocking interconnection network based switch using input queueing as $N \to \infty$. Table 6.1 illustrates the saturation throughput as a function of N (see To Look Further)

Table 6.1	
N	Saturation Throughput
1	1.0000
2	0.7500
3	0.6825
4	0.6553
5	0.6399
6	0.6302
7	0.6234
8	0.6184
∞	0.5858

Table 6.1 is based on a table appearing in [BHAN]. This is confirmed by simulation results [KARO 86] plotted in Figure 6.12.

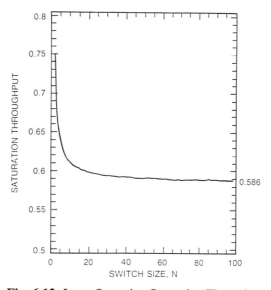

Fig. 6.12: Input Queueing Saturation Throughput
© 1987 IEEE

6.7 Case Study II: Queueing on a Single-buffered Banyan Network

6.7.1 Introduction

As has been mentioned in the previous section, there are many interconnection networks besides that of the crossbar architecture. One popular architecture is that of the Banyan network. A 4-stage (16x16) Banyan network is illustrated in Figure 6.13.

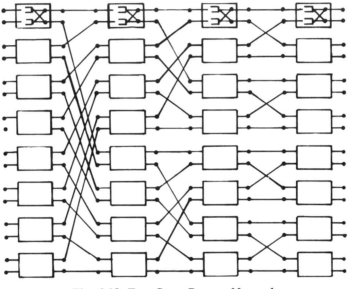

Fig. 6.13: Four Stage Banyan Network
© 1983 IEEE

Rather than place queues at the inputs or outputs of the switch, buffers are placed within each of the switching elements (boxes). Each switching element is a 2×2 cross-bar switch.

The Banyan network has an important self-routing property. Label the outputs with binary numbers in ascending order, from the top output to the bottom output. These are the output addresses. When a packet arrives at the input to the Banyan network, the first switching element routes the packet to its upper output if the first bit of the destination address is a 0 and to the lower output if it is a 1. The second switching element that the packet enters uses the same routing rule for the second bit in the destination address and so on. Using this routing rule a packet will find its way to the correct output no matter which input it enters. The self-routing property is important as it allows routing to be done distributively.

In what follows we will discuss the performance evaluation of a Banyan network where each queue, as in Figure 6.13, has a single buffer. This material is based on work published in 1983 [JENQ] by Y.C. Jenq.

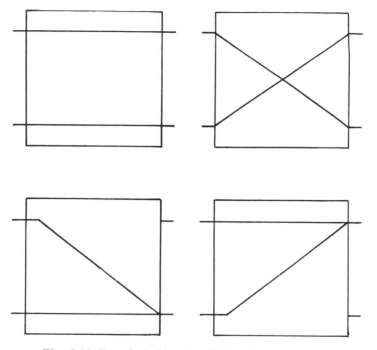

Fig. 6.14: Four Switching Possibilities for 2x2 Switch

Each 2×2 switching element can switch in four ways. These are illustrated in Figure 6.14. If there is a packet in each buffer, then two of these switching arrangements lead to conflict and two do not. That is, conflict occurs when both packets are destined for the same output. What happens is that one packet is randomly selected to go to the next stage and the other one remains in its buffer. However, a packet can only move forward if the buffer at the next stage is empty or if it is full but that packet is able to move forward. As it is put in [JENQ]: "Thus the ability for a packet to move forward depends on the state of the entire portion of the network succeeding the current stage".

In terms of implementation, this again is a synchronous system. Let a slot be τ seconds long and be composed of $\tau_1 + \tau_2$ periods. During τ_1, control signals propagate from the last stage back to the first so that it is determined whether each packet can move forward or must stay in its buffer because it is blocked. During τ_2 the packets that can, move.

6.7.2 The Model Assumptions

How many states would a Markov model of the single-buffered Banyan network have? Let n be the number of stages. Let each element have single buffers. Then a little thought will show that the number of states is $2^{n \cdot 2^n}$. Thus the number of states grows exponentially with the number of stages. For instance, the four-stage Banyan network of Figure 6.13 has 1.8×10^{19} states.

In order to make progress with such a model, Jenq made several simplifying assumptions.

First of all, loading is assumed to be balanced. That is, arriving packets are destined for each output with equal probability. The load on each input is $0 \leq q(1) \leq 1$.

With a balanced load the state of each switching element in stage k should be statistically the same. Thus Jenq looked at only a single switching element at stage k.

Matters are also simplified if the states of the two buffers within a switching element are assumed to be statistically independent. On one hand this assumption can be justified by noting that packets entering the inputs of a switching element originate from disjoint and independent network inputs. On the other hand, packets within the same switching element do interfere with one another. In [JENQ] models are constructed with and without this assumption and the results are shown to be similar. In what follows we will make use of this assumption.

6.7.3 The Model and Solution

Following [JENQ] closely, we now make some definitions:

$p_0(k,t)=$
   the probability that a switching element buffer at stage k is empty at the beginning of the tth clock period.

$p_1(k,t)=1-p_0(k,t)$

$q(k,t)=$
   the probability that a packet is ready to enter a switching element buffer at stage k during the tth clock period.

$r(k,t)=$
   the probability that a packet in a switching element buffer at stage k is able to move (forward) into the next stage during the tth clock period.

The basic idea behind Jenq's performance evaluation will be to write a series of probabilistic equations, recursive in the stage number and in time, for the above quantities.

First of all we have

$$q(k,t) = .75 \times p_1(k-1,t) p_1(k-1,t) \qquad (6.124)$$

$$+ 0.5 \times p_0(k-1,t)p_1(k-1,t) + 0.5 \times p_1(k-1,t)p_0(k-1,t),$$

$$k = 2,3,4,....n.$$

This equation relates $q(k,t)$ to $p_{0,1}(k-1,t)$. For instance, the first term multiplies out the probability that both buffers in the k-1 stage are filled with the 0.75 probability that one of these two packets enters the input line of the kth stage switching element (see Figure 6.14). This term is followed by two terms corresponding to only one of the buffers in the k-1 stage switching element being filled and routing its packet, with probability 0.5, to the kth stage switching element.

Next comes an equation for $r(k,t)$:

$$r(k,t) = [p_0(k,t) + 0.75p_1(k,t)] \qquad (6.125)$$

$$\times [p_0(k+1,t) + p_1(k+1,t)r(k+1,t)],$$

$$k = 1,2,3,...n-1,$$

$$r(n,t) = [p_0(n,t) + 0.75p_1(n,t)]. \qquad (6.126)$$

The first term in the brackets is the probability that, with one buffer in the kth stage switching element full, either the other buffer is empty or is full but does not interfere. The second term is the probability that the buffer in the $(k+1)$st stage switching element will be empty by the time the packet from the kth stage arrives.

Finally we have

$$p_0(k,t+1) = [1-q(k,t)][p_0(k,t) + p_1(k,t)r(k,t)], \qquad (6.127)$$

$$p_1(k,t+1) = 1-p_0(k,t+1). \qquad (6.128)$$

The first term in the brackets is the probability that there is no incoming packet to a kth stage switching element buffer. The second term in brackets is the probability that the kth stage switching element buffer is empty or empties out.

This set of equations models the dynamics of the system. Notice that they are indexed by k, slot time. If there is an equilibrium solution, these quantities should converge to time-independent values for $q(k)$, $r(k)$, $p_0(k)$, and $p_1(k)$. The equations can, in fact, be solved iteratively for the equilibrium values.

The two performance measures of most interest, as usual, are throughput and delay. The normalized throughput, $\overline{Y}$, or the average number of output packets per output link per slot is

$$\overline{Y} = p_1(k)r(k), \qquad k=1,2,3,\ldots n. \qquad (6.129)$$

The normalized average delay τ_{norm}, is

$$\tau_{norm} = \frac{1}{n}\sum_{k=1}^{n}\frac{1}{r(k)}. \qquad (6.130)$$

For instance, if $r(k)=0.5$ for some stage k, then the delay in moving through that stage is $1/0.5=2$ slots. The normalized average delay is computed by summing each stage's average delay and dividing by n, the number of stages. The minimal delay is thus 1.0. This corresponds to a delay of one time slot for each stage.

In Figure 6.15 the average throughput is plotted versus the independent input load parameter $q(1)$. Below $q(1)=0.4$ the throughput grows linearly with input load as little blocking is occurring internal to the Banyan network. Beyond $q(1)=0.4$ the throughput saturates (to 0.45 for $n=10$). That throughput saturates is understandable because of the blocking that occurs when two packets in a switching element attempt to access the same output link or when a packet cannot move forward.

In Figure 6.16 the normalized average delay is plotted as a function of $q(1)$. From this graph it appears that normalized delay is in the range of 1.0 to 1.55.

6.8 Case Study III: DQDB Erasure Station Location

6.8.1 Introduction

A trend that became apparent during the 1980s was the dominant influence that high capacity fiber optic transmission systems would have on new computer network architectures. One example of this is the Distributed Queue Dual Bus (DQDB) system . DQDB is a network architecture and protocol that can serve as

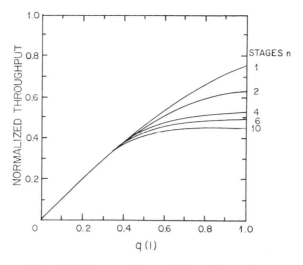

Fig. 6.15: Banyan Network Mean Throughput
© 1983 IEEE

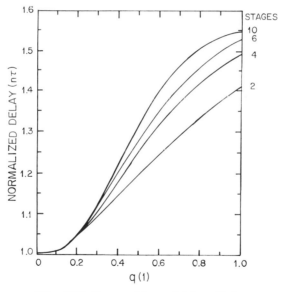

Fig. 6.16: Banyan Network Mean Delay
© 1983 IEEE

either a high-capacity local area network or as a metropolitan area network. A metropolitan area network is a high-capacity network spanning, perhaps, 50 km, that interconnects local area networks. The IEEE 802.6 standards committee has developed DQDB as a standard for Metropolitan Area Networks. DQDB is derived from the earlier QPSX [NEWM], which was developed at the University of Western Australia in conjunction with Telecom Australia.

DQDB consists of N stations interconnected by two fiber optic buses transmitting information in opposite directions at 150 Mbps per bus (Figure 6.17). Stations tap onto both buses and transmit information on the bus that leads to the receiving station. A clever protocol, described below, controls access to the buses.

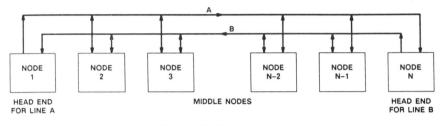

Fig. 6.17: DQDB Network Architecture
© 1990 IEEE

Taps can be either passive or active. While passive taps are highly reliable, they sap light energy from the fiber, and with current technology only several tens of taps are possible on a single fiber. Active taps regenerate the light signal.

In DQDB fixed length slots are generated at each head end and travel down the buses. Each slot contains a "busy bit" and a "request bit". DQDB really consists of two symmetrical transmission systems, one for each bus.

Whenever a station writes into a slot, the busy bit is set to one and prevents downstream stations from overwriting into the same slot. Whenever a station has a packet to send, it sets the first available request bit on the *opposite* bus.

DQDB is intended to implement a distributed queueing protocol. Suppose we consider information transmission on bus A. Each station counts the difference between the number of requests flowing past it, from right to left, on bus B and the number of empty (nonbusy) slots flowing past it, from left to right, from upstream on bus A. When a packet is ready for transmission, a station waits a number of empty slots equal to the current difference before transmitting.

If it were not for the non-zero propagation delays encountered by the request bits and the finite bandwidth, DQDB would distributively emulate a FIFO queue, networkwide. That is, packets would be transmitted onto the bus(es) in the order of their arrivals. Because there is nonzero propagation delay and bandwidth is finite, FIFO emulation is only an approximation and there are discrepancies in the "fairness" with which access by each station is treated [FILI].

In a purely passive implementation of DQDB, when a busy slot passes a receiver the slot cannot be reused by stations further downstream. An active station may "erase" busy slots that have already reached their destinations so that stations further downstream may reuse such slots. However, there is a penalty - delay is introduced by an active station as it must read the packet's address and decide whether to pass it or erase it. Therefore, a compromise has been proposed

[GARR] of placing a limited number of "erasure stations" in the network that would perform this function in order to boost the network's capacity.

In what follows, the optimal locations for placing erasure stations and the associated increase in network capacity is calculated. This material is reprinted from a paper [GARR] by S.-Q. Li and M. Garrett. It is reprinted with the permission of the IEEE (copyright 1990 IEEE). In 6.8.2 the network will be modeled continuously. A discrete formulation is also possible [GARR]. This is a good example of a successful attempt at tackling a distributed queueing problem.

6.8.2 Optimal Location of Erasure Nodes

by Mark W. Garrett and San-Qi Li

The first question regarding erasure nodes is: Where do they go? We simplify the problem for the moment by assuming that load is uniformly distributed among all source/destination pairs. For a dual bus network we need only analyze one side; the other behaves identically by symmetry. In one direction, the load offered by any station is proportional to the number of destination stations downstream. Therefore, the offered load is highest at the first station and decreases linearly to the last station, which sends zero because all of its traffic is sent on the opposite bus. For convenience we model the network in a continuous fashion, as if there were a station at every point on the line segment from zero to unity.

We first consider the case with no erasure nodes and no reuse of slots. Traffic is generated from any point, x, on the network according to the unidirectional transmission density function,

$$f_T(x) = 2\gamma(1-x), \quad 0 \le x \le 1, \tag{6.131}$$

where γ is the maximum throughput or capacity for the single bus normalized to the case without slot reuse. The probability that a slot at point x is occupied is given by the distribution

$$F_O(x) = \int_0^x f_T(\chi)d\chi = \gamma x(2-x). \tag{6.132}$$

For the network to be fully occupied means that all slots measured at the end of a bus are full, or, $F_O(1)=1$. This yields $\gamma=1$, i.e., without slot reuse the network capacity is the nominal value.

If we now consider destination release (DR), we have a reception density function along the unit segment,

$$f_R(x) = 2\gamma x, \qquad 0 \le x \le 1, \tag{6.133}$$

and the occupancy distribution at any point is given by

$$F_{O,DR}(x) = \qquad (6.134)$$

$$\int_0^x f_T(\chi) - f_R(\chi)d\chi = 2\gamma x(1-x).$$

Setting the first derivative to zero shows that the network is now most congested in the center rather than at the far end. The maximum throughput is found by setting $F_{O,DR}(1/2)=1$, yielding $\gamma=2$, i.e., continuous destination release results in a doubling of the network capacity.

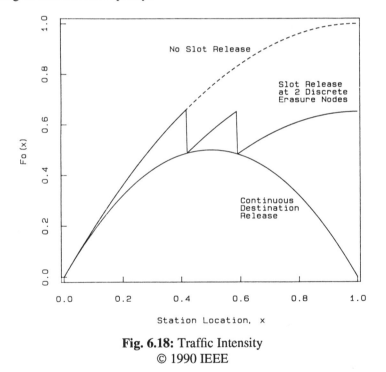

Fig. 6.18: Traffic Intensity
© 1990 IEEE

If we now consider adding discrete erasure nodes, it is clear that traffic builds up as with no reuse, equation (6.132), until the first erasure node, as shown in Figure 6.18. The traffic level is then reduced to the value of $F_{O,DR}(x)$ with continuous destination release, equation (6.134), and again increases monotonically until the next erasure node or the end of the network. The problem is to maximize γ by optimally locating n_e erasure nodes, subject to the constraint that the probability of occupancy be not more than 100% at all points. The local maxima in $F_O(x)$ are clearly just before each erasure node, i.e. at $x=L_k^-$, and at the end of the network. The maximum γ occurs when these maxima are just equal to unity. We now have n_e+1 constraints

$$F_O(L_k^-)=1, \quad F_O(1)=1, \quad 1 \le k \le n_e, \tag{6.135}$$

and the same number of variables $L_1, \cdots L_{n_e}, \gamma$, where L_k denotes the location of the kth erasure node. We solve for γ and the L_k's by identifying some relationships from which we can use a simple numerical method.

From the network head end to the first erasure node, the occupancy follows equation (6.132). Setting $F_O(L_1^-)=1$ we have

$$L_1^2 - 2L_1 + 1/\gamma = 0, \tag{6.136}$$

yielding with $0 \le L_1 \le 1$,

$$\boxed{L_1 = 1 - \sqrt{1-1/\gamma}. \quad (6.137)}$$

Generally, just beyond any erasure node the occupancy is given by the DR expression, equation (6.134),

$$F_O(L_k^+)=F_{O,DR}(L_k)=2\gamma L_k(1-L_k), \tag{6.138}$$

and between erasure nodes, using equations. (6.132) and (6.138),

$$F_O(x)=2\gamma L_k(1-L_k)+\int_{L_k}^{x} 2\gamma(1-\chi)d\chi,$$

$$F_O(x)=-\gamma(x^2-2x+L_k^2), \quad L_k<x<L_{k+1}. \tag{6.139}$$

At $x=1$ this gives the constraint on L_{n_e} as

$$F_O(1) = 1 = \gamma-\gamma L_{n_e}^2, \tag{6.140}$$

yielding

$$L_{n_e} = \sqrt{1-1/\gamma}. \quad (6.141)$$

Equations (6.137) and (6.141) give the locations of the first and last erasure nodes as functions of γ and are shown in Figure 6.19. For $n_e=1$, the single erasure node is optimally located at the point of intersection, x =0.5 and γ=1.33. For more than one erasure node we use equation (6.139) with $F_O(L_{k+1})=1$ to get

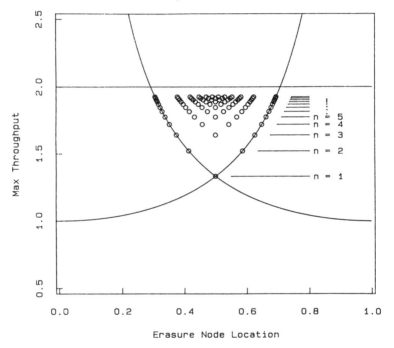

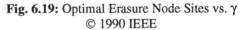

Fig. 6.19: Optimal Erasure Node Sites vs. γ
© 1990 IEEE

$$1/\gamma = 2L_{k+1}-L_{k+1}^2-L_k^2, \quad 1 \le k \le n_e, \quad (6.142)$$

yielding

$$L_{k+1} = 1 - \sqrt{1 - 1/\gamma - L_k^2}. \quad (6.143)$$

This set of equations, with the boundary equations (6.137) and (6.141), is easily solved numerically by searching for a value of γ for which the value of L_{n_e} given by equation (6.141) matches the value obtained by iterating equation (6.143) $n-1$ times, from L_1 given by equation (6.137). Figure 6.19 shows the resulting locations for up to 12 erasure nodes. A substantial throughput gain is possible for up to five or six erasure nodes, with only marginal gains thereafter [GARR].

These results show that erasure nodes are quite non-uniformly spaced. Indeed, with continuous reuse, equation (6.134), traffic is concentrated in the center of the network, and so erasure nodes are optimally located near the center also. We note that as $n_e \to \infty$, $L_1 \to 1 - \sqrt{1/2}$. Equation (6.132) shows that this is the first location where traffic builds up to 100% occupancy when $\gamma = 2$, which is the limit of γ as $n_e \to \infty$.

To Look Further

Discrete time Geom/D/1 and Geom/D/1/n queues are examined in [GRAV].

In [KARO 86] an interesting point is made that the same asymptotic throughput as for input queueing packet switches has also been obtained in the different context of memory interference in synchronous multiprocessor systems [BASK 76], [BHAN]. Several recent ideas for improving the performance of nonblocking switches with input buffering appear in [KARO 92]. The effect of reduced speedups in a nonblocking switch is examined in [OIE]. A nonblocking switch carrying two classes of traffic is studied in [GUER]. A wide variety of work on performance evaluation in high-speed switching appears in [GANZ] and [ROBE 93].

Jenq's 1983 paper also examines switch delay when there are infinite sized buffers at the Banyan network inputs. In [KIM] probabilistic equations are used to account for nonuniform traffic patterns in a Banyan network with single or multiple buffers. Multiple buffers in a Banyan network are also examined in [YOON]. Some methods for improving Banyan network performance are described in [ANIO] and [KIM 90b].

An alternative look at erasure node placement in DQDB appears in [RODR]. A survey of the DQDB literature appears in [MUKH].

Problems

The summations of section A.7 in the appendix are useful for a number of these problems.

6.1 For the system portrayed in Figures 6.1 and 6.2:

a) Write an expression for the utilization of buffer A.

b) If in slot t both buffers are empty, what is the probability at slot $t+1$ that both buffers are full? That one buffer is full? That both buffers are empty?

c) If in slot t both buffers are empty, what is the probability in slot $t+2$ that both buffers are full?

6.2 Consider a two-state (silence(0) and talkspurt(1)) discrete time Markov chain modeling the stochastic generation of packets in a TASI like [SCHW 87] packet voice system. Here time alternates between periods of packet generation (talkspurt) and silence. Let:

a: Probability of making a transition from silence to talkspurt.

s: Probability of making a transition from talkspurt to silence.

The probabilities of making a transition from a state to itself are 1-a and 1-s for the silence and talkspurt state, respectively.

a) Draw the state transition diagram.

b) Find the equilibrium probability of silence and of a talkspurt.

c) Using algebra prove that the rate at which transitions are made from silence to talkspurt equals the rate at which transitions are made from talkspurt to silence.

d) Write out the transition matrix P of this system.

e) Let $a=0.6$ and $s=0.8$ and the initial state probabilities at slot 0 be $[\pi_0, \pi_1] = [0.5, 0.5]$. Calculate the state probabilities at the third slot using equation (6.6).

f) Compare the result of (e) with the result using the method of (b).

6.3 a) Write out the state transition matrix, P, for the system of Figure 6.1 and Figure 6.2.

b) For such a matrix the entries of each row must sum to one. Why?

c) Check this for the first row.

6.4 a) Consider a system like that in Figure 6.1 with 5 stations. Draw the state transition graph when the state variable is the total number of packets in the buffers (one per station) across the network.

b) Relate the structure of the graph to that found in section 3.4.

6.5 Compare the computational requirements of equations (6.5) and (6.6) for solving the equilibrium state probabilities of discrete time queueing systems. Note that solving N general linear equations requires time proportional to the cube of N.

6.6 **Computer Project:** Write and execute a computer program to solve the system of Figure 6.2 for the equilibrium state probabilities using equation (6.6). Assign initial probabilities to each state uniformly. Include code to renormalize the state probability vector after each iteration. Have the program calculate the performance measures of equations (6.8) through (6.11). Produce graphs of these performance measures for "a" from 0.1 to 1.0 in steps of 1.0 with s=0.25,0.50,0.75. Hand in a report consisting of a program flow chart, the code listing, and graphs.

6.7 Consider a stream of traffic consisting of voice or data packets. Time is slotted. Voice packets, data packets and idle slots all have the same width. There are no "collisions". The probability of a voice packet arrival, p_v, the probability of a data packet arrival, p_d, and the probability of an idle slot, p_{idle}, are such that:

$$p_v+p_d+p_{idle}=1.0$$

These probabilities are independent of one another.

(a) Suppose we observe the following sequence of packets: a voice packet followed by a data packet, followed by two consecutive voice packets followed by an idle slot. Sketch this sequence of packets.

(b) Write an expression for the probability of the "exact sequence" of (a) occurring (that is a voice packet, a data packet, two voice packets and an idle slot).

(c) Let a voice "packet train" consist of i consecutive voice packets. Write an expression for the probability of a voice packet train of i consecutive packets occurring.

(d) Find the mean (average) length of the packet train of part (c).

6.8 Consider a Bernoulli process modeling packet arrivals to a buffer. Let p be the probability of a packet arrival in a slot. Let a number of packets arriving in consecutive slots be called a "packet train" [XION].

a) *Given that a packet train arrives*, write out an expression for the probability of i packets in a packet train.

b) Find the mean number of packets in a packet train as a function of p.

c) Plot the result of (b).

6.9 Consider a different packet train model for the previous problem where p is the probability of the arrival of the first packet after a period of empty slots (really the probability of the occurrence of the train itself) and q is the probability of each succeeding packet. With this model packet trains may occur infrequently (small p) but when they do occur they can consist of a large number of packets (q close to 1).

a) *Given that a packet train arrives*, write out the probability of i packets in the packet train.

b) Find the mean number of packets in a packet train as a function of q.

c) Compare the mean number of packets in a packet train if $p=.01$ and $q=0.9$ for the model of this problem and the model of problem 6.8.

6.10 From the discussion in the text it can be deduced that the mean throughput (mean number of packets transmitted per slot) of the packet train model of problem 6.8 is p.

a) Find the mean throughput of the packet train model of problem 6.9 (Hint: First develop an expression for the mean length of an "empty slot train").

b) Plot it for $q=0.9$ and $0 \le p \le 1.0$.

c) Show that the expression of (a) reduces to p if $p=q$ (problem 6.8's model).

6.11 a) Under a Bernoulli process for what value of p is the variance in the number of arrivals maximized (see equation (6.22))?

b) Why is it maximized at this value?

6.12 Consider a discrete time Bernoulli arrival process as discussed in the text. If $p=0.8$ and ten empty slots have passed since the last arrival, what is the expected number of slots until the next arrival?

6.13 Consider the situation described in the previous problem. If ten empty slots have indeed passed since the last arrival what is the probability of ten *or more* further empty slots? Is it

$$1 - \sum_{i=1}^{10} Prob\,(Arrival\ in\ Slot\ i) \quad (a)$$

or:

$$1 - \sum_{i=1}^{20} Prob\,(Arrival\ in\ Slot\ i). \quad (b)$$

6.14 From (6.27), for a geometric distribution one has

$$Prob\,(Arrival\ in\ Slot\ i) = (1-p)^{i-1}p, \quad i=1,2,3.$$

a) Determine the cumulative distribution function. That is, find the probability of an arrival *by* slot i.

b) Plot it versus i for $p=0.2$ and $p=0.5$.

6.15 Consider a discrete time model of a packet switch with N inputs. Let p be the probability of an arrival in a slot on an input.

a) Find the cumulative distribution function of the time by which at least one packet has arrived on an input.

b) Find the cumulative distribution function of the time by which at least one packet has arrived on each of the N inputs.

c) Find the probability density function of (b).

d) Let $p=0.5$ and $N=16$. Plot the cumulative distribution function of (b).

6.16 Consider a rectangular crossbar switch similar to the one in Fig. 6.7 but with a inputs and b outputs. Sketch the crossbar.

In each time slot 0 or 1 packets arrive at each input. Each input's packet arrival process is Bernoulli. A packet arrives on an input with probability p and no packet arrives on an input with probability 1-p. Individual packets are (uniformly) equally likely to go to any output. There are no buffers so that if more than one packet arrives at the inputs in one slot for the same destination (output), one packet is randomly selected to go to the destination (output) and the other packets are lost.

(a) Focus on a single output (called the "tagged" output). Find the probability that the tagged output receives a packet in a given time slot from a specific input.

(b) Using the result of (a), express the mean (average) throughput of the switch.

(c) Let $p=1$ (heavy load) and $a=b=\infty$ (very large switch). Find the mean switch throughput.

6.17 Consider a 2×2 switching element (2 inputs, 2 outputs). Each arriving packet prefers one of the two outputs for routing purposes. If two arriving packets (one on each input line) prefer the same output line one packet is chosen randomly for transmission on that output line, and the other is "dropped" (lost). If each of the two packets prefer different outputs, both are routed accordingly. The probability of a packet arriving on a given input line during a slot is p. Each packet prefers an output with equal probability.

a) Calculate the probability of a contention event, that is two packets preferring the same output, in a slot.

b) For a *given* arriving packet calculate the probability of being dropped. Why is this different from the answer of (a)?

c) Calculate the mean switch throughput as a function of p. Normalize the throughput as the maximal number of packets that can be delivered to the outputs each slot is 2.0.

d) Plot the mean throughput vs. p.

6.18 Consider two packet switch "modules", each with 4 inputs. Both modules comprise a switching system. Sketch the complete system.

Time on each input is slotted into equi-width slots. One packet exactly fits into one slot. The inputs are synchronized so that in each slot one packet may or may not arrive on each input. Each input's packet arrival process is Bernoulli. In a slot a packet arrives at a given input with independent probability p and a packet may not arrive ("empty slot") with probability 1-p.

(a) Write an expression for the probability of one or more arrivals in one slot to the entire switching "system" (both modules combined).

(b) Write an expression for the probability of one or more arrivals in one slot at "each" switch module during the same slot.

(c) Write an expression for the probability that exactly 2 packets arrive in a slot to the entire switching system.

6.19 Consider a packet switching element with 3 inputs and 2 outputs. Inputs and outputs are slotted. One packet can fit in one slot. Each input's packet arrival process is Bernoulli. The probability of a packet arriving at a given input in a given slot is p. The probability of an empty slot is 1-p. Sketch the switching element.

With 3 inputs and 2 outputs if 1 or 2 packets arrive to the switch in the same slot there is no problem. The 1 or 2 packets are transmitted on the outputs. If 3 packets arrive on the inputs in the same slot, 2 packets are then randomly selected for transmission on the outputs and the third is dropped (erased).

(a) Find the probability a packet is dropped.

(b) Find the probability of "no" arrivals to the switch in a given slot.

(c) Find the mean switch throughput (average number of packets transmitted on the output links). Hint: Use a weighted sum.

6.20 Consider a network of three switching elements, A, B, and C. All three switching elements have 2 inputs and one output. The outputs of A and B are the inputs of switching element C. Sketch this system.

Time is slotted. Packets are fixed in size. Each input's packet arrival process is Bernoulli. In each slot a packet arrives on each of the inputs of A and B with independent probability p and no packet arrives on each input with independent probability 1-p. For elements A and B if one packet arrives on either of their two inputs in a slot then it is sent to the element output. If two packets arrive to either element A or B in a slot, one is randomly chosen to go to the output and the other is dropped (erased). Element C also has one output and can pass only one packet from input to output.

(a) Find the probability of a packet on the output link of A.

(b) Find the probability of at least one arrival to element C in a slot.

(c) Find the mean throughput (packets per slot transmitted) of element C.

6.21 Consider a network of three identical 4 input : 2 output switching elements, A, B, and C. The outputs of A and B are the inputs of switching element C.

Each input's packet arrival process (for A and B) is Bernoulli. That is, time is slotted on each input and the probability of an arrival on an input is p and the probability of no arrival is (1-p). For each switching element in each slot up to 2 arriving packets are sent to the switching element's outputs and additional packets are dropped (erased).

(a) Find an expression for the mean throughput of switching of element A.

(b) Consider a link from switching element A to C. Find q, the probability of a packet on this link in a slot.

(c) Using the results of part (a) and (b), find an expression for the mean throughput of switching element C.

6.22 Consider a packet switching element with N inputs and either 1 or 2 outputs.

The packet arrival process on each input is Bernoulli. That is, time is slotted on each input and the probability of an arrival on an input is p. The probability of no arrival is 1-p. In an N:1 switching element in each slot an arriving packet is selected randomly and sent to the output, additional packets are dropped (or erased). In an N:2 switching element in each slot up to 2 arriving packets are chosen randomly and sent to the outputs and additional packets are dropped.

(a) For an N:1 switching element, develop an expression for the mean "throughput", including N and p in the expression.

(b) Do the same for an N:2 switching element.

(c) Develop an expression for the mean number of packets dropped by an N:2 switching element during a slot.

6.23 Consider a 2 input: 2 output switching element.

Each of the two inputs receives a synchronized Bernoulli process (p is the probability of arrival in a slot, 1-p is the probability of no arrival). There is a buffer for each output. Sketch the switching element. Label the inputs A and B and the outputs O1 and O2. For each buffer, if it is not empty, there is one departure per slot. Assume uniform traffic: p is the same for both inputs, and a packet is equally and independently likely to go to any buffer.

Also, the packets are placed in the buffers so that each buffer contains packets for its output only. There is no limit to the size of the buffers.

(a) What is the probability that 2 arriving packets (1 on each input) are placed in the same buffer? Hint: Create a table(s) of all possibilities and associated probability mass.

(b) Draw the discrete time Markov chain for one of the buffers and label it with the transition probabilities. Do this for states 0,1,2 and 3 only. Assume that a packet(s) arriving to an empty buffer must wait at least one slot before leaving the buffer. Hint: As a necessary (but not sufficient) condition check that the sum of the transition probabilities leaving each state should sum to 1.0.

6.24 Consider a 2 input: 3 output switching element.

Each of the two inputs receives a synchronized Bernoulli process (p is the probability of arrival in a slot, 1-p is the probability of no arrival). There is a buffer for each output. Sketch the switching element. Label the inputs A and B and the outputs O1, O2, and O3. For each buffer, if it is not empty, there is one departure per slot. Assume uniform traffic: p is the same for both inputs, and a packet is equally and independently likely to go to any buffer.

Also, the packets are placed in the buffers so that each buffer contains packets for its output only. There is no limit to the size of the buffers.

(a) What is the probability that 2 arriving packets (1 on each input) are placed in the same buffer? Hint: Create a table(s) of all possibilities and associated probability mass.

(b) Draw the discrete time Markov chain for one of the buffers and label it with the transition probabilities. Do this for states 0,1,2 and 3 only. Assume that a packet(s) arriving to an empty buffer must wait at least one slot before leaving the buffer. Hint: As a necessary (but not sufficient) condition check that the sum of the transition probabilities leaving each state should sum to 1.0.

6.25 Consider a 2 input: 3 output switching element.

Label the inputs A and B and the outputs O1, O2, and O3. Arrivals are a (synchronized) Bernoulli process. There are no buffers in this model. Sketch the switching element. If more than one input packet is destined for the same output, one packet is selected randomly to go to the output and the other is dropped (lost). Traffic can be assumed to be uniform (p is the independent probability of arrival in a slot on each input, packets are equally and independently likely to go to any output).

(a) Find an expression for the mean throughput of the switching element, as a function of p.

6.26 Consider the 2 input: 3 output switching element, without buffers again as in the previous problem.

Assume uniform traffic, as in the previous problem. The inputs are, again, a synchronized Bernoulli process. If there is more than 1 packet for an output, a packet is randomly chosen to be deflected to an idle output.

(a) Find the probability that a packet is deflected, as a function of p.

6.27 Consider a 3 input: 2 output switching element. Label the inputs as A, B and C and the outputs as O1 and O2.

Arrivals are a (synchronized) Bernoulli process. There are no buffers in this model. Sketch the switching element. If more than one input packet is destined for the same output, one packet is selected randomly to go to the output and the other(s) are dropped (lost). Traffic can be assumed to be uniform (p is the probability of an arrival in a slot on each input, packets are equally and independently likely to go to any output).

(a) Find an expression for the mean throughput of the switching element, as a function of p.

6.28 Consider the 3 input: 2 output switching element, without buffers again as in the previous problem.

Assume uniform traffic, as in the previous problem. The inputs are, again, a synchronized Bernoulli process. If there is more than 1 packet for an output, a packet is randomly chosen to be deflected to an idle output, "if possible". Packets that do not get their first choice of output, and are not deflected, are lost (dropped).

(a) Find the probability that a packet is "deflected", as a function of p.

(b) Find the probability that a packet is "lost (dropped)", as a function of p.

6.29 Consider a binomial distribution for $N=4$ and $p=0.6$.

a) Plot it.

b) Calculate its mean and variance.

6.30 Consider a discrete time model of an input line carrying three classes of traffic: voice, video and data. Let the probability of the arrival of a packet be p. For a *given* packet the probability that it is voice is q_v, the probability that it is video is q_{tv}, and the probability that it is data is q_d.

a) Calculate, for each class of traffic, the probability of i arrivals in N slots.

b) Calculate, for each class of traffic, the mean number of arrivals in N slots.

c) Calculate the probability of i combined arrivals of either voice or data packets.

6.31 **Computer Project:** Write software that includes an evaluation of the binomial distribution. For a given p and N the software should calculate the values of i 1) below which 25% of the arrivals occur (25% quantile) 2) between which and centered at the mean 50% of the arrivals occur and 3) above which 25% of the arrivals occur (75% quantile). Use this software to compute these quantities for the arrivals, during one slot, to a packet switch with 16 and 64 inputs as a function of p. Hand in a report consisting of a program flow chart, the code listing, and graphs.

6.32 Consider N stations that have access to M discrete time slotted buses. In a given slot the probability that a station has a single packet to transmit is p. One packet fills one slot. A station chooses a bus randomly and attempts to transmit in the current slot. If more than one station attempts to transmit on the same bus during the same slot, none of the packets are successfully transmitted (the overlapped transmissions are garbled). The system is memoryless so that whether a station has a packet in one slot and whether it is successful does not influence whether it will have a packet in a future slot.

a) What is the probability that a station attempts to transmit on a given bus during a slot?

b) Calculate the probability of a "collision" event on a given bus (more than one simultaneous transmission).

c) Calculate the mean throughput of a single bus.

d) Calculate the total mean throughput of all M buses.

e) Calculate an expression for p that maximizes the mean throughput of (d).

f) Plot the optimal mean throughput of (d) and (e) versus N for $M=1,3,5$.

g) Find the mean optimal throughput as $N \to \infty$. Compare with (f).

6.33 Consider bus networks with N stations, M discrete time slotted buses and R connections per station such as these examples:

Example 1:

There are four buses and four stations. Station 1 is connected to buses 3 and 4, station 2 is connected to buses 2 and 3, station 3 is connected to buses 1 and 2 and station 4 is connected to buses 1 and 4. Sketch this situation.

Example 2:

There are 6 buses and 3 stations. Station 1 is connected to buses 3,4,5,6; station 2 is connected to buses 1,2,3,4 and station 3 is connected to buses 1,2,5,6. Sketch this situation.

Naturally RN/M should be an integer. This is the number of stations connected per bus. Let the probability that a station attempts to transmit in a slot on a given connected bus be p $(0<p<1)$. Note that this definition is somewhat different from problem 6.32. Here a station may transmit on more than one bus in a slot with different packets,

(a) Develop an expression for the system throughput. We are assuming that collisions result if more than two stations transmit on the same bus in the same slot.

(b) Find an expression for p that maximizes throughput. Show all steps.

Note: Do a reality check by substituting in the parameters of the two illustrated examples into the result of (b) and seeing if the value(s) of p is reasonable.

6.34 Consider two Ethernet style buses operating in a "slotted" multiple access mode. They are connected by a repeater.

Bus 1 has N stations and one file server on it.

Bus 2 has M stations and one file server on it.

Assume that the transmissions by each station are a Bernoulli process. Each station transmits with probability p in a slot. The file server (FS) on bus 1 transmits with probability q in a slot and the file server on bus 2 transmits with probability r in a slot. Sketch this system.

(a) Assume that the repeater forwards all traffic between the two buses so that they are effectively one bus. Assume that the slots on each bus are synchronized. Find an expression for the probability of a successful transmission in a slot. Use the usual discrete time multiple access model where if two or more devices transmit during the same slot a "collision" occurs and no packet gets through. For a successful transmission only one device may transmit in a slot.

(b) Suppose now that the repeater is replaced by a bridge that forwards only packets destined for the other bus. Find an expression for the probability of a successful transmission on bus 1. Let x be the probability that a bus 2 packet has a destination on bus 1. Use the usual discrete time multiple access model. State any assumptions.

6.35 Consider a Bernoulli packet arrival process. Let $N=20$ and $p=.01$.

a) Calculate $b(i,N,p)$ for $i=0,1,2$ using the binomial distribution.

b) Calculate the same quantities using the Poisson approximation to the binomial distribution of section 6.3.4. Compare with the results of (a).

6.36 Consider a Geom/Geom/2/4 discrete time queueing system where a packet may enter an empty queue and leave during the same slot.

a) Draw the state transition diagram, and from first principles, label each transition with the corresponding transition probabilities.

b) Compare the result of (a) with equations (6.83-6.85).

6.37 Consider a Geom/Geom/2/4 discrete time queueing system where a packet entering an empty queue may *not* leave during the same slot.

a) Draw the state transition diagram, and from first principles, label each transition with the corresponding transition probabilities.

b) Compare the result with equations (6.83) and (6.86-6.87).

6.38 Consider a Geom/Geom/1/N queueing system (one server and a capacity of N) where a customer arriving to an empty queue *cannot* enter and leave during the same slot.

a) Draw and label the state transition diagram. Verify the expressions for the transition probabilities suggested in this chapter.

b) Develop an expression for P_n, the equilibrium state probability, as a function of s,p and P_0. Hint: Draw vertical boundaries between adjacent states in the state transition diagram, and equate the flow of probability flux from left to right to the flow from right to left.

c) Find a closed form expression for P_0. Hint: Make use of one of the useful summations in the appendix A.7.

d) Develop an expression for the mean throughput of the queueing system. Plot mean throughput versus p.

6.39 Consider a Geom/Geom/1/N queueing system (one server and a capacity of N) where a customer arriving to an empty queue *can* enter and leave during the same slot (virtual cut through policy). Draw and label the state transition diagram. Verify the expressions in the chapter for the transition probabilities (6.94).

6.40 Consider a Geom/Geom/1 discrete time, slotted queueing system.

Let p be the arrival probability and s be the service completion probability. Assume that a customer arriving at an empty queue can be serviced in the same slot. Also assume that if the queue is full an arriving customer can enter the system during the same slot that a customer leaves the server--- leading to no net change in state.

(a) Draw and label the state transition diagram of this queueing system.

(b) Using algebra explain why the utilization of the queueing system has the following behavior: utilization is a monotonically increasing function of p and utilization is a monotonically decreasing function of s.

6.41 **Computer Project:** Write software for a discrete event simulation of an input queueing nonblocking space division packet switch with a uniform traffic load as described in Case Study I. Write the software so that one of the inputs is N, the size of the switch. The program output should include mean throughput as a function of input load, p. Check your mean throughput results under saturation with Table 6.1.

6.42 **Computer Project:** Write software to implement the iterative equations for the blocking Banyan network of Case Study II, and solve them to produce the curves of Figures 6.15 and 6.16. Note that one is solving for equilibrium values so t is not an argument but k is. That is, one needs one set of equations for each stage of the switch, k. To produce an iterative solution start with an initial guess of the system variables and use the equations to produce a new guess that is re-substituted into the equations and so on. Continue iterating until the variables converge to constant values.

6.43 **Computer Project:** Write software to solve for the optimal DQDB erasure station locations as in Figure 6.19. Use the procedure outlined in the discussion following equation (6.143).

Chapter 7: Network
Traffic Modeling

7.1 Introduction

The first six chapters of this book have given you, the reader, the basics of queueing theory and performance evaluation for computer networks and computer systems. However this is only a beginning. There are a plethora of techniques to model traffic and evaluate network performance. Overviews of these are given in section 7.2 (for continuous time models) and section 7.3 (for discrete time models). Solution methods for these models are discussed in section 7.4.

There has been much interest this decade in investigating bursty traffic sources. Section 7.5 provides a generic introduction to characterizing burstiness. The special paradigm of self-similar traffic and long-range dependence is examined in Section 7.6

7.2 Continuous Time Models

In a continuous time model packet arrivals or departures may occur at arbitrary instants of time. That is, at least in terms of mathematical theory, times of events are real numbers of infinite precision. In the following the taxonomy of Kuehn [KUEH] is followed in classifying continuous time models. In particular these models have applications as arrival models.

7.2.1 Poisson Process (PP or M)

This is the simplest and oldest arrival model. It was used by Erlang in the early part of the twentieth century to model arrivals to telephone exchanges. It is a good model when there are a large number of independent users, each of which is generating traffic according to a Poisson process and no source dominates the traffic generation. As mentioned in Chapter 2, a time invariant Poisson process is simply one where every time instant is equally likely/unlikely to have an event (call). This is true of individuals placing calls to a telephone network over a short enough time period so that time of day variations are not a factor.

Two other important properties of the Poisson process are that it is memoryless and inter-arrival times are negative exponentially distributed. An extensive

333

discussion of Poisson processes appears in section 2.2. The Poisson process is also an important component of several of the following traffic arrival models.

7.2.2 Generally Modulated Poisson Process (GMPP)

Here the Poisson arrival rate as a function of time, $\lambda(t)$, is determined by a second stochastic process with a finite state space. If this second "modulating" process (or "generator") is a Markov chain, one has a Markov Modulated Poisson Process (see below). Other modulating processes are possible though. A GMPP is what is called a doubly stochastic process: it has two stochastic dimensions. The GMPP is also known as a switched Poisson process.

7.2.3 Markov Modulated Poisson Process (MMPP)

As mentioned, this is a special and important case of the GMPP. The modulating process is a finite Markov chain where q_{ij} is the transition rate from state i to state j (and i is not equal to j). The modulating Markov chain runs like any continuous time Markov chain. When the modulating chain is in state i ($i=1,2,...,n$), the MMPP arrival rate is λ_i.

The MMPP has been frequently used for traffic modeling. For instance, Schwartz [SCHW 96, pg. 48] considered using a one dimensional Markov chain as the modulating process to model N multiplexed and packetized voice sources:

$$q_{i,i+1} = (N-i)\lambda, \tag{7.1}$$

$$q_{i+1,i} = (i+1)\alpha, \tag{7.2}$$

$$i = 0,1,2 \cdots N.$$

Here λ is the Poisson talkspurt (see section 7.2.5) generation rate for one user and α is the individual talkspurt termination rate (inverse of the holding time). Here the MMPP traffic arrival rate is $i\beta$ for state i. Naturally β is the packet generation rate for one call while in talkspurt (assuming packetized voice).

MMPP's can also be used to model video traffic [SCHW 96].

It should be noted that in general the equilibrium state probabilities of the modulating process are:

$$\underline{p} = [p_1, p_2, p_3,...,p_N]. \tag{7.3}$$

The sum of these probabilities is one (normalization). The probabilities also satisfy the matrix equation:

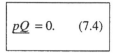

$$\underline{p}\,\underline{Q} = 0. \qquad (7.4)$$

Here $\underline{Q}$ is a square matrix where the ij th element is q_{ij}:

$$\underline{Q} = [q_{ij}]. \qquad (7.5)$$

Here also:

$$q_{ii} = -\sum_{j \neq i}^{N} q_{ij} \qquad i=1,2,...,N. \qquad (7.6)$$

The matrix $\underline{Q}$ is known as the infinitesimal generating matrix for the modulating Markov chain.

A renewal stochastic process is one where inter-arrival times are independent and identically distributed. A MMPP is not a renewal process.

Markov modulation is a technique that has been widely used in high-speed network performance evaluation. Michiel and Laevens [MICH] list the following degrees of freedom that characterize a Markov modulated source:

→ The number of states in the modulating Markov chain.

→ The modulating chain topology and transition rates.

→ The stochastic process invoked by each state in the modulating Markov chain. These could be Poisson processes (as above) in continuous time, Bernoulli processes in discrete time (see below), or deterministic processes. Usually the same type of process is invoked by all states in a modulating Markov chain.

→ The time evolution of the modulating Markov chain. This includes the choice of using either continuous or discrete time models and variations of these.

The book by Schwartz [SCHW 96] provides a good introduction to Markov modulated processes.

7.2.4 Switched Poisson Process (SPP)

The switched Poisson process consists of two states and is a Markov modulated process. The modulating two-state Markov chain has transition rates q_{01} and q_{10}. The arrival rates of the SPP are λ_0 when the Markov chain is in state 0 and λ_1 when the chain is in state 1.

One can describe the modulating Markov chain by the equations:

$$p_1 = \frac{q_{01}}{q_{10}} p_0, \tag{7.7}$$

$$p_0 + p_1 = 1. \tag{7.8}$$

Here p_i is the equilibrium probability of the i th state.

Solving, one has:

$$p_1 = \frac{q_{01}}{q_{01} + q_{10}}, \tag{7.9}$$

$$p_0 = \frac{q_{10}}{q_{01} + q_{10}}. \tag{7.10}$$

The average arrival rate is then:

$$\lambda_{AVG} = \lambda_0 \frac{q_{10}}{q_{01} + q_{10}} + \lambda_1 \frac{q_{01}}{q_{01} + q_{10}}. \tag{7.11}$$

7.2.5 Interrupted Poisson Process (IPP)

This is a special case of a MMPP and SPP. It is a SPP where $\lambda_0 = 0$ and λ_1 is not equal to zero.

The IPP can be used to model a packetized voice conversation. Human speech consists of alternating periods of speech and silence. A typical model would have state one indicate a talkspurt (ON) and state zero indicate a silence (OFF). The presence of a significant amount of silence in speech has been used as a basis for compression. One such system for analog telephone conversations is TASI (time assigned speech interpolation). As described above, the probability a speaker is active (ON) is $q_{01}/q_{01} + q_{10}$. The inverse of this is the "TASI advantage", the factor by which more circuits can be accommodated.

Also note the average arrival rate of packets for an IPP is:

$$\overline{\lambda}_{AVG} = \lambda_1 \frac{q_{01}}{q_{01} + q_{10}}. \tag{7.12}$$

The interrupted Poison process is a renewal process.

7.2.6 Markovian Arrival Process (MAP)

A Markovian arrival process also uses a finite state Markov chain. The transition matrix Q is as above. However here at each transition that is transited an arrival occurs with some fixed probability. A MAP with batch arrivals is a BMAP. Work on MAPs originated with Neuts [NEUT][NEUT 89]. Lucatoni [LUCA] devised a related model that was shown to be equivalent to the model of Neuts. Feldman and Whitt [FELD 98] suggest the use of MAPs and BMAPs to model complex network behavior. Asmussen and Koole have proved that any stationary point process can be approximated arbitrarily closely by a Markovian arrival process [ASMU].

7.2.7 Autoregressive Moving Average Model (ARMA)

The time series based ARMA (autoregressive moving average) model has been used to model video sources. One can have [KUEH]:

$$X_i = g(\alpha Z_{i-m} + Y_i + V_i). \tag{7.13}$$

Here X_i is a discrete state ARMA process modeling the number of cells in the nth interval. The random variables Y_i and Z_i are series of zero mean correlated Gaussian random variables representing frame and scene correlation, respectively. Also α is a constant between 0 and 1. The random variable V_i is a series of zero mean uncorrelated Gaussian random variables (a form of white noise). Finally $g(\,)$ is a non-linear and memoryless function transforming the sums of Gaussian random variables to the desired distribution $f(x)$.

7.2.8 Fluid Flow Approximation Model (FFA)

In a fluid flow model the fine packet stream is replaced by a coarse continuous valued flow to provide a more tractable analysis. The flow rate may be modulated by a Markov chain. Fluid flow modeling has been used for studying statistical multiplexing, admission control, video multiplexing, and leaky-bucket analysis [SCHW 96].

7.2.9 Self-Similarity Source Model (SSS)

Since 1993 there has been evidence that important types of network traffic exhibit self-similarity: statistically similar behavior over many time scales. Self-similar models are discussed in section 7.6.

7.2.10 Renewal Processes (RP,GI)

In a renewal process arrivals have inter-arrival times that are independently and identically distributed (called "i.i.d."). A renewal process can be seen to be a generalization of a Poisson process. A renewal process is sometimes called a general independent point process.

7.2.11 Semi-Markov Processes (SMP)

Semi-Markov processes [MICH] are a generalization of renewal processes and Markov models. A semi-Markov process achieves this flexibility by allowing more freedom with respect to the time between transitions. This is accomplished by allowing the probability density function of the time τ in state m before transitioning to state n to be $f_{mn}(\tau)$. Note that this probability density function is a function of both m, the current state, and n, the next state.

A semi-Markov process is a continuous process associated with a discrete state space. Because a semi-Markov process requires more knowledge to predict the future than simply the current state, it is not a Markov process [MICH]. If the random time between transitions, or sojourn times, depend only on the state m, one has a special semi-Markov process.

7.3 Discrete Time Models

In a discrete time model, time is slotted. The most common model assumes fixed size slots with one packet fitting exactly in one slot. ATM (asynchronous traffic mode) is a key technology for which this is a fundamental model. The slot width in seconds is a basic parameter of ATM and other similar technologies.

Again, the taxonomy of Kuehn [KUEH] is followed in classifying discrete time models. While most of the following discrete time models have continuous time analogues, the details of modeling are somewhat different.

7.3.1 Deterministic Process (DP)

In a deterministic process packets arrive at fixed multiples of a basic slot (i.e., every slot, every other slot, every third slot). While not a stochastic process itself, deterministic processes are often modeled together with stochastic processes. One example is a GEOM/D/1 queue where arrivals follow a Bernoulli process (see below) and service completions are deterministic. Deterministic processes model certain communication protocol actions well.

7.3.2. Bernoulli Process (BP)

In a Bernoulli process (see section 6.3.1) the probability of one arrival in a slot is p and the probability of no arrival in a slot is $1-p$. Arrivals are independent from slot to slot so that the Bernoulli process is memoryless (the discrete time analogue of the continuous time Poisson process). Note that the Bernoulli process is the only discrete time memoryless process just as the Poisson process is the only continuous time memoryless process. The Bernoulli process is a good model for the aggregate flow of many independent sources, each of which follows a Bernoulli process and none of which dominates and for which self-similarity is not present.

The inter-arrival times of a Bernoulli process follow a geometric distribution (see section 6.3.2). The geometric distribution is thus the discrete time analogue of the continuous time negative exponential distribution for Poisson process inter-arrival times.

The Bernoulli process is a good model for a first study of discrete time systems. Its simplicity aids tractability. It has been used often in the performance evaluation of switching fabrics [ROBE 93].

In a batch Bernoulli process (BBP) up to N packets may arrive in each slot where the number of packets in each slot is binomially distributed (see section 6.3.3) One may think of a BBP as a type of superposition of independent and identically distributed Bernoulli processes.

7.3.3 Generally Modulated Deterministic Process (GMDP)

This is a doubly stochastic discrete time point process. The deterministic packet arrival rate is a function of a modulating stochastic process with a finite state space. That is, the current deterministic arrival rate is a function of the current state. The modulating process is sometimes called a generator process.

7.3.4 Markov Modulated Deterministic Process (MMDP)

This is a special case of the GMPP with a discrete time finite Markov chain (see section 6.2) as the modulating process.

For instance, a video multiplexer arrival model may use a one dimensional Markov chain as the modulating process with packets generated at deterministic rate iR packets/sec in state i where R is a constant.

7.3.5. Switched Deterministic Process (SDP)

The switched deterministic process is a MMDP with two states. Deterministic arrival rates λ_0 and λ_1 are non-zero.

7.3.6 Interrupted Deterministic Process (IDP)

This is a special case of a MMDP and SDP where there is a two-state discrete time modulating Markov chain. Let state 1 be the "on" state and state 0 be the "off" state. Deterministic arrival rate λ_0 (when in state 0) is equal to zero and deterministic arrival rate λ_1 (when in state 1) is non-zero. The transition probabilities of the modulating Markov chain are:

$$a_{11} = 1-\alpha, \tag{7.14}$$

$$a_{10} = \alpha, \tag{7.15}$$

$$a_{00} = 1-\beta, \tag{7.16}$$

$$a_{01} = \beta. \tag{7.17}$$

Here α is the transition probability of going from state 1 to state 0. Also β is the transition probability of going from state 0 to state 1. Naturally α and β are between 0 and 1. Thus, the equilibrium state probability for this two-state Markov chain can be found as:

$$\pi_1 = \frac{\beta}{\alpha}\pi_0, \tag{7.18}$$

$$\pi_0 + \pi_1 = 1. \tag{7.19}$$

Solving:

$$\pi_1 = \frac{\beta}{\alpha+\beta}, \tag{7.20}$$

$$\pi_0 = \frac{\alpha}{\alpha + \beta}. \tag{7.21}$$

The distribution of time spent in each state is geometric. For state 0:

$$f_0(i) = \beta(1-\beta)^{i-1}. \tag{7.22}$$

Also for state 1:

$$f_1(i) = \alpha(1-\alpha)^{i-1}. \tag{7.23}$$

The mean amount of time spent in state 0 is $1/\beta$ and the mean amount of time spent in state 1 is $1/\alpha$ when the slot length is normalized.

The two-state IDP model is sometimes called an On/Off source, a burst/silence model, or talkspurt/silence model [KUEH].

7.3.7 Discrete Time Markovian Arrival Process (DMAP)

The discrete time Markovian arrival process is generated by a discrete time finite state Markov chain. There is a transition matrix where the ijth element is a_{ij} and where some transited transitions cause a packet arrival and the remainder do not. One can generalize the DMAP to include batch arrivals upon transiting certain transitions [BLON][BLON B].

7.3.8 Discrete Renewal Process (DRP)

Here inter-arrival times are independent and identically distributed. This is a generalization of the Bernoulli process.

7.4 Solution Methods

There are, what are by now, a number of standard techniques for calculating the statistics of models developed for broadband networks. The first four techniques listed below [MICH] are useful for calculating equilibrium state probabilities. In section 7.4.5 transient solutions are discussed.

7.4.1 Simulation

As discussed in section 4.6, simulation is a robust tool for producing performance results. Advantages of simulation include the ability to capture complex protocol detail and relative ease of implementation. Disadvantages include scalability problems to the billions of packets and rare events (10^{-11} and smaller) that are characteristic of today's protocols and networks. Another problem is that simulation may not give the meaningful insight into a model's behavior that an analytical model, if possible, may produce.

7.4.2 Linear Equation Solution

Most queueing models can be solved in theory by solving the associated linear flow balance equations. This is true of both continuous time and discrete time models. Such flow balance equations are often written in matrix form.

The applicability of this solution method is limited by the state space explosion problem, the large number of states, and hence equations, necessary for even modestly sized models. This problem is further compounded by the super-linear computational complexity of linear equation set solution methods. Some matrices have special structure that can be exploited for computational purposes. The reader interested in more detailed information on this topic can find it in [KRIE], books by Neuts [NEUT] [NEUT 89], and in this book.

7.4.3 Probability Generating Function

The basic idea here is to transform a sequence of unknown numbers (a probability distribution) into a frequency like domain where it is sometimes easier to make analytical progress. The final result is transformed back as a known probability distribution.

Using this technique requires good analytical skills. Michiel and Laevens [MICH] point out that many early investigations were satisfied with only low-order moments, which are usually easily found. Today there is much interest in low probability events (such as buffer overflow) so that it is currently of more interest to compute complete probability distributions.

7.4.4 Fluid Flow Approximation

The approach here is to model buffer occupancy as a continuous random variable that well describes the gross behavior of the effect of many packets passing through a buffer. Arrivals to the buffer follow some stochastic model and departures often are modeled as occurring at a constant rate.

A classic paper by Anick, Mitra and Sondhi [ANIC] [SCHW 96] first demonstrated the usefulness of the fluid flow approach in the context of a FIFO queue with identical negative exponentially distributed ON-OFF sources with packet arrival rates proportional to the number of ON sources. Anick et al. found closed form expressions for the problem eigenvalues and eigenvectors as well as for the buffer occupancy distribution.

The fluid flow approach can be used for both infinite and finite sized buffers. The fluid flow approach has thus far been used to model statistical multiplexing but also for modeling admission control, video multiplexing, and leaky-buckets [SCHW 96].

7.4.5 Transient Effect Models

Transient effect models capture the behavior of a system in a limited time period, usually after some initiating event such as an arrival process being turned on. Even an M/M/1 queue transient model results in the use of differential equations and a relatively complex solution (see section 2.11). Thus typically solutions are more difficult to achieve and more complex in format for transient models compared to equilibrium models. Simulation can provide solutions for transient models.

7.5 Burstiness

A "bursty" traffic source is one that generates traffic in (random) clusters. Certainly Poisson and Bernoulli traffic have "clusters" of traffic and are more bursty than a deterministic (periodic) source. However because this perceived burstiness can be smoothed through aggregation (unlike self-similar sources), Poisson and Bernoulli traffic are not considered to be truly bursty. Many actual data sources exhibit burstiness beyond that of Poisson or Bernoulli traffic. This presents new problems in attempting to aggregate (multiplex) traffic streams and optimally sizing packet buffers. The understanding of bursty traffic sources and their implications for network design is currently a major research area in teletraffic modeling.

In this section following the taxonomy of Michiel and Laevens [MICH] some common methods of characterizing burstiness are discussed. In section 7.6 a special type of bursty traffic, self-similar traffic, is examined.

7.5.1 Ratio of Peak Rate to Mean Rate

The ratio of peak packet arrival rate to mean (average) packet arrival rate is appealing in its simplicity as a burstiness measure. However, as Michiel and Laevens describe [MICH], it suffers from some difficulties.

One problem is that the time period over which to measure the "peak" rate is ambiguous. Moreover the peak to mean ratio does not provide information on the length of the ON period of a source. This is a problem as mean burst length is a key parameter in designing buffers but one can double both the ON and OFF periods of a source yet the peak to mean ratio is identical for both cases. Finally, the peak to mean ratio is different for aggregated ON/OFF sources generating deterministic or Bernoulli traffic but studies have shown a large degree of similarity between the two types of aggregated streams.

7.5.2 Coefficient of Variation of Traffic Load

The standard definition of the coefficient of variation of a random variable is the ratio of the random variable's standard deviation to its mean. Note that the standard deviation of a random variable is the square root of its variance. Thus, if N is the number of packets transmitted per packet time,

$$\bar{N} = \rho, \tag{7.24}$$

$$VAR\,(N) = \sigma_N^2 = E\,[(N - \bar{N})^2], \tag{7.25}$$

$$SCV = (\sigma_N / \rho)^2. \tag{7.26}$$

Here SCV is the "squared" coefficient of variation.

One problem with the coefficient of variation of the traffic load is that it is a single number measuring a form of normalized variance rather than measuring burstiness directly. Michiel and Laevens [MICH] feel that the following two measures, the index of dispersion and spectral functions, better capture burstiness information.

7.5.3 Index of Dispersion

One can study a point process either in terms of the inter-arrival times or the number of events in a time interval. Thus there are two indices of dispersion, the index of dispersion for intervals (IDI) and the index of dispersion for counts (IDC) [GUSE][MICH].

IDI

The IDI is given by:

$$J_n = \frac{VAR\left[\sum_{k=1}^{n} X_{i+k}\right]}{nE^2[X]} \qquad n=1,2,3,\dots \ . \qquad (7.27)$$

Here the X_i are the inter-arrival intervals. The index of dispersion for intervals is thus a function of n, the number of intervals used. Among its properties are that the SCV of intervals equals J_1 and for a Poisson process $J_n=1$ for $n=1,2,3,\dots$[MICH].

A few words should be said about the conditions under which the IDI exists and is stable [MICH]. It is not required to assume that the X_i are independent and identically distributed. Under weak stationarity conditions $E[\sum_{k=1}^{n} X_{i+k}]$ and $VAR[\sum_{k=1}^{n} X_{i+k}]$ have definite values that are independent of time. Also, let the covariance function be:

$$COVAR[X,Y] = E[(X-\bar{X})(Y-\bar{Y})]. \qquad (7.28)$$

Then, if the X_i are uncorrelated $VAR[\sum_{k=1}^{n} X_{i+k}]$ is simply $n \times VAR[X]$.

However, if this is not the case, as is more likely, then:

$$VAR[\sum_{k=1}^{n} X_{i+k}] = n \times VAR[X] + 2\sum_{j=1}^{n-1}\sum_{k=1}^{j} COVAR[X_j,X_{j+k}]. \qquad (7.29)$$

As can be seen the IDI is influenced by the correlation between successive interarrival times.

IDC

One can also count the number of arrivals in an interval (of length t), N_t. The index of dispersion for counts is:

$$I(t) = \frac{VAR[N_t]}{E[N_t]}. \qquad (7.30)$$

As can be seen, the IDC is the ratio of the variance of the number of arrivals in an interval of length t to the mean number of arrivals, both being taken over the same interval.

Also here t is a continuous parameter. For a Poisson process $I(t)$ is 1 for all t (see section 2.2.3).

Using [GUSE]:

$$\lim_{n \to \infty} J_n = \lim_{t \to \infty} I(t) = \frac{VAR\,[X]}{\overline{X}^2} \left[1 + 2 \sum_{j=1}^{\infty} \frac{C(j)}{VAR\,[X]}\right]. \qquad (7.31)$$

Here $C(j)$ is the autocovariance of the inter-arrival times:

$$C(j) = COVAR\,[X_n, X_{n+j}] = \qquad (7.32)$$

$$= E\,[(X_n - \overline{X})(X_{n+j} - \overline{X})] = E\,[X_n X_{n+j}] - \overline{X}^2.$$

Thus in the limit the IDI and IDC are equal. The IDI and IDC are discussed further in [GUSE].

7.5.4 Spectral Characteristics

Some researchers such as S.Q. Li [LI A][LI B] and R. Grunenfelder [GRUN] have examined bursty traffic and its influence on queues in terms of its spectral (frequency) properties. This leads to an examination of input traffic second, third and fourth order statistics. Thus one examines the correlations and power spectra of (t_1, t_2), (t_1, t_2, t_3), and (t_1, t_2, t_3, t_4).

The general conclusion of this work is that knowledge of the input spectrum is useful for buffer analysis. Moreover it was found that low frequency input power has much more of an influence on buffer behavior than high frequency input power.

Michiel and Laevens [MICH] stress that the key factor in the use of spectral techniques is the use of second order statistics. They point out that one can show for the autocovariance spectra at zero frequency:

$$S(f) = \sum_{k=-\infty}^{\infty} C(k)e^{-j2\pi f k}. \qquad (7.33)$$

Letting $f = 0$:

$$S(0) = \sum_{k=-\infty}^{\infty} C(k), \qquad (7.34)$$

$$S(0) = C(0) + 2 \sum_{k=1}^{\infty} C(k), \qquad (7.35)$$

$$S(0) = VAR[X] + 2\sum_{k=1}^{\infty} C(k),\qquad(7.36)$$

$$S(0) = VAR[X]\left[1 + 2\sum_{k=1}^{\infty}\frac{C(k)}{VAR[X]}\right].\qquad(7.37)$$

Here X is the number of arrivals per unit time.

Comparing this with equation (7.31) one can see that:

$$\lim_{n\to\infty} J_n = J_\infty = \frac{S(0)}{X^2}.\qquad(7.38)$$

Thus one can conclude that J_∞ and $S(0)$ are almost identical. In the sense that J_∞ contains second order information, the IDI (and the IDC as well) are in some sense spectral burstiness measures.

7.5.5 Some Other Techniques

Other techniques for characterizing burstiness include [MICH] capturing statistical parameters of buffer underload and overload periods [YEGE] [JABB] and the stochastic process entropy rate [PLOT].

7.5.6 Queueing Performance under Burstiness

Let's consider a multiplexer [MICH]. Assume it is being fed by a number of non-bursty sources. One can plot the probability that the buffer occupancy is greater than n, a number of packets in the multiplexer, on a vertical log scale. The horizontal axis is simply n, the number of packets. In this case the curve is a decreasing straight line (with negative slope). As one moves to larger values of n, the probability that the buffer occupancy is greater than n decreases.

Suppose now that the multiplexer is fed by a number of bursty sources. The same curve now has two parts. In the first region (for small n) there is again an almost linear decrease in the probability that buffer occupancy is greater than n as n is increased. This first part of the curve is the cell region where arrivals are relatively random (Poisson-like) in nature. After a certain point though the negative slope transitions (at a curve "knee") to a smaller (in magnitude) and relatively constant slope value. This is the burst region. Here several sources are supplying bursts of packets at the same time. The load is thus larger and queue length increases. The curve in the burst region is indeed falling as the probability of very large queue sizes continues to diminish but at a slower rate compared to the cell region.

The key result of this is that a buffer of size B may be sufficient to achieve a certain probability of packet loss if traffic is non-bursty but would be too small for bursty traffic. Thus larger buffers will be needed if sources are bursty in order to achieve the same buffer occupancy performance level. This point is returned to below in discussing network performance under self-similar traffic.

7.6 Self-Similar Traffic

There was some unfocused sense in the performance evaluation community prior to 1993 that there were modeling problems with using Markovian statistics to describe data traffic. However the issue was crystallized in 1993 with the publication of "On the Self-Similar Nature of Ethernet Traffic" [LELA] by Leland, Taqqu, Willinger and Wilson. Leland and his co-authors reported on a statistical analysis of hundreds of millions of Ethernet measurements made between 1989 and 1992. They found that the traffic behavior in their experiment was fractal-like in nature and exhibited "self-similarity". That is, the statistical behavior was similar across many different time scales (i.e., tenths of seconds, seconds, tens of seconds, hundreds of seconds, etc.).

This sort of behavior is in fact very different from Markovian models. For instance, aggregating (or multiplexing) streams of traffic that are truly Markovian results in a "smoothed" (and less bursty) stream. But an aggregation of self-similar streams results in a bursty aggregate stream. This difference in behavior has potential implications for buffer design, backbone network design, and other networking problems.

Other work on self-similar traffic followed such as Paxson and Floyd's "Wide Area Traffic: The Failure of Poisson Modeling" in 1995 [PAXS]. As a result of all this some researchers felt it was necessary to completely rewrite the theory of network traffic modeling. Others felt that while Poisson models might not always be appropriate, more sophisticated models using conventional techniques could handle network traffic modeling and performance evaluation.

The paper by Leland and his coauthors was significant for not only pointing out what might be wrong with the current model, but also for suggesting an alternative: the self-similarity paradigm. Because of this paradigm's importance in ongoing research on network traffic modeling, this section discusses self-similarity.

7.6.1 Some Basics

Self-similarity models have a history that precedes their introduction into the networking field. The term "self-similar" was introduced by B. Mandelbrot, a pioneer in the study of fractals. Original applications included hydrology and geophysics. Other applications and citations to papers on the statistical theory of self-similar processes appear in [TAQQ 85].

A few basics first. The autocorrelation function of a continuous time process X is:

$$R_X(t_1, t_2) = E[X_{t_1} X_{t_2}].$$
(7.39)

Let $\bar{X}$ be the mean value of the process X. Then the autocovariance function of the process X is:

$$C_X(t_1, t_2) = E[(X_{t_1} - \bar{X})(X_{t_2} - \bar{X})].$$
(7.40)

In terms of a discrete time process the autocorrelation function is:

$$R_X(i, j) = E[X_i X_j].$$
(7.41)

The discrete time autocovariance function is:

$$C_X(i, j) = E[(X_i - \bar{X})(X_j - \bar{X})].$$
(7.42)

The statistical properties of a stationary process do not change over time. A process is stationary up to order m, if for any legitimate $t_1, t_2, t_3, \ldots, t_n$ and for any lag k and for any positive integers $m_1, m_2, m_3, \ldots, m_n$ satisfying $m_1 + m_2 + m_3 + \cdots + m_n \leq m$:

$$E[X_{t_1}^{m_1} X_{t_2}^{m_2} \cdots X_{t_n}^{m_n}] = E[X_{t_1+k}^{m_1} X_{t_2+k}^{m_2} \cdots X_{t_n+k}^{m_n}].$$
(7.43)

This definition of stationarity [CHAN][PRIE] is weaker than complete stationarity. A completely stationary process is one where the joint probability distribution for $(X_{t_1}, X_{t_2}, X_{t_3}, \ldots, X_{t_n})$ is identical to the joint probability distribution for $(X_{t_1+k}, X_{t_2+k}, X_{t_3+k}, \ldots, X_{t_n+k})$ for all k larger than zero.

A process X is wide sense (or covariance) stationary if its mean is a constant and the autocorrelation function depends only on the time difference $k = j - i$ (in terms of a discrete time process). A process that is stationary up to order two is also wide sense stationary if the second moments exist. However, the converse is not necessarily true [THOM]. Thus [MICH]:

$$R(k) = E[X_n X_{n+k}],$$
(7.44)

$$C(k) = E[(X_n - \bar{X})(X_{n+k} - \bar{X})].$$ (7.45)

It can be easily shown that:

$$R(k) = C(k) + \bar{X}^2.$$ (7.46)

7.6.2 Self-Similarity

Following Leland et. al. [LELA], let $X = (X_t : t = 0,1,2,3,...)$ be a covariance stationary stochastic process with mean $\bar{X}$, variance σ^2, and autocorrelation function $r(k)$ for positive k. Assume that:

$$\lim_{k \to \infty} r(k) \approx k^{-\beta} L(t).$$ (7.47)

Here $0 < \beta < 1$. Also L is "slowly varying" at infinity. That is, $\lim_{t \to \infty} L(tx)/L(t) = 1$ for all $x > 0$.

Now for each $m = 1,2,3,...$ let $X^{(m)} = (X_k^{(m)} : k = 1,2,3,...)$ be a new (covariance stationary) series created by averaging nonoverlapping blocks of size m. Let the time series have autocorrelation function $r^{(m)}$. That is, for each positive integer m and $k \geq 1$:

$$X_k^{(m)} = \frac{1}{m}(X_{km-m+1} + X_{km-m+2} + \cdots + X_{km}).$$ (7.48)

The process X is exactly second order self-similar if for $m = 1,2,3,...$:

$$VAR(X^{(m)}) = \sigma^2 m^{-\beta}.$$ (7.49)

$$r^{(m)}(k) = r(k), \qquad k \geq 0.$$ (7.50)

This self-similar process has self-similarity (Hurst) parameter H:

$$\boxed{H = 1 - \beta/2.} \quad (7.51)$$

The process X is called asymptotically second order self-similar if for all k large enough,

$$\lim_{m \to \infty} r^{(m)}(k) \to r(k). \qquad (7.52)$$

Here $r(k)$ is described by equation (7.47).

It should be noted that it is asymptotic self-similarity, not exact self-similarity, that is often encountered in practice. Almost all of the following performance results depend on asymptotic self-similarity. The reason behind this is that exact self-similarity (and finite variance) implies that the traffic process is fractional Gaussian noise. This model is too restrictive in practice. As an example, fractional Gaussian noise has a normal marginal distribution at all time scales. This is in fact not the case for traffic processes on lightly loaded links at fine time scales.

Self-similarity has a number of equivalent effects [LELA][MICH]:

→ Variances that decay slowly. In other words, the variance of the sample mean decreases more slowly than the more usual reciprocal of the sample size. That is:

$$\lim_{m \to \infty} VAR[X_k^{(m)}] \approx cm^{-\beta} \quad where \quad 0 < \beta < 1. \qquad (7.53)$$

Here c is a constant.

→ Hyperbolic decay of autocorrelation rather than exponential decay. Thus, the autocorrelation function is nonsummable ($\sum_k r(k) = \infty$) or long range dependent. That is, $\lim_{k \to \infty} r(k) \approx k^{-\beta}$ or equation (7.47) is satisfied.

→ The power spectral density $S(f)$ behaves like $1/f$ noise (power law) near the origin. That is, $\lim_{f \to 0} S(f) \approx f^{-(1-\beta)}$.

Self-similarity implies long range dependence (again, $\sum_k r(k) = \infty$). A process for which this summation is finite is called short-range dependent. A short range dependent process is different from a long range dependent process in that [MICH]:

→ Variances decay more quickly ($VAR[X_k^{(m)}]$ decays as m^{-1}).

→ $r(k)$ decays in an exponential manner ($\lim_{k \to \infty} r(k) \approx \gamma^{-k}$).

$\rightarrow$ A finite and approximately constant power spectral density occurs about the frequency origin.

$\rightarrow$ The process is similar to pure noise in a second order sense for $\lim\limits_{m \to \infty} r^{(m)}(k) \to 0$ where k is not equal to zero.

Leland et. al. [LELA] describe the most striking intuitive aspect of both an exact and an asymptotically second order self-similar process is that they both have a nondegenerate correlation structure in the limit as m approaches infinity. This is in contrast to short range dependent processes where the correlation structure (function) goes to zero for large m.

7.6.3 The Hurst Effect

The Hurst parameter takes on values between 0.5 and 1.0. For short range dependent processes $H = 0.5$. Many naturally occurring self-similar processes have H at about 0.7. Leland et al. [LELA] present estimates of the Hurst parameter for Ethernet traffic of about 0.85 to 0.95.

Hurst observed an empirical relation known as Hurst's law or the Hurst effect. Assume a set of data $X_1, X_2, X_3, \ldots, X_n$ with sample mean $\overline{X}(n) = \dfrac{1}{n} \sum\limits_{k=1}^{n} X_k$ and sample variance $S^2(n) = \dfrac{1}{n-1} \sum\limits_{k=1}^{n} (X_k - \overline{X}(n))^2$. The rescaled adjusted range statistic is:

$$\frac{R(n)}{S(n)} = \frac{1}{S(n)} [\max(0, W_1, W_2, W_3, \ldots, W_n) \tag{7.54}$$
$$-\min(0, W_1, W_2, W_3, \ldots, W_n)].$$

Here for $k = 1, 2, 3, \ldots, n$:

$$W_k = (X_1 + X_2 + X_3 + \cdots + X_k) - k\overline{X}(n). \tag{7.55}$$

Here W_k is a measure of the deviation of the process X_k from its mean value and $R(n)$ is a measure of the "record" value of this deviation [LELA][MICH].

For naturally occurring self-similar processes Hurst determined that $\lim\limits_{n \to \infty} E[R(n)/S(n)] \approx cn^H$ where H is the Hurst parameter and c is a constant. Typically H is about 0.7 for such processes. Again for short range dependent processes H is 0.5 and the constant may be different.

7.6.4 Roots of Self-Similarity

Crovella and Bestavros, in a 1997 paper titled "Self-Similarity in World Wide Web Traffic: Evidence and Possible Causes" [CROV], examined reasons why network traffic is self-similar. In an experiment they recorded the statistics of over half a million Web (URL) requests and 130,000 transferred files over a 37-workstation local area network. They found the distribution of transmission times may be heavy tailed (because of the distribution of file sizes). They also found that silent times (due to user thinking) were also heavy-tailed.

If one considers an individual user to be modeled as such a simple ON/OFF source, the question is what happens to network traffic flow if many such ON/OFF users are aggregated? Willinger, Taqqu, Sherman and Wilson in a 1997 paper [WILL] found that if the individual ON/OFF sources were described by heavy-tailed infinite variance distributions (i.e., such as a Pareto distribution), the aggregate flow is self-similar or long-range dependent.

When comparing ON and OFF times the distribution with the heavier tail has been found to be the determinant of the actual network traffic self-similarity levels [CROV] [WILL]. Thus since Crovella and Bestavros found the ON time distribution to be heavier tailed than the OFF time distribution, they concluded that the distribution of Web file sizes was the main cause of Web traffic self-similarity. Related to this, Park, Kim and Crovella [PARK] showed that it is sufficient to draw file sizes from a heavy-tailed distribution to generate self-similar network traffic.

There have been reports of observed self-similarity in Ethernet traffic [LELA], World Wide Web traffic [CROV], and TCP, FTP, and TELNET traffic [PAXS]. Other environments where self-similar traffic has been observed are summarized in [STAL].

7.6.5 Detecting Self-Similarity

There are a number of useful techniques to detect the presence of self-similarity. Among these are [MICH] [SAHI]:

→ *Visual Observation*: One can examine traffic plots on different time scales. Naturally self-similar traffic appears visually "similar" across many time scales whereas short range dependent traffic smooths out when sufficiently aggregated.

It is important to realize that when visually inspecting traffic on different time scales, the Y axis must be rescaled by the same factor as the X (time) axis. Otherwise it is possible to (incorrectly) expand the Y axis to match the observed variability, which can superficially resemble a self-similar situation. Being somewhat subjective, visual observation should be used in conjunction with other detection techniques.

→ *R/S Plot*: One can plot $\log[R(n)/S(n)]$ (see equation (7.54)) versus $\log(n)$. One can find the Hurst parameter through linear regression. That is, for self-similar traffic, the R/S statistic grows according to a power law as a function of n with exponent H. Thus H can be estimated from a log-log plot of R/S versus n [SAHI].

→ *Variance Time Plot*: One plots $\log(VAR[X_k^{(m)}])$ versus $\log(m)$. On this log-log plot a line with slope $(-\beta)$ greater than -1 indicates self-similarity. One can estimate $\beta=2(1-H)$.

→ *Spectral Power Density*: One can plot $\log|S(f)|$ versus $\log f$ to estimate $1-\beta$.

Actually these different techniques may provide significantly different values of the Hurst parameter H. Between this and the need to process large amounts of data, some have questioned the practicality of the Hurst parameter when used as a traffic descriptor [PARU] [MICH]. A very good paper discussing the advantages and disadvantages of various H estimation techniques is [TAQQ 95]. Also, a useful tool for scaling analysis, including H estimation, is the discrete wavelet transform [ABRY].

7.6.6 Network Performance

The effect of self-similar traffic on network performance is the subject of ongoing research. To some extent it appears that using self-similarity models is sometimes appropriate and sometimes is not [STAL].

In theory, the presence of self-similar traffic should necessitate larger buffers to meet cell loss requirements than if traffic were more Poisson like (see section 7.5.6). However it should be noted that asymptotic self-similarity is a statement about correlations over long time scales. Because of this, it may not be relevant in certain situations. One such situation is the use of small buffers. As an example, buffering is often kept small in the transmission of VBR video in order to minimize delays. For this case only short range correlations are important for predicting buffer overflow as demonstrated by [HEYM 96]. A second example is ATM networks where fairly small per-flow buffering is often used. It has been shown in [LIN] that for the ATM AAL5 frame reassembly operation both the usual Markovian traffic model and long range dependent traffic model yield similar queue behavior.

On the other hand Erramilli, Narayan and Willinger reported in 1996 that "...long range dependence has considerable impact on queueing performance, and is a dominant characteristic for a number of packet traffic engineering problems" [ERRA]. Through additional research the areas where self-similar traffic plays an important role should become more evident with time.

To Look Further

Aside from the references in this chapter's discussion, the theory and modeling of self-similar and long range dependent traffic is also explored in [ANDE 97], [ANDE 98], [ASMU], [AVRE], [CHAN], [CHOE], [DAIG], [DANI], [DIAM], [GORD 96], [GRUN], [IZQU], [JELE] [KRIE], [MAYO], [PAXS B], [TAQQ 97], [TAQQ 97B], and [TSYB]. Simulation and performance evaluation of self-similar traffic is discussed in [FELD 98], [KRIS], [LAMB], and [TAQQ 95]. Measurement and performance evaluation for an ATM local network appears in [LI]. A study of self-similar traffic for LEO satellite systems is in [PAPA]. An excellent introduction to the topic of self-similar traffic appears in [STAL]. Excellent tutorials on the topics of this chapter appear in [KUEH], [MICH], and [SAHI].

Appendix: Probability Theory Review

A.1 Probability

Probability theory is ultimately concerned with the results of statistical "experiments". An experiment will be assumed to have a number of outcomes that can not be further decomposed [FELL]. For instance, if we pick a card from a randomly shuffled deck of cards, there are 52 possible outcomes. Each such outcome is a **sample point**. The collection of such sample points is the **sample space**, S. It is assumed that the sample points are mutually exclusive.

Sample points can be aggregated into subsets of the sample space called **events**. For instance, in picking a card in the previous example, two potential events are the selection of a red card and the selection of a black card.

A **probability measure** defined on S is an association of real numbers to the events, which satisfies the following axioms [THOM], [KLEI 75]:

$$P(S) = 1.0, \tag{A.1}$$

$$0 \le P(A) \le 1 \quad \text{for } any \; event \; A, \tag{A.2}$$

and if A and B are mutually exclusive:

$$P(A + B) = P(A) + P(B). \tag{A.3}$$

These equations are the axiomatic way to define probability [PAPO]. One can also define it, intuitively, in terms of relative frequency. If an event A occurs n_A times out of n tries, then

$$P(A) = \lim_{n \to \infty} \frac{n_A}{n}. \tag{A.4}$$

Let $P(A|B)$ be the **conditional probability** of event A occurring in an experiment given that B has occurred. Let $P(AB)$ be the probability of the intersection of events A and B. Then:

$$P(AB) = P(A)P(B \mid A) = P(B)P(A \mid B). \tag{A.5}$$

357

Two events are **statistically independent** if

$$P(B \mid A) = P(B) \text{ and } P(A \mid B) = P(A). \tag{A.6}$$

This, together with (A.5), implies that for two statistically independent events

$$P(AB) = P(A)P(B). \tag{A.7}$$

For instance, let A be the event of drawing a diamond and B be the event of drawing a red card. Clearly, $P(A)$=0.25, $P(B)$=0.5, $P(A|B)$=0.5, $P(B|A)$=1.0, $P(AB)$=0.25. However, $P(AB){\neq}P(A)P(B)$ as these events are *not* independent.

In general, for independent events A,B,C...

$$P(ABC...) = P(A)P(B)P(C).... \tag{A.8}$$

$P(ABC...)$ is known as the **joint probability** and $P(A)$ is an example of a **marginal probability**.

A.2 Densities and Distribution Functions

Suppose that we are interested in a continuous random variable, X, that assumes real values. Then the **cumulative distribution function** [THOM] is $F(a) = P(X{\leq}a)$ for argument a. Among the properties of the cumulative distribution function are

$$F(a) \geq 0, \tag{A.9}$$

$$F(-\infty) = 0, \quad F(\infty) = 1. \tag{A.10}$$

If $a_1 < a_2$ then

$$F(a_2) - F(a_1) = P(a_1 < x < a_2). \tag{A.11}$$

Finally, the cumulative distribution function is nondecreasing as a function of a:

$$F(a+\Delta) \geq F(a) \quad \text{if } \Delta > 0. \tag{A.12}$$

A **probability density function** is related to the cumulative distribution function by

$$f(x) = \frac{dF(X)}{dx} \qquad (A.13)$$

or

$$F(x) = \int_{-\infty}^{x} f(y)dy. \qquad (A.14)$$

For a continuous probability density function

$$f(x) \geq 0, \qquad (A.15)$$

$$\int_{-\infty}^{\infty} f(x)dx = 1, \qquad (A.16)$$

$$P[a_1 < x \leq a_2] = \int_{a_1}^{a_2} f(x)dx. \qquad (A.17)$$

Now consider a probability density function on a discrete state space. Then

$$f(x_i) \geq 0, \qquad (A.18)$$

$$\sum_{x_i \varepsilon S} f(x_i) = 1, \qquad (A.19)$$

$$P[a_1 \leq x \leq a_2] = \sum_{i=a_1}^{a_2} f(x_i). \qquad (A.20)$$

Example: The negative exponential density is used throughout this text:

$$f(x) = \mu e^{-\mu x}, \qquad x \geq 0.$$

The cumulative distribution function is

$$F(x) = \int_0^x \mu e^{-\mu y} dy = 1 - e^{-\mu x}, \quad x \geq 0.$$

A.3 Joint Densities and Distributions

The **joint distribution function** of random variables X_1 and X_2 is

$$F(x_1, x_2) = P[X_1 \leq x_1, X_2 < x_2] \tag{A.21}$$

for arguments x_1 and x_2 and the **joint density function** is:

$$f(x_1, x_2) = \frac{d^2 F(x_1, x_2)}{dx_1 dx_2}. \tag{A.22}$$

These two expressions can be extended, in the natural way, to situations involving more than two random variables.

The **marginal density function** is related to the joint density function through

$$f(x_1) = \int_{-\infty}^{\infty} f(x_1, x_2) dx_2. \tag{A.23a}$$

That is, if the joint density function is integrated out over x_2, what is left is $f(x_1)$. For a joint distribution of n variables, one integrates out over all but the ith variable to obtain $f(x_i)$. For random variables (say two) over a discrete state space a summation can be used to compute a marginal density function from the joint density function:

$$f(x_i) = \sum_j f(x_i, x_j). \tag{A.23b}$$

The **conditional probability density** is related to the joint and marginal probability densities through

$$f(x_1 | x_2) = \frac{f(x_1, x_2)}{f(x_2)}. \tag{A.24}$$

A collection of n random variables $X_1, X_2, \cdots X_n$ is said to be **independent** if the joint probability density function is equal to the product of the marginal probability density functions:

$$p(x_1, x_2, \cdots x_n) = p(x_1)p(x_2) \cdots p(x_n).$$ (A.25)

A.4 Expectations

The **expectation** of a random variable, X, is:

$$E(X) = \int_{-\infty}^{\infty} xf(x)dx.$$ (A.26)

Intuitively, the expectation is an integration (averaging) of x weighted by the probability of x occurring, $f(x)$. It is simply the mean or average value of a random variable, X. For the discrete case:

$$E(X) = \sum_i x_i f(x_i).$$ (A.27)

In general

$$E(X_1 + X_2 + \cdots + X_n) =$$ (A.28)
$$E(X_1) + E(X_2) + \cdots + E(X_n).$$

More specifically, if there are n independent random variables, $X_1, X_2, \cdots X_n$, then [DEGR]:

$$E(\prod_{i=1}^{n} X_i) = \prod_{i=1}^{n} E(X_i).$$ (A.29)

This is a consequence of $f(x_1, x_2, \cdots x_n) = f(x_1)f(x_2) \cdots f(x_n)$.

The **nth moment** of a probability density is

$$E(X^n) = \int_{-\infty}^{\infty} x^n f(x)dx$$ (A.30)

for the continuous case and

$$E(X^n) = \sum_i x_i^n f(x_i) \qquad (A.31)$$

for the discrete case. Moments can be used to characterize a probability density function [THOM]. Usually, though, only the lower moments play a significant role. For instance, the **variance** of a random variable, σ^2, is a measure of the "spread" of a density about its mean:

$$\sigma^2 = E[(X-E(X))^2], \qquad (A.32)$$

$$\sigma^2 = \int_{-\infty}^{\infty} (x-E(X))^2 f(x)dx, \qquad (A.33)$$

or for the discrete case:

$$\sigma^2 = \sum_i (x_i-E(x_i))^2 f(x_i). \qquad (A.34)$$

The variance is related to the 2nd moment through

$$\sigma^2 = E(X^2)-(E(X))^2. \qquad (A.35)$$

A.5 Convolution

Suppose $Y=X_1+X_2$ where the random variables X_1 and X_2 are independent with marginal densities $f_1(x_1)$ and $f_2(x_2)$. Then [PAPO], [KLEI 75] the density of Y, $f_Y(y)$, is given by

$$f_Y(y_1) = \int_{-\infty}^{\infty} f_1(y_1-y_2)f_2(y_2)dy_2. \qquad (A.36)$$

This function is called the **convolution** of f_1 and f_2. Convolution operations are used in circuit theory, signal processing, and queueing theory.

A.6 Combinatorics

The number of combinations of n elements taken m at a time is:

$$\begin{bmatrix} n \\ m \end{bmatrix} = \frac{n!}{(n-m)!m!}.$$ (A.37)

A.7 Some Useful Summations

The following summations and closed form expressions, taken from this text, are often useful:

$$\sum_{n=0}^{\infty} x^n = \frac{1}{1-x}, \quad 0 \le x < 1,$$ (A.38)

$$\sum_{n=0}^{N} x^n = \frac{1-x^{N+1}}{1-x},$$ (A.39)

$$\sum_{n=0}^{\infty} nx^n = \frac{x}{(1-x)^2}, \quad 0 \le x < 1,$$ (A.40)

$$\sum_{n=0}^{\infty} n^2 x^n = \frac{x(1+x)}{(1-x)^3}, \quad 0 \le x < 1,$$ (A.41)

$$\sum_{n=0}^{\infty} \frac{1}{n!} x^n = e^x.$$ (A.42)

A.8 Useful Moment-generating Function Identities

The following identities are often useful in solving problems involving moment-generating functions:

$$\sum_{n=1}^{\infty} p_{n-1} z^n = zP(z),$$ (A.43)

$$\sum_{n=0}^{\infty} p_{n+1} z^n = z^{-1}[P(z) - p_0],$$ (A.44)

$$p(1) = \sum_{n=0}^{\infty} p_n = 1.0, \tag{A.45}$$

$$\frac{dP(z)}{dz}\Big|_{z=1} = \sum_{n=0}^{\infty} np_n = E(n), \tag{A.46}$$

$$\frac{d^2P(z)}{dz^2}\Big|_{z=1} = E(n^2) - E(n). \tag{A.47}$$

References

[ABAT]

Abate, J. and Whitt, W., "Calculating Time-Dependent Performance Measures for the M/M/1 Queue", *IEEE Transactions on Communications*, Vol. 37, No. 10, Oct. 1989, pp. 1102–1104.

[ABRY]

Abry, P. and Veitch, D., "Wavelet Analysis of Long Range Dependent Traffic", *IEEE Transactions on Information Theory*, Vol. 44, 1998, pp. 2–15.

[ACAM]

Acampora, A.S. and Karol, M.J., "An Overview of Lightwave Packet Networks", *IEEE Network*, Jan. 1989, pp. 29–41.

[ACKR]

Ackroyd, M.H., "M/M/1 Transient State Occupancy Probabilities Via the Discrete Fourier Transform", *IEEE Transactions on Communications*, Vol. COM-30, No. 3, March 1982, pp. 557–559.

A discussion of this technique relative to that of [JONE] appears in [CHIE].

[ADAS]

Adas, A., "Traffic Models in Broadband Networks", *IEEE Communications Magazine*, Vol. 35, July 1997, pp. 82–89.

[ALBI]

Albin, S.A., "Approximating a Point Process by a Renewal Process II: Superposition Arrival Process to Queues", *Operations Research*, Vol. 32, No. 2, Sept./Oct. 1984, pp. 1133–1162.

[AMMA]

Ammar, H.H. and Liu, R.W., "Analysis of the Generalized Stochastic Petri Nets by State Aggregation", *International Workshop on Timed Petri Nets*, Torino, Italy, IEEE Computer Society Press, July, 1985, pp. 88–95.

[ANDE 97]

Andersen, A.T. and Nielsen, B.F., "An Application of Superpositions of Two State Markovian Sources to the Modeling of Self-Similar Behaviour", *Proceedings of IEEE INFOCOM'97*, Kobe, Japan, 1997, pp. 196–204.

[ANDE 98]

Andersen, A.T. and Nielsen, B.F., "A Markovian Approach for Modeling Packet Traffic with Long Range Dependence", *IEEE Journal on Selected Areas in Communications*, Vol. 16, No. 5, June 1998, pp. 719–732.

[ANIC]

Anick, D., Mitra, D., Sondhi, M.M., "Stochastic Theory of a Data-Handling System with Multiple Sources", *Bell System Technical Journal*, Vol. 61, No. 8, Oct. 1982, pp. 1871–1894.

[ANIO]

Aniodo, G.J. and Seeto, A.W., "Multipath Interconnection: A Technique for Reducing Congestion within Fast Packet Switching Fabrics", *IEEE Journal on Selected Areas in Communications*, Vol. 6, No. 9, Dec. 1988, pp. 1480–1488.

[ASMU]

Asmussen, S. and Koole, G., "Marked Point Processes as Limit of Markovian Arrival Streams", *Journal of Applied Probability*, Vol. 30, 1993, pp. 365–372.

[ATAL]

Atalay, V., Gelenbe, E. and Yalabik, N., "Image texture generation with the random neural network model", *International Conference on Artificial Neural Networks'91*, Helsinki, Finland, 1991, North-Holland, T. Kohonen, editor.

[AVRE]

Avresky, D.R., Shurbanov, V., Horst, R. and Mehra, P., "Performance Evaluation of the Server Net^R SAN under Self-Similar Traffic", *Proceedings of the 13th International Parallel Processing Symposium and 10th Symposium on Parallel and Distributed Processing*, San Juan, Puerto Rico, IEEE Computer Society Press, Los Alamitos CA, 1999, pp. 143–147.

[BADR]

Badr, H.G., Gelernter, D. and Podar, S., "An Adaptive Communications Protocol for Network Computers", *Performance Evaluation*, Vol. 6, 1986, pp. 35–51.

[BAE]

Bae, J.J. and Suda, T., "Survey of Traffic Control Schemes and Protocols in ATM Networks", *Proceedings of the IEEE*, Vol. 79, No. 2, Feb. 1991, pp. 170–189.

An excellent survey of ATM related issues.

[BAIL]

Bailey, N.T.J., "A Continuous Time Treatment of a Single Queue Using Generating Functions", *Journal of the Royal Statistical Society*, Series B, Vol. 16, pp. 288–291.

[BARD]

Bard, Y., "Some Extensions to Multiclass Queueing Network Analysis", in *Performance of Computer Systems*, ed. by M. Arato, A. Butrimenko and E. Gelenbe, North-Holland, Amsterdam, 1979.

This paper introduced the idea of approximate mean value analysis.

[BASK 75]

Baskett, F., Chandy, K.M., Muntz, R.R. and Palacios, F., "Open, Closed and Mixed Networks of Queues with Different Classes of Customers", *Journal of the ACM*, Vol. 22, No. 2, April 1975, pp. 248–260.

This paper generalizes the earlier work of Jackson, and Gordon and Newell, in describing classes of queueing networks with product form solution. It is often referred to as the "BCMP paper".

[BASK 76]

Baskett, F. and Smith, A.J., "Interference in Multiprocessor Computer Systems with Interleaved Memory", *Communications of the ACM*, Vol. 19, No. 6, June 1976, pp. 327–334.

[BERG]

Berger, A.W., Naldi, M., DeGiovanni, L. and Villen-Altamirano, "Standardization of Traffic Measurements and Models for Broadband Networks: Open Issues", *Computer Networks and ISDN Systems*, Vol. 30, 1998, pp. 1327–1340.

[BHAN]

Bhandarkar, D.P., "Analysis of Memory Interference in Multiprocessors", *IEEE Transactions on Computers*, Vol. C-24, No. 9, Sept. 1975, pp. 897–908.

[BHAT]

Bhatia, S., Mouftah, H.T. and Ilyas, M., "Hybrid Modeling Techniques for Complex Computer Networks", *IEEE International Conference on Communications 1987*, Seattle WA., June 1987, pp. 1311–1355.

[BIRM]

Birman, A. and Kogan, Y., "Asymptotic Evaluation of Closed Queueing Networks with Many Stations", *IBM Research Report RC 17559 (77264)*, Dec. 19, 1991.

[BIRM 89]

Birman, A., Chang, P.-C., Chen, J.S.-C. and Guerin, R., "Buffer Sizing in an ISDN Frame Relay Switch", *Technical Report RC 14836, IBM Research Report*, Aug. 1989.

[BLON]

Blondia, C., "A Discrete Time Batch Markovian Arrival Process as B-ISDN Traffic Model", *Belgian Journal of Operations Research, Statistics and Computer Science*, Vol. 32, No. 3, 1992, pp. 3–23.

[BLON B]

Blondia, C., and Casals, O., "Statistical Multiplexing of VBR Sources: A Matrix Analytic Approach", *Performance Evaluation*, Vol. 16, 1992, pp. 5–20.

[BLUM]

Blum, A., et al., "Experiments with Decomposition of Extended Queueing Network Models", *Modeling Techniques and Tools for Performance Analysis*, D. Potier, Ed., North-Holland, Amsterdam, The Netherlands, 1985 pp. 623–640.

[BODN]

Bodnar, B.L. and Liu, A.C., "Modeling and Performance Analysis of Single-Bus Tightly-Coupled Multiprocessors", *IEEE Transactions on Computers*, Vol. 38, No. 3, March 1989, pp. 464–470.

[BOUC]

Boucherie, R.J. and van Dijk, N.M., "Product Forms for Queueing Networks with State Dependent Multiple Job Transactions", *Advances in Applied Probability*, Vol. 23, 1991, pp. 152–187.

[BRAU]

Brauer, W., (editor), *Net Theory and Applications*: Proceedings of the Advanced Course on General Net Theory of Processes and Systems, Hamburg, 1979, Springer-Verlag, 1980.

[BREM]

Bremaud, P., *Point Processes and Queues: Martingale Dynamics*, Springer-Verlag, New York, 1981.

[BRUE 78]

Bruell, S.C., *On Single and Multiple Job Class Queueing Network Models of Computing Systems*, Ph.D. Thesis, Computer Sciences Dept., Purdue University, Dec. 1978.

Later condensed into [BRUE 80].

[BRUE 80]

Bruell, S.C. and Balbo, G., *Computational Algorithms for Closed Queueing Networks*, North-Holland, N.Y., 1980.

A very accessible treatment of the convolution algorithm for single and multi-class networks.

[BRUE 86]

Bruell, S.C. and Ghanta, S., "Throughput Bounds for Generalized Stochastic Petri Net Models", *International Workshop on Timed Petri Nets*, Torino, Italy, IEEE Computer Society Press, July 1985, pp. 250–261.

[BRUM]

Brumelle, S.L., "On the Relation Between Customers and Time Averages in Queues", *Journal of Applied Probability*, Vol. 8, No. 3, pp. 508–520.

[BURK]

Burke, P.J., "The Output of a Queueing System", *Operations Research*, Vol. 4, 1956, 699–704.

[BURM]

Burman, D.Y., Lehoczky, J.P., and Lim, Y., "Insensitivity of Blocking Probabilities in a Circuit-Switching Network", *Journal of Applied Probability*, Vol. 21, 1984, pp. 850–859.

[BUZE 76]

Buzen, J.P., "Operational Analysis: The Key to the New Generation of Performance Prediction Tools", *Proceedings of IEEE COMPCON*, Sept. 1976.

[BUZE 73]

Buzen, J.P., "Computational Algorithms for Closed Queueing Networks with Exponential Servers", *Communications of the ACM*, Vol. 16, No. 9, Sept. 1973, pp. 527–531.

[CANT]

Cantrell, P.E., "Computation of the Transient M/M/1 Queue cdf, pdf and Mean with Generalized Q-Functions", *IEEE Transactions on Communications*, Vol. COM-34, No. 8, August 1986, pp. 814–817.

[CASS]

Cassandras, C.G. and Strickland, S.G., "Perturbation Analysis Techniques for Communication Networks", *IEEE Global Telecommunications Conference*, New Orleans, LA, Dec. 1985, pp. 13.5.1–13.5.5.

[CHAN]

Chang, H.-C., *Congestion Analysis in ATM Networks with Correlated Traffic*, Ph.D Thesis, Dept. of Applied Mathematics and Statistics, University at Stony Brook, Oct. 1998. Advisor: H. Badr.

This thesis shows that for a deterministic, single server system with finite buffer capacity and a semi-Markov process input (SMP/D/1/s), the output is also SMP. Numerical results show that the input correlation structure is "remarkably persistent" in the output traffic. There is also a simulation study demonstrating that short term correlation effects tend to play the dominant role in determining congestion levels, even in the presence of self-similar traffic.

[CHAN 75]

Chandy, K.M., Herzog, U. and Woo, L., "Parametric Analysis of Queueing Networks", *IBM Journal of Research and Development*, Vol. 19, 1975, pp. 43–49.

[CHAN 79]

Chandy, K.M., Holmes, V., and Misra, J., "Distributed Simulation of Networks", *Computer Networks*, Vol. 3, 1979, pp. 105–113.

[CHAN 80]

Chandy, K.M. and Sauer, C.H., "Computational Algorithms for Product Form Queueing Networks", IBM Res. Rep. RC7950, IBM Thomas J. Watson Research Center, Yorktown Heights, NY (1980) (also in the Proceedings Supplement, Performance 80, ACM Press, Toronto, Ont.)

Describes the LBANC algorithm.

[CHAN 81]

Chandy, K.M. and Misra, J., "Asynchronous Distributed Simulation via a Sequence of Parallel Computations", *Communications of the ACM*, Vol. 24, April 1981, pp. 198–206.

[CHAN 82]

Chandy, K.M. and Neuse, D., "Linearizer: A Heuristic Algorithm for Queueing Network Models of Computing Systems", *Communications of the ACM*, Vol. 25, No. 2, 1982, pp. 126–133.

[CHAO]

Chao, X. and Pinedo, M., "On G-Networks: Queues with Positive and Negative Arrivals", *accepted for Probability in the Engineering and Information Sciences*.

[CHEN]

Chen, W., *Solution Manual for Telecommunications Networks: Protocol, Modeling and Analysis*, Addison-Wesley, 1987.

[CHEN b]

Chen, T.M., Walrand, J. and Messerschmitt, D.G., "Dynamic Priority Protocols for Packet Voice", *IEEE Journal on Selected Areas of Communications*, Vol. 7, No. 5, June 1989, pp. 632–643.

[CHEN c]

Chen, J.S.-C., Guerin, R., and Stern, T.E., "Markov-Modulated Flow Model for the Output Queues of a Packet Switch", *IEEE Transactions on Communications*, Vol. 40, No. 6, June 1992, pp. 1098–1110.

[CHIE]

Chie, C.M., "Comments on "M/M/1 Transient State Occupancy Probabilities Via the Discrete Fourier Transform", *IEEE Transactions on Communications*, Vol. COM-31, No. 2, Feb. 1983, pp. 289–290.

[CHIO]

Chiola, G., "A Software Package for the Analysis of Generalized Stochastic Petri Net Models", *International Workshop on Timed Petri Nets*, July 1985, Torino, Italy, IEEE Computer Society Press, 1986, pp. 136–143.

[CHIP]

Chipalkatti, R., Kurose, J.F. and Towsley, D., "Scheduling Policies for Real Time and Non-Real Time Traffic in a Statistical Multiplexer", *Proceedings of IEEE INFOCOM'89*, Ottawa, Canada, April 1989, pp. 774–783.

[CHOE]

Choe, J. and Shroff, N.B., "Queueing Analysis of High Speed Multiplexers including Long-Range Dependent Arrival Processes", *Proceedings of IEEE INFOCOM'99*, New York, NY, 1999, pp. 617–624.

[CIAR]

Ciardo, G., Muppala, J. and Trivedi, K., "SPNP: Stochastic Petri Net Package", *Proceedings of the Petri Nets and Performance Models*, Kyoto, Japan, Dec. 1989, IEEE Computer Society Press, pp. 142–151.

[CIAR 91]

Ciardo, G., Muppala, J. and Trivedi, K., "On the Solution of GSPN Reward Models", *Performance Evaluation*, Vol. 12, No. 4, July 1991, pp. 237–253.

[CIDO]

Cidon, I., Gopal, I., Gopal, P.M., Guerin, R., Janniello, J., and Kaplan, M., "The plaNET/ORBIT High Speed Network", *Technical Report RC 18270, ISM Research Report*, Aug. 1992.

[CIDO 88]

Cidon, I., Gopal, I., Grover, G. and Sidi, M., "Real Time Packet Switching: A Performance Analysis", *IEEE Journal on Selected Areas in Communications*, Vol. SAC-6, No. 9, Dec. 1988, pp. 1576–1586.

[CIDO 93]

Cidon, I., Guerin, R., Khamisy, A. and Sidi, M., "Analysis of a Correlated Queue in a Communication System", *IEEE Transactions on Information Theory*, Vol. 39, No. 2, March 1993.

[CLAR]

Clark, D., Davie, B., Farber, D., Gopal, I., Kadaba, B., Sincoskie, D., Smith, J. and Tennnenhouse, D., "An Overview of the Aurora Gigabit Testbed", *Proceedings of IEEE INFOCOM'92*, Florence Italy, 1992, pp. 569–581.

[COBH]

Cobham, A., "Priority Assignment in Waiting Line Problems", *Operations Research*, Vol. 2, 1959, pp. 70–76.

[COHE]

Cohen, J.W., *The Single Server Queue*, North-Holland Publishing Co., 1969.

[COMF]

Comfort, J.C., "Distributed Simulation on a Network of Transputers", *Proceedings of the European Simulation Multiconf. Simulation Comput. Integrated Manufacturing*, Vienna, Austria, July 1987, pp. 25–29.

[CONW 86]

Conway, A. and Georganas, N. "RECAL—A New Efficient Algorithm for the Exact Analysis of Multiple-Chain Queueing Networks", *Journal of the ACM*, Vol. 33, No. 4, Oct. 1986, pp. 768–791.

A polynomial algorithm for models with multiple routing chains.

[CONW 89]

Conway, A.E., De Souza E., Silva, E., and Lavenberg, S.S., "Mean Value Analysis by Chain of Product Form Queueing Networks", *IEEE Transactions on Computers*, Vol. 38, No. 3, March 1989, pp. 432–441.

[CONW 89b]

Conway, A.E. and Georganas, N.D., *Queueing Networks—Exact Computational Algorithm: A Unified Theory Based on Decomposition and Aggregation*, The MIT Press, Cambridge, Mass., 1989.

Includes a discussion of the RECAL algorithm.

[COOP]

Cooper, R.B., *Introduction to Queueing Theory*, North-Holland Publishing Co., New York, 1981.

A solid introduction to classical queueing theory.

[COUR 77]

Courtois, P.-J., *Decomposibility: Queueing and Computer System Applications*, Academic Press, New York, 1977.

[COUR 85]

Courtois, P.-J., "On Time and Space Decomposition of Complex Structures", *Communications of the ACM*, Vol. 28, No. 6, June 1985, pp. 590–603.

Models of large and complex systems can often be reduced to smaller submodels through decomposition. This paper discusses certain criteria for successful decompositions.

[COX]

Cox, D.R. and Smith, W.L., *Queues*, Chapman and Hall, London, 1961.

[COX 66]

Cox, D.R. and Lewis, P.A.W., *The Statistical Analysis of Series of Events*, Methuen, London, 1966.

[COX 78]

Cox, D.R. and Miller, H., *The Theory of Stochastic Processes*, Chapman and Hall, London, 1978.

[COX 80]

Cox, D.R. and Isham, V., *Point Processes*, Chapman and Hall, London, 1980.

[COX 84]

Cox, D.R., "Long Range Dependence: A Review", in *Statistics: An Appraisal*, David H.A., and Davis, H.T., eds., The Iowa State University Press, Ames, Iowa, 1984.

[CRAN]
Crane, M. and Lemoine, A., *An Introduction to the Regenerative Method for Simulation Analysis*, Springer-Verlag, New York, 1977.

[CROS]
Crosby, S. and Krzesinski, A.E., "Product Form Solutions for Multiserver Centres with Concurrent Classes of Customers", *Performance Evaluations*, Vol. 11, No. 4, Nov. 1990, pp. 265–281.

[CROV]
Crovella, M.E. and Bestavros, A., "Self-Similarity in World Wide Web Traffic: Evidence and Possible Causes", *IEEE/ACM Transactions on Network*, Vol. 5, No. 6, Dec. 1977, pp. 835–846.

An empirical traffic study showing that the properties of the World Wide Web traffic are consistent with self-similarity.

[DAIG]
Daigle, J.N. and Roughan, M., "Queue-Length Distributions for Multi-Priority Queueing Systems", *Proceedings of IEEE INFOCOM'99*, New York, NY, 1999, pp. 641–647.

[DANI]
Daniels, T. and Blondia, C., "Asymptotic Behavior of a Discrete-Time Queue with Long Range Dependent Input", *Proceedings of IEEE INFOCOM'99*, New York, NY, 1999, pp. 633–640.

[DAVI]
Davidson, D. and Reynolds, P., "Performance Analysis of a Distributed Simulation Algorithm Based on Logical Processes", *Proceedings of the Winter Simulation Symposium*, Arlington, VA, Dec. 1983, pp. 267–268.

[DECE]
Decegama, A., "Parallel Processing Simulation of Large Computer Networks", *Proceedings of the Symposium on Simulation of Computer Networks*, Colorado Springs, CO, Aug. 1987, pp. 51–63.

[DEGR]
DeGroot, M.H., *Probability and Statistics*, Addison-Wesley, Reading, Mass., 1975.

[DENN]
Dennis, J., (editor), *Record of the Project MAC Conference on Concurrent Systems and Parallel Computation*, ACM, New York, June 1970.

[DEPR]
M. de Prycker, *Asynchronous Transfer Mode: Solution for Broadband ISDN*, Ellis Horwood, Simon and Schuster, 1991.

A solid and practical introduction to ATM. Largely non-mathematical.

[DESO]
De Souza E Silva, E. and Lavenberg, S.S., "Calculating Joint Queue-

Length Distributions in Product-Form Queueing Networks", *Journal of the Association of Computing Machinery*, Vol. 36, No. 1, Jan. 1989, pp. 194–207.

[DIAM]

Diamond, J.E. and Alfa, A.S., "Matrix Analytical Model of an ATM Output Buffer with Self-Similar Traffic", *Performance Evaluation*, Vol. 31, 1998, pp. 201–210.

[DIAZ]

Diaz, M., "Modeling and Analysis of Communication and Cooperation Protocols Using Petri Net Based Models", *Protocol Verification Specification, Testing and Verification*, C. Sunshine, editor, North-Holland, New York 1982.

[DIDO]

Di Donato, A.R. and Jarnagin, M.P., "A Method for Computing and Circular Coverage Function", *Math. Comput.*, Vol. 16, July 1962, pp. 347–355.

[DIJK]

Dijkstra, E.W., "Cooperating Sequential Processes", *Programming Languages*, F. Genvys, editor, Academic Press, New York, 1968, pp. 43–112.

[DISN]

Disney, R.L., *Traffic Processes in Queueing Networks: A Markov Renewal Approach*, The Johns Hopkins University Press, Baltimore, Md., 1987.

[DOYL]

Doyle, P.G. and Snell, J.L., *Random Walks and Electric Networks*, The Carus Mathematical Monographs No. 22, Published by the Mathematical Association of America, 1984.

A tutorial introduction to the fascinating relation between electric circuits and random walks.

[DUGA]

Dugan, J.B., Ciardo, G., Bobbio, A., and Trivedi, K., "The Design of a Unified Package for the Solution of Stochastic Petri Net Models", *International Workshop on Timed Petri Nets*, Torino, Italy, IEEE Computer Society Press, July 1985, pp. 6–13.

[EAGE]

Eager, D.L., and Lipscomb, J.N., "The AMVA Priority Approximation", *Performance Evaluation*, North-Holland, Vol. 8, No. 3, 1988, pp 173–193.

The MVA Priority Approximation is proposed as a relatively inexpensive and accurate approximation technique for queueing networks with priority scheduled service centers.

376

[EILO]

Eilon, S., "A Simpler Proof of $L = \lambda W$", *Operations Researc* 1969, pp. 915–916.

[EMER]

Emer, J.S., and Ramakrishnan, K., "Performance Consideration: ..ui Distributed Services—A Case Study: Mass Storage", *Proceedings of the 8th International Conference on Distributed Computing Systems*, June 1988, pp. 289–297.

[ERDE]

Erdelyi, A., *Asymptotic Expansions*, Dover Publications, New York, 1956. See also *Asymptotic Expansions* by E.T. Copson, Cambridge University Press, 1965, Chapters 5,6 and see also *Introduction to Asymptotics and Special Functions* by F.W.J. Oliver, Academic Press, New York, 1974, Chapter 3.

[ERRA]

Erramilli, A., Narayan, O. and Willinger, W., "Experimental Queueing Analysis with Long-Range Dependent Packet Traffic", *IEEE/ACM Transactions on Networking*, Vol. 4, No. 2, April 1996, pp. 209–223.

[FELD]

Feldmeier, D.C., "Traffic Measurements on a Token Ring Network", *Proceedings of the Computer Networking Symposium*, Washington D.C., 1986, IEEE Computer Society Press, pp. 236–243.

Describes data traffic measurements made on a ten megabit token ring network at M.I.T. A comparison is made with the Ethernet findings of Shoch and Hupp.

[FELD 98]

Feldman, A. and Whitt, W., "Fitting Mixtures of Exponentials to Long-Tail Distributions to Analyze Network Performance Models", *Performance Evaluation*, Vol. 31, 1998, pp. 245–279.

[FELL]

Feller, W., *An Introduction to Probability Theory and Its Applications*, Wiley, New York, 1950.
This is a classic introductory graduate level work on probability theory.

[FEND]

Fendick, K.W., Saksena, V.R. and Whitt, W., "Dependence in Packet Queues", *IEEE Transactions on Communications*, Vol. COM-37, No. 11, Nov. 1989, pp. 1173–1183.

[FILI]

Filipiak, J., "Access Protection for Fairness in a Distributed Queue Dual Bus Metropolitan Area Network", *Proceedings of the IEEE International Conference on Communications*, 1989, pp. 635–639.

[FILM]

Filman, R.E. and Friedman, D.P., *Coordinated Computing, Tools and Techniques for Distributed Software*, McGraw-Hill, New York, 1984.

[FIRB]

Firby, P.A. and Gardiner, C.F., *Surface Topology*, published by Ellis Horwood Limited, Chichester and distributed by Halsted Press: a division of John Wiley & Sons, New York, 1982.

A very readable introduction to surface topology.

[FISH 78a]

Fishman, G.S., *Principles of Discrete-Event Simulation*, Wiley, New York 1978.

[FISH 78b]

Fishman, G.S., "Grouping Observations in Digital Simulation", *Management Science*, Vol. 24, 1978, pp. 510–521.

[FLOR]

Florin, G. and Natkin, S., "Evaluation des performances d'un protocole de communication a l'aide des reseaux de Petri et des processus stochastiques", *J. AFCET Multiprocesseurs et Multiordinateurs en Temps Re'el*, 1978, pp. E0–E26.

This is the first stochastic Petri network paper. A list of early papers by these authors appears in the IEEE Transactions on Computers, Vol. C-32, No. 9, Sept. 1983, pg. 880. Another early, independent, appearance of stochastic Petri nets is in [MOLL 82].

[FLOR 91]

Florin, G. and Natkin, S., "Generalization of Queueing Network Product Form Solutions to Stochastic Petri Nets", *IEEE Transactions on Software Engineering*, Vol. 17, 1991, pp. 99–107.

[FORS]

√ Frost, V. and Melamed, B., "Traffic Modeling for Telecommunications Networks", *IEEE Communications Magazine*, Vol. 32, March 1994, pp. 70–81.

[FRAN]

Franklin, M.A., "A VLSI Performance Comparison of Banyan and Crossbar Communication Networks", *IEEE Transactions on Computers*, Vol. C-30, NO. 4, April 1981, pp. 283–290.

[FREN 87]

Frenkel, K.A., "Profiles in Computing: Alan L. Scherr, Big Blue's Time-Sharing Pioneer", *Communications of the ACM*, Vol. 30, No. 10, Oct. 1987, pp. 824–828.

Scherr discusses his preference for analysis over simulation.

[FROS 86]

Frost, V.S. and Shanmugan, K.S., "Hybrid Approaches to Network Simulation", *IEEE International Conference on Communications 1986*, Toronto, Ontario, June 1986, pp. 228–234.

[FROS 88]

Frost, V.S., Larue, W., and Shanmugan, K.S., "Efficient Techniques for the Simulation of Computer Networks", *IEEE Journal of Selected Areas in Communications*, Vol. 6, No. 1, January 1988, pp. 146–157.

[FROS 92]

Frosch, D. and Natarajan, K., "Product Form Solutions for Closed Synchronized Systems of Stochastic Sequential Processes", *Proceedings of the 1992 International Computer Symposium*, Taichung, Taiwan ROC, Dec. 1992, pp. 392–402.

[FROS 93]

Frosch-Wilke, D., "Exact Performance Analysis of a Class of Product Form Stochastic Petri Nets", in *Computer and Telecommunications Performance Engineering*, M.E. Woodward et. al. eds., Pentech Press, Nov. 1993.

[GANZ]

Ganz, A., Grillo, D., Takagi, H. and Tantawi, A., eds., "Special Issue on Performance Modeling of High Speed Telecommunication Systems", *Performance Evaluation*, Vol. 16, No. 1–3, Nov. 1992.

A special issue on the performance evaluation of such network technologies as ATM, ISDN, DQDB, and WDM.

[GARR]

Garrett, M.W. and Li, S.-Q., "A Study of Slot Reuse in Dual Bus Multiple Access Networks", *Proceedings of IEEE INFOCOM'90*, San Francisco, CA, June 1990.

See also [RODR].

[GELE 90]

Gelenbe, E. and Shassberger, R., "Note on the Stability of G-networks", *to appear in Probability and Applications in Engineering and the Information Sciences*, Cambridge University Press, 1992.

[GELE 91a]

Gelenbe, E., "Product Form Networks with Negative and Positive Customers", *Journal of Applied Probability*, Vol. 28, 1991, pp. 656–663.

[GELE 91b]

Gelenbe, E., Glynn, P. and Sigman, K., "Queues with Negative Arrivals", *Journal of Applied Probability*, Vol. 28, 1991, pp. 245–250.

[GELE 91c]

Gelenbe, E., Stafylopatis, A. and Likas, A., "Associative Memory Operation of the Random Network Model", *International Conference on Artifi-*

cial Neural Networks, Helsinki, Finland, 1991, North-Holland, T. Kohonen, editor.

[GERL]

Gerla, M. and Kleinrock, L., "On the Topological Design of Distributed Computer Networks", *IEEE Transactions on Communications*, vol. COM-25, No. 1, Jan. 1977, pp. 48–60.

[GHAN]

Ghanta, S. and Bruell, S.C., "Throughput Bounds for Generalized Stochastic Petri Net Models", *International Workshop on Timed Petri Nets*, Torino, Italy, July 1985, IEEE Computer Society Press, 1986, pp. 250–262.

[GIDR]

Gidron, R., and Temple, A., "TeraNet: A Multihop Multichannel ATM Lightwave Network", *Proceedings of the IEEE International Conference on Communications '91*, Denver Colorado, June 1991, pp. 602–608.

[GLYN 86a]

Glynn, P. and Sanders, J., "Monte Carlo Optimization of Stochastic Systems: Two New Approaches", *Proceedings of the 1986 ASME Computer Engineering Conference*, 1986.

[GLYN 86b]

Glynn, P., "Optimization of Stochastic Systems", *Proceedings of the Winter Simulation Symposium*, Washington, D.C., Dec. 1986, pp. 52–59.

[GOOD]

Goodman, M.S., "Multiwavelength Networks and New Approaches to Packet Switching", *IEEE Communications Magazine*, Oct. 1989, pp. 27–35.

[GORD]

Gordon, W.J. and Newell, G.F., "Closed Queueing Systems with Exponential Servers", *Operations Research*, Vol. 15, 1967, pp. 254–265.

[GORD 96]

Gordon, J.J., "Long Range Correlations in Multiplexed Pareto Traffic", *Proceedings of the International IFIP-IEEE Conference on Broadband Communications*, Montreal, Quebec, Chapman and Hall, London and New York, 1996, pp. 28–39.

[GRAV]

"On the Geo/D/1 and Geo/D/1/n Queues", *Performance Evaluation*, Vol. 11, No. 2, July 1990, pp. 117–125.

[GREE]

Greenberg, A.G. and McKenna, J., "Solution of Closed, Product Form, Queueing Networks Via The RECAL and Tree-RECAL Methods on a Shared Memory Multiprocessor", *Proceedings of the 1989 ACM SIGMETRICS and Performance'89 International Conference on Measurement and Modeling of Computer Systems*, Berkeley CA, May 1989, pp. 127–135.

[GRES]

Gressier, E. "A Stochastic Petri Net Model for Ethernet", *International Workshop on Timed Petri Nets*, Torino, Italy, July 1985, IEEE Computer Society Press, 1986, pp. 296–306.

[GROS]

Gross, D. and Harris, C.M., *Fundamentals of Queueing Theory*, Wiley, New York, 1974, 1985.

A very thorough and excellent introduction to classical queueing theory.

[GRUN]

Grunenfelder, R. and Robert, S., "Which Arrival Law Parameters Are Decisive for Queueing System Performance?", *Proceedings of the 14th International Teletraffic Congress*, Elsevier, 1994, pp. 191–200.

[GUER]

Chen, J.S.-C. and Guerin, R., "Performance Study of an Input Queueing Packet Switch", *IEEE Transactions on Communications*, Vol. 39, No. 1, Jan. 1991, pp. 117–126.

[GUSE]

Gusella, R., "Characterizing the Variability of Arrival Processes with Indexes of Dispersion", *IEEE Journal on Selected Areas in Communications*, Vol. 9, No. 2, 1991, pp. 203–211.

[HAAS]

Haas, P.J. and Shedler, G.S., "Regenerative Simulation of Stochastic Petri Nets", *International Workshop on Timed Petri Nets*, Torino, Italy, IEEE Computer Society Press, July 1985, pp. 14–21.

[HAMI 86a]

Hamilton, Jr., R.L., *Modeling and Analysis of Multihop and Priority Random Access Computer Networks*, Ph.D Thesis, School of Electrical Engineering, Purdue University, 1986.

[HAMI 86b]

Hamilton, Jr., R.L. and Coyle, E.J., "A Two-Hop Packet Radio Network with Product Form Solution", *Proceedings of the 1986 Information Sciences and Systems Conference*, Princeton University, Princeton N.J., March 1986, pp. 871–876.

[HAMI 89]

Hamilton, Jr., R.L. and Yu, C., "A Buffered Two Node Packet Radio Network with Product Form Solution", *Proceedings of IEEE INFOCOM'89*, Ottawa, Canada, April 1989, pp. 520–528. Journal version in *IEEE Transactions on Communications*, Vol. 39, No. 1, Jan. 1991, pp. 62–75.

[HAMM]

✓   Hammond, J.L. and O'Reilly, P.J.P., *Performance Analysis of Local Computer Networks*, Addison-Wesley, Reading, Mass., 1986.

A self-contained and highly readable treatment of this topic.

[HAVI]

Haviv, M., *Personal Communication*, April 1992.

[HEID 81]

Heidelberger, P. and Welch, P.D., "A Spectral Method for Confidence Interval Generation and Run Length Control in Simulation", *Communications of the ACM*, Vol. 24, 1981, pp. 233–245.

[HEID 84]

Heidelberger, P., and Lavenberg, S.S., "Computer Performance Evaluation Methodology", *IEEE Transactions on Computers*, Vol. C-33, Dec. 1984, pp. 1195–1220.

[HEID 85]

Heidelberger, P., "Statistical Analysis of Parallel Simulation", *Proceedings of the Winter Simulation Symposium*, Washington D.C., Dec. 1985, pp. 290–295; also in IBM Corp., Yorktown Heights, NY, Tech. Rep. RC12733.

[HEID 86]

Heidelberger, P., "Limitations of Infinitesimal Perturbation Analysis", IBM Thomas J. Watson Research Center, Yorktown Heights, NY, *Tech. Rep.*, RC11891, May 1986.

[HEND]

Henderson, W. and Taylor P.G., "Embedded Processes in Stochastic Petri Nets", *IEEE Transactions on Software Engineering*, Vol. 17, 1991, pp. 108–116.

[HEND 89]

Henderson, W. and Taylor P.G., "Aggregation Methods in Exact Performance Analysis of Stochastic Petri Nets", *Proceedings of the Third International Workshop on Petri Nets and Performance Models*, Kyoto, Japan, IEEE Computer Society Press, 1989.

[HEND 90]

Henderson, W. and Taylor, P.G., "Product Form In Networks of Queues with Batch Arrivals and Batch Services", *Queueing Systems*, Vol. 6, No. 1, April 1990, pp. 71–88.

[HEND 92]

Henderson, W. and Taylor, P.G., "Insensitivity in Discrete Time Queues with a Moving Server", *Queueing Systems*, Vol. 11, No. 3, Sept. 1992, pp. 273–297.

[HEND 93]

Henderson, W. and Lucic, D., "Aggregation and Disaggregation through

Insensitivity in Stochastic Petri Nets", *Performance Evaluation*, Vol. 17, No. 2, March 1993, pp. 91–114.

[HERZ]

Herzog, U., Woo, L. and Chandy, K.M., "Solution of Queueing Problems by a Recursive Technique", *IBM Journal of Research and Development*, May 1975, pp. 295–300.

This is the first paper to systematically look at recursive solutions for nonproduct form models.

[HEYM]

Heyman, D.P. and Stidham Jr., S., "The Relation Between Customer and Time Averages in Queues", *Operations Research*, Vol. 28, No. 4, 1980, pp. 983–994.

[HEYM 96]

Heyman, D.P. and Lakshman, T.V., "What are the Implications of Long-Range Dependence for VBR-Video Traffic Engineering", *IEEE/ACM Transactions on Networking*, Vol. 4, No. 3, June 1996, pp. 301–317.

[HO 83a]

Ilo, Y.C. and Cassandras, C., "A New Approach to the Analysis of Discrete Event Dynamic Systems", *Automatica*, Vol. 19, 1983, pp. 149–167.

[HO 83b]

Ho, Y.C. and Cao, X., "Perturbation Analysis and Optimization of Queueing Networks", *Journal of Optimization Theory and Applications*, Vol. 40, 1983, pp. 554–582.

[HO 84]

Ho, Y.C., et al., "Perturbation Analysis and Optimization of Large Multiclass (Non-Product Form) Queueing Networks", *Large Scale Systems*, Vol. 7, 1984, pp. 165–180.

[HOLT]

Holt, A., Saint, H., Shapiro, R. and Warshall, S., Final Report of the Information System Theory Project, *Technical Report RADC-TR-68-305*, Rome Air Development Center, Griffiss Air Force Base, New York, 1968, 352 pages.

[HOWA]

Howard, R.A., *Dynamic Probabilistic Systems: Vol. I: Markov Models*, John Wiley & Sons, New York, 1971.

[HSIA]

Hsiao, M.-T. T. and Lazar, A.A., "An Extension to Norton's Equivalent", *Queueing Systems*, J.C. Baltzer A.G. Scientific Publishing Co., Vol. 5, 1989, pp. 401–412.

[HUAN]

Huang, H.-Y., and Robertazzi, R.G., "Recursive Solutions for Discrete Time Queues with Applications to High Speed Switching Fabrics", *Pro-*

ceedings of the 1992 Conference on Information Sciences and Systems, Princeton, N.J., March 1992, pp. 256–261.

[HUI]

Hui, J.Y. and Arthurs, E., "A Broadband Packet Switch for Integrated Transport", *IEEE Journal on Selected Areas in Communications*, Vol. SAC-5, No. 8, Oct. 1987, pp. 1264–1273.

This paper was originally presented at the International Conference on Communications, ICC'87, at Seattle WA, June 1987.

[HWAN]

Hwang, K. and Briggs, F.A., *Computer Architecture and Parallel Processing*, McGraw-Hill, New York, 1984.

A thorough treatment.

[HYMA 91]

Hyman, J.M., Lazar A.A. and Pacifici, G., "Real-Time Scheduling with Quality of Service Constraints", *IEEE Journal on Selected Areas in Communications*, Nov. 1991.

[HYMA 92]

Hyman, J.M., Lazar A.A. and Pacifici, G., "Joint Scheduling and Admission Control for ATS-based Switching Nodes", *IEEE Journal on Selected Areas in Communications*, Jan. 1992.

[IBE]

Ibe, O.C. and Trivedi, K.S., "Stochastic Petri Net Analysis of Finite Population Vacation Queueing Systems", *Queueing Systems*, Vol. 8, No. 2, April 1991, pp. 111–128.

[IZQU]

Izquierdo, M.R. and Reeves, D.S., "A Survey of Statistical Source Models for Variable Bit Rate Compressed Video", *Multimedia Systems*, Vol. 7, 1999, pp. 199–213.

[JABB]

Jabbari, B. and Yegenoglu, F., "An Efficient Method for Computing Cell Loss Probability for Heterogeneous Bursty Traffic in ATM Networks", *International Journal of Digital and Analog Communication Systems*, Vol. 5, 1992, pp. 49–48.

[JACK 57]

Jackson, J.R., "Networks of Waiting Lines", *Operations Research*, Vol. 5, 1957, pp. 518–521.

The first product form solution paper.

[JACK 64]

Jackson, J.R., "Job Shop Like Queueing Systems", *Management Sciences*, Vol. 10, No. 1, 1964, pp. 131–142.
Generalizes the work of [JACK 57].

[JAIN]

Jain, R. and Turner, R., "Workload Characterization Using Image Accounting", *Proceedings of the Computer Performance Evaluation Users Group 18th Meeting*, Oct. 1982, pp. 111–120.

[JAIN]

Jain, R., *The Art of Computer Systems Performance Analysis: Techniques for Experimental Design, Measurement, Simulation and Modeling*, Wiley, New York, 1991.

An excellent and comprehensive treatment of the subject.

[JEFF]

Jefferson, D., "Virtual Time", *ACM Transactions on Programming Languages and Systems*, Vol. 7, July 1985, pp. 404–425.

[JELE]

Jelenkovic, P.R., "Network Multiplexer with Truncated Heavy-Tailed Arrival Streams", *Proceedings of IEEE INFOCOM'99*, New York, NY, 1999, pp. 625–631.

[JENQ]

Jenq, Y.-C., "Performance Analysis of a Packet Switch Based on Single-Buffered Banyan Network", *IEEE Journal on Selected Areas in Communication*, Vol. SAC-1, No. 6, Dec. 1983, pp. 1014–1021.

[JEWE]

Jewell, W.S., "A Simple Proof of $L = \lambda W$", *Operations Research*, Vol. 15, 1967, pp. 1109–1116.

[JONE]

Jones, S.K., Cavin III, R.K. and Johnston, D.A., "An Efficient Computational Procedure for the Evaluation of the M/M/1 Transient State Occupancy Probabilities", *IEEE Transactions on Communications*, Vol. COM-28, No. 12, Dec. 1980, pp. 2019–2020.

For a discussion of the merits of this technique relative to that in [ACKR] see [CHIE].

[KARO 86]

Karol, M.J., Hluchyj, M.G. and Morgan, S.P., "Input Vs. Output Queueing on a Space-Division Packet Switch", *IEEE Global Communications Conference*, Houston TX, Dec. 1986, pp. 659–665.

[KARO 87]

Karol, M.J., Hluchyj, M.G. and Morgan, S.P., "Input Vs. Output Queueing on a Space-Division Packet Switch", *IEEE Transactions on Communications*, Vol. COM-35, No. 12, Dec. 1987, pp. 1347–1356.

[KARO 88]

Hluchyj, M.G. and Karol, M..G, "Queueing in Space-Division Packet Switching", *IEEE INFOCOM '88*, New Orleans La., April 1988, pp. 334–343.

[KARO 92]

Karol, M.J., Eng, K.Y. and Obara, H., "Improving the Performance of Input-Queued ATM Packet Switches", *Proceedings of IEEE INFO-COM'92*, Florence, Italy, May 1992, pp. 110–115.

[KAUF]

Kaufman, L., Gopinath, B., and Wunderlich, E.F., "Analysis of Packet Network Congestion Control Using Sparse Matrix Algorithms", *IEEE Transactions on Communications*, Vol. COM-29, No. 4, April 1981, pp. 453–465.

[KELL]

Kelly, F.P., *Reversibility and Stochastic Networks*, John Wiley & Sons, New York, 1979.

A very good introduction to the concept of reversibility. A great many interesting and unusual stochastic models are presented.

[KELL 75]

Kelly, F.P., "Networks of Queues with Customers of Different Types", *Journal of Applied Probability*, Vol. 12, 1975, pp. 542–554.

[KELL 76]

Kelly, F.P., "Networks of Queues", *Advances in Applied Probability*, Vol. 8, 1976, pp. 416–432.

[KEND]

Kendall, D.G., "Some Problems in the Theory of Queues", *Journal of the Royal Statistical Society*, Series B, Vol. 13, No. 2, 1951, pp. 151–185.

The simple imbedded Markov chain proof of the Pollaczek-Khinchin mean value formula used in this book first appears in this paper. Contains an extensive discussion section.

[KERM]

Kermani, P. and Kleinrock, L., "Virtual Cut-Through: A New Computer Communication Switching Technique", *Computer Networks*, Vol. 3, 1979, pp. 267–286.

[KHIN]

Khinchin, A.Y., "Mathematisches uber die Erwartung vor einem offentlichen Schalter" or "Mathematical Theory of Stationary Queues", *Matem. Sbornik*, Vol. 39, 1932, pp. 73–84.

This paper is one of two early analysis of the M/G/1 queueing system. It is in Russian with a German summary. See also [POLL] and [KEND].

[KIM]

Kim, H.S. and Leon-Garcia, A., "Performance of Buffered Banyan Networks Under Nonuniform Traffic Patterns", *IEEE Transactions on Communications*, Vol. 38, No. 5, May 1990, pp. 648–658.

[KIM 90b]

Kim, H.S. and Leon-Garcia, A., "A Self-Routing Multistage Switching Network for Broadband ISBN", *IEEE Journal on Selected Areas in Communications*, Vol. 8, No. 3, April 1990, pp. 459–466.

[KLEI 64]

Kleinrock, L., *Communication Nets*, McGraw-Hill, New York, 1964, Reprinted: Dover, New York, 1972.

[KLEI 65]

Kleinrock, L., "A Conservation Law for a Wide Class of Queueing Systems", *Naval Research Logistics Quarterly*, Vol. 12, 1965, pp. 181–192.

[KLEI 75]

Kleinrock, L., *Queueing Systems, Vol. I: Theory*, Wiley, New York, 1975. A very thorough and excellent introduction to classical queueing theory. Vol. II deals with applications.

[KOBA]

Kobayashi, H., *Modeling and Analysis: An Introduction to System Performance Evaluation*, Addison-Wesley, Reading, Mass. 1978.

[KOGA]

Kogan, Ya.A. and Boguslavsky, L.B., "Asymptotic Analysis of Memory Interference in Multiprocessors with Private Cache Memories", *Performance Evaluation*, Vol. 5, 1985, pp. 97–104.

[KRIE]

Krieger, U., Muller-Clostermann, B. and Sczittnick, M., "Modeling and Analysis of Communication Systems Based on Computational Methods for Markov Chains", *IEEE Journal on Selected Areas in Communications*, Vol. 8, 1990, pp. 1630–1648.

[KRIS]

Krishnan, K-R. and Meempat, G., "Long-Range Dependence in VBR Video Streams and ATM Traffic Engineering", *Performance Evaluation*, Vol. 30, 1997, pp. 45–56.

Further examines issues raised in [HEYM 96]

[KRIT]

Kritzinger, P.S., "A Performance Model of the OSI Communication Architecture", *IEEE Transactions on Communications*, Vol. COM-34, No. 6, 1986, pp. 554–563.

[KUEH]

Kuehn, P.J., "Reminder on Queueing Theory for ATM Network", *Telecommunication Systems*, J.C. Baltzer AG, Science Publishers, Vol. 5, 1996, pp. 1–24.

An excellent survey of stochastic processes useful for traffic modeling.

[KURO]

Kurose, J.F. and Mouftah, H.T., "Computer-Aided Modeling, Analysis, and Design of Communication Networks", *IEEE Journal on Selected Areas in Communications*, Vol. 6, No. 1, Jan. 1988, pp. 130–145.

This is the paper in which Chapter 4's discussion on simulation first appeared.

[LAMB]

√ Lambadaris, I., Devetsikiotis, M. et al., "Traffic Modelling and Design Methodologies for Broadband Networks", *Canadian Journal of Electrical and Computer Engineering*, Vol. 20, No. 3, 1995, pp. 105–115.

[LAVE 80]

Lavenberg, S.S. and Reiser, M., "Stationary State Probabilities of Arrival Instants for Closed Queueing Networks with Multiple Types of Customers", *Journal of Applied Probability*, Vol. 17, 1980, pp. 1048–1061.

This article presents the *Arrival Theorem* and deals with a generalization of BCMP type networks.

[LAVE 83]

Lavenberg, S.S. (ed.), *Computer Performance Modeling Handbook*, Academic Press, New York, 1983.

Includes a particularly strong summary of analytical queueing results and extended discussions of simulation.

[LAW 82]

Law, A.M. and Kelton, W.D., *Simulation Modeling and Analysis*, McGraw-Hill Book Co., New York, 1982.

A solid introductory text.

[LAW 83]

Law, A., "Statistical Analysis of Simulation Output Data", *Operations Research*, Vol. 31, Nov. 1983, pp. 983–1029.

[LAZA 84a]

Lazar, A.A. and Robertazzi, T.G., "The Geometry of Lattices for Markovian Queueing Networks", *Columbia University Tech. Rep., 1984, SUNY at Stony Brook CEAS Tech. Rep. 471.*

[LAZA 84b]

Lazar, A.A. and Robertazzi, T.G., "The Geometry of Lattices for Multiclass Markovian Queueing Networks", *Proceedings of the 1984 Information Sciences and Systems Conference*, Princeton University, Princeton N.J., March 1984, pp. 164–168.

[LAZA 84c]

Lazar, A.A., "An Algebraic Topological Approach to Markovian Queueing Networks", *Proceedings of the 1984 Information Sciences and Systems Conference*, Princeton University, Princeton N.J., March 1984, pp. 437–442.

[LAZA 86]

Lazar, A.A. and Robertazzi, T.G., "The Algebraic and Geometric Structure of Markovian Petri Network Lattices", *Proceedings of the Twenty-Fourth Annual Allerton Conference on Communication, Control and Computing*, University of Illinois, Urbana-Champaign Ill., 1986, pp. 834–843.

[LAZA 87]

Lazar, A.A. and Ferrandiz, J.M., "Geometric Analysis of Quasi-Birth and Death Processes by Flow Redirection", *Proceedings of the 1987 Conference on Information Sciences and Systems*, The Johns Hopkins University, Baltimore, MD, March 1987, pp. 865–870.

[LAZA 87b]

Lazar, A.A. and Robertazzi, T.G., "Markovian Petri Net Protocols with Product Form Solution", *IEEE INFOCOM'87*, San Francisco, CA, 1987, pp. 1054–1062. Journal version appears in *Performance Evaluation*, Vol. 12, 1991, pp. 67–77 under the same title.

[LAZA 90]

Lazar, A.A., Temple, A.T. and Gidron, R. "An Architecture for Integrated Networks that Guarantees Quality of Serivice", *International Journal of Digital and Analog Communication Systems*, Vol. 3, April–June 1990, pp. 229–238.

[LAZA 91]

Lazar, A.A. and Pacifici, G., "Control of Resources in Broadband Networks with Quality of Service Guarantees", *IEEE Communications Magazine*, Vol. 29, No. 10, Sept. 1991, pp. 66–73.

[LAZO]

Lazowska, E., Zahorjan, J., Cheriton, D., and Zwaenepoel, W., "File Access Performance of Diskless Workstations", *ACM Transactions on Computer Systems*, Vol. 4, No. 3, Aug. 1986.

[LELA]

Leland, W.E., Taqqu, M.S., Willinger, W. and Wilson, D.V., "On the Self-Similar Nature of Ethernet Traffic (Extended Version)". *IEEE/ACM Transactions on Networking*, Vol. 2, No. 1, Feb. 1994, pp. 1–15.

The first paper to show network traffic to be self-similar.

[LEON]

Leon Garcia, A., *Probability and Random Processes for Electrical Engineering*, Addison-Wesley, Reading, Mass., 1989.

An excellent introduction for students.

[LI]

Li, J.-S., Wolist, A. and Popescu-Zeletin, R., "Measurement and Performance Evaluation of NFS Traffic in ATM Network", *Computer Communications*, Vol. 22, 1999, pp. 101–109.

[LI A]

Li, S.-Q. and Hwang, C.-L., "Queue Response to Input Correlation Functions: Discrete Spectral Analysis", *IEEE/ACM Transactions on Networking*, Vol. 1, No. 5, 1993, pp. 522–533.

[LI B]

Li, S.-Q. and Hwang, C.-L., "Queue Response to Input Correlation Functions: Continuous Spectral Analysis", *IEEE/ACM Transactions on Networking*, Vol. 1, No. 6, 1993, pp. 678–692.

[LIGG]

Liggett, T.M., *Interacting Particle Systems*, Springer-Verlag, New York, 1985.

[LIM]

Lim, Y. and Kobza, J., "Analysis of a Delay-Dependent Priority Discipline in a Multi-Class Traffic Packet Switching Node", *Proceedings of IEEE INFOCOM'88*, New Orleans, LA, April 1988, pp. 889–898.

[LIN]

Lin, G.C. and Suda, T., "On the Impact of Long-Range Dependent Traffic in Dimensioning ATM Network Buffer", *Proceedings of IEEE INFOCOM'98*, San Francisco, 1998, pp. 1317–1324.

[LITT]

Little, J.D.C., "A Proof of the Queueing Formula $L = \lambda W$", *Operations Research*, Vol. 9, 1961, pp. 383–387.

[LITW]

Littelwood, M., Gallagher, I.D. and Adams, J.L., "Evolution Towards an ATD Multi-Service Network", *British Telecommunications Technology Journal*, Vol. 5, No. 2, April 1987, pp. 52–62.

A good discussion of the advantages of ATM technology.

[LIU]

Liu, Z. and Nain, P., "Sensitivity Results in Open, Closed and Mixed Product Form Queueing Networks", *Performance Evaluation*, Vol. 13, No. 4, Nov. 1991, pp. 237–251.

[LUBA]

Lubachevsky, B., and Ramakrishnan, K., "Parallel Time-Driven Simulation of a Network using Shared Memory MIMD Computer", *Modeling Tools and Techniques for Performance Analysis*, D. Potier, Ed., North-Holland, Amsterdam, 1985.

[LUCA]

Lucatoni, D., "New Results on the Single Server Queue with a Batch Markovian Arrival Process", *Stochastic Models*, Vol. 7, 1991, pp. 1–46.

[MARC]

Marcum, J.I., "A Statistical Theory of Target Detection by Pulsed Radar",

IRE Transactions on Information Theory, Vol. IT-6, pp. 59–267, April 1960.

[MARS]
Marsan, M. Ajmone, Balbo, G., Conte G., *Performance Models of Multi-processor Systems*, The MIT Press, Cambridge, Mass., 1986.
An excellent treatment of modeling multiprocessors using queueing and Petri Nets.

[MARS 83]
Marsan, M.J., Balbo, G., Conte, G. and Gregoretti, F., "Modeling Bus Contention and Memory Interference in a Multiprocessor System", *IEEE Transactions on Computers*, Vol. C-32, No. 1, 1983, pp. 60–72.
Presents the original multiprocessor models upon which the modified product form models of chapter 5 are based.

[MARS 84]
Marsan, M., Balbo, G. and Conte G., "A Class of Generalized Stochastic Petri Nets for the Performance of Evaluation of Multiprocessor Systems", *ACM Transactions on Computer Systems*, May 1984.
Introduces immediate transitions for stochastic Petri Nets.

[MARS 84b]
Marsan, M., Balbo, G., Chiola, G. and Donatelli, S., "On the Product Form Solution of a Class of Multiple Bus Multiprocessor System Models", *Proceedings of the International Workshop of the Modeling and Performance Evaluation of Parallel Systems*, Grenoble, France, 1984.

[MARS 85]
Marsan, M., Balbo, G., Bobbio, A., Chiola, G., Conte, G., and Cumani, A., "On Petri Nets with Stochastic Timing", *International Workshop on Timed Petri Nets*, Torino, Italy, IEEE Computer Society Press, July 1985, pp. 80–87.

[MARS 86]
Marsan, M., Balbo, G., Chiola, G. and Donatelli, S., "On The Product Form Solution of a Class of Multiple Bus Multiprocessor System Models", *Journal of Systems and Software*, Vol. 6, 1986, pp. 117–124.

[MARS 90]
Marsan, M., Donatelli, S. and Neri, F., "GSPN Models of Markovian Multiserver Multiqueue Systems", *Performance Evaluation*, Vol. 11, No. 4, Nov. 1990, pp. 227–240.

[MASS]
Massey, W., *Algebraic Topology: An Introduction*, Springer-Verlag, New York, 1977.

[MAYO]
Mayor, G. and Silvester, J., "Time Scale Analysis of an ATM Queueing System with Long-Range Dependent Traffic", *Proceedings of IEEE IN-FOCOM'97*, Kobe, Japan, 1997, pp. 205–212.

[McKEN 81]

McKenna, J., Mitra, D., and Ramakrishnan, K.G., "A Class of Closed Markovian Queueing Networks: Integral Representations, Asymptotic Expansions, and Generalizations", *The Bell System Technical Journal*, Vol. 60, No. 5, May–June 1981, pp. 599–641.

[McKEN 82]

McKenna, J. and Mitra, D., "Integral Representations and Asymptotic Expansions for Closed Markovian Queueing Networks: Normal Usage", *The Bell System Technical Journal*, Vol. 61, No. 5, May–June 1982, pp. 661–683.

[McKEN 84]

McKenna, J. and Mitra, D., "Asymptotic Expansions and Integral Representations of Moments of Queue Lengths in Closed Markovian Networks", *Journal of the ACM*, Vol. 31, No. 2, April 1984, pp. 346–360.

[McKEN 86]

McKenna, J. and Mitra, D., "Asymptotic Expansions for Closed Markovian Networks with State-Dependent Service Rates", *Journal of the ACM*, Vol. 33, No. 3, July 1986, pp. 568–592.

[McKEN 87]

McKenna, J., "Asymptotic Expansions of the Sojourn Time Distribution Functions of Jobs in Closed, Product-Form Queueing Networks", *Journal of the ACM*, Vol. 34, No. 4, Oct. 1987, pp. 985–1003.

[MICH]

Michiel, H. and Laevens, K., "Teletraffic Engineering in a Broad-Band Era", *Proceedings of the IEEE*, Vol. 85, No. 12, Dec. 1997, pp. 2007–2033

An excellent survey of teletraffic modeling

[MISR]

Misra, J., "Distributed Discrete Event Simulation", *ACM Computing Surveys*, Vol. 18, No. 1, March 1986, pp. 39–66.

[MIYA]

Miyazawa, M., "Time and Customer Processes in Queues with Stationary Inputs", *Journal of Applied Probability*, Vol. 14, No. 2, pp. 349–357.

[MOLL]

Molloy, M.K., *Fundamentals of Performance Modeling*, Macmillan Publishing Co., New York, 1989.

An excellent introduction to the basics of performance evaluation theory.

[MOLL 82]

Molloy, M.K., "Performance Analysis Using Stochastic Petri Nets", *IEEE Transactions on Computers*, Vol. C-31, No. 9, 1982, pp. 913–917.

References

[MOLL 84]
Molloy, M.K., "Modeling and Analysis of LAN Protocols Using Labeled Petri Nets", *Technical Report 84-15*, Dept. of Computer Science, The University of Texas at Austin, Austin, TX, 1984.

[MOLL 86]
Molloy, M.K., "Fast Bounds for Stochastic Petri Nets", *International Workshop on Timed Petri Nets*, Torino, Italy, IEEE Computer Society Press, July 1985, pp. 244–249.

[MUEL]
Mueller, B. and Reinhardt, J., *Neural Networks: An Introduction*, Springer-Verlag 1990.

An excellent graduate level introduction to neural networks. Particularly strong in the physics and mathematical aspects of these nets. A floppy diskette containing neural networks programs is included.

[MUKH]
Mukherjee, B. and Bisdikian, C., "A Journey Through the DQDB Network Literature", *Performance Evaluation*, North-Holland Publishing Co., Dec. 1992.

[NEUT]
Neuts, M.F., *Matrix-Geometric Solutions in Stochastic Models: An Algorithmic Approach*, The Johns Hopkins University Press, Baltimore Md., 1981.

[NEUT 89]
Neuts, M.F., *Structured Stochastic Matrices of M/G/1 Type and Their Applications*, Marcel Dekker, New York, 1989.

[NEWE]
Newell, G.F., *Applications of Queueing Theory*, Chapman and Hall, London, 1971.

[NEWM]
Newman, R.M. and Hullet, J.L., "Distributed Queueing: A Fast and Efficient Packet Access Protocol for QPSX", *Proceedings of the International Conference on Computer Communications* (ICCC), 1986.

[NICH]
Nichols, W.G. and Emer, J.S., "Design and Implementation of the VAX Distributed File Service", *Digital Technical Journal*, No. 9, June 1989, pp. 16–28.

[OIE]
Oie, Y., Murata, M., Kubota, K. and Miyahara, H., "Effect of Speedup in Nonblocking Packet Switch", *Proceedings of IEEE International Conference on Communications'89*, 1989, pp. 410–414.

[ONVU 90]
Onvural, R.O., "Closed Queueing Networks with Blocking", in Takagi,

391

tic Analysis of Computer and Communication Systems, 1990,

O. and Akyildiz, I., eds., *Proceedings of the 2nd International on Queueing Networks with Blocking*, Research Triangle Park 992, Elsevier, 1993.

[OREI]

O'Reilly, P., and Hammond, J.L., "An Efficient Technique for Performance Studies of CSMA/CD Local Networks", *IEEE Journal of Selected Areas in Communications*, Vol. SAC-2, Jan. 1984, pp. 238–249.

[OUST]

Ousterhout, J., et al., "A Trace-driven Analysis of the UNIX 4.2 BSD File System", *Proceedings of the 10th ACM Symposium on Operating Systems Principles*, Dec. 1985.

[PAPA]

Papapetrou, E., Gragopoulos, I. and Pavlidou, F.-N., "Performance Evaluation of LEO Satellite Constellations with Inter-Satellite Links under Self-Similar and Poisson Traffic", *International Journal of Satellite Communications*, Vol. 17, 1999, pp. 51–64.

[PAPO]

Papoulis, A., *Probability, Random Variables and Stochastic Processes*, McGraw-Hill, New York, 1965.

[PAPO 84]

Papoulis, A., *Probability, Random Variables and Stochastic Processes*, 2nd Edition, McGraw-Hill Book Company, New York, 1984.

[PARE]

Pareek, S. and Robertazzi, T.G., "An Algorithm for the Exact Decomposition of a Class of Non-Product Form Queueing Models", *Proceedings of the 1990 Conference on Information Sciences and Systems*, Princeton N.J., March 1990.

[PARK]

Park, K., Kim, G.T. and Crovella, M.E., "On the Relationship Between File Sizes, Transport Protocols and Self-Similar Network Traffic", *Proceedings of the 4th International Conference on Network Protocols (ICNP'96)*, Oct. 1996, pp. 171–180.

[PARU]

Parulekar, M. and Makowski, A., "Tail Probabilities for a Multiplexer with Self-Similar Traffic", *Proceedings of the IEEE INFOCOM'96*, San Francisco, 1996, pp. 1452–1459.

[PARZ]

Parzen, E., *Stochastic Processes*, Holden-Day, San Francisco, 1962.

[PAXS]

Paxson, V. and Floyd, S., "Wide Area Traffic: The Failure of Poisson Modeling", *IEEE/ACM Transactions on Networking*, Vol. 3, No. 3, June 1995, pp. 226–244.

[PAXS B]

Paxson, V., "Fast Approximate Synthesis of Fractional Gaussian Noise for Generating Self-Similar Network Traffic", *ACM Computer Communication Review*, Vol. 27, Oct. 1997, pp. 5–18.

[PEAC]

Peacock, J., Wong, J., and Manning, E., "Distributed Simulation using a Network of Processors", *Computer Networks*, Vol. 3, 1970, pp. 44–56.

[PERR 84]

Perros, H.G., "Queueing Networks with Blocking: A Bibliography", *Performance Evaluation Review*, Vol. 12, Aug. 1984, pp. 8–12.

[PERR 89]

Perros, H.G. and Altiok, T., ed., *Proceedings of the First International Workshop on Queueing Networks with Blocking*, North Carolina State University, Raleigh N.C., May 1988, Elsevier North-Holland, 1989.

[PERR 89b]

Perros, H.G. and Akyildiz, I., eds., Special Issue on Queueing Networks with Blocking, *Performance Evaluation*, Vol. 10, 1989.

[PERR 90]

Perros, H.G., "Approximate Algorithms for Open Networks with Blocking", in Takagi, ed., *Stochastic Analysis of Computer and Communication Systems*, Elsevier, 1990, pp. 451–498.

[PETE]

Peterson, J.L., *Petri Net Theory and the Modeling of Systems*, Prentice Hall, Englewood Cliffs N.J., 1981.

An excellent introductory text on Petri Nets.

[PETR]

Petri, C.A., *Kommunikation mit Automaten*, Ph.D Thesis, Univ. of Bonn, Fed. Rep. of Germany, 1962.

This is the original work on Petri Nets.

[PLOT]

Plotkin, N. and Varaiya, P., "The Entropy of Traffic Streams in ATM Virtual Circuits", *Proceedings of IEEE INFOCOM 1994*, 1994, pp. 1038–1045.

[POLL]

Pollaczek, F., "Uber eine Aufgabe dev Wahrscheinlichkeitstheorie", *Math. Zeitschrift*, Vol. 32, 1930, pp. 64–100, 729–750.
One of the two early analysis of the M/G/1 queueing system. See also [KHIN] and [KEND].

[PRIE]

Priestly, M.D., *Spectral Analysis and Time Series*, Academic Press, 1981.

[RAMA]

Ramakrishnan, K.G. and Mitra, D., "An Overview of PANACEA, a Software Package for Analyzing Markovian Queueing Networks", *The Bell System Technical Journal*, Vol. 61, No. 10, Dec. 1982, pg. 2849.

[RAMA 86]

Ramakrishnan, K. and Emer, J., "A Model of File Server Performance for a Heterogeneous Distributed System", *Proceedings of the ACM SIGCOMM '86 Symposium*, August 1986, pp. 338–347.

[RAMA 89]

Ramakrishnan, K. and Emer, J., "Performance Analysis of Mass Storage Service Alternatives for Distributed Systems", *IEEE Transactions on Software Engineering*, vol. 15, no. 2, 1989.

[REED]

Reed, D., Malony, A., and McCredie, B., "Parallel Discrete Simulation: A Shared Memory Approach", *Performance Evaluation Review*, Vol. 15, May 1987, pp. 36–39.

[REIB]

Reibman, A., Smith, R. and Trivedi, K., "Markov and Markov Reward Model Transient Analysis: An Overview of Numerical Approaches", *European Journal of Operational Research*, Vol. 40, 1989, North-Holland, pp. 257–267.

[REIC]

Reich, E., "Waiting Times when Queues are in Tandem", *Annals of Mathematical Statistics*, Vol. 28, 1957, pp. 768–773.

[REIM]

Reiman, M., and Weiss, A., "Sensitivity Analysis via Likelihood Ratios", *Proceedings of the Winter Simulation Symposium*, Washington D.C., Dec. 1986, pp. 285–289; extended version available as unpublished report.

[REIS 73]

Reiser, M. and Kobayashi, H., "Recursive Algorithms for General Queueing Networks with Exponential Servers", *IBM Research Report RC 4254*, Yorktown Heights N.Y., 1973.

[REIS 75]

Reiser, M. and Kobayashi, H., "Queueing Networks with Multiple Closed Chains: Theory and Computational Algorithms", *IBM Journal of Research and Development*, Vol. 19, 1975, pp. 283–294.

[REIS 79]

Reiser, M., "A Queueing Network Analysis of Computer Communications Networks with Window Flow Control", *IEEE Transactions on Communications*, Vol. COM-27, No. 8, August 1979, pp. 1199–1209.

[REIS 80]
Reiser, M. and Lavenberg, S.S., "Mean-value Analysis of Closed Multi-chain Queueing Networks", *Journal of the ACM*, Vol. 27, No. 2, April 1980, pp. 313–322.

[REIS 81]
Reiser, M., "Mean-value Analysis and Convolution Method for Queue-Dependent Servers in Closed Queueing Networks", *Performance Evaluation*, Vol. 1, 1981, pp. 7–18.

The discussion here relates MVA to the convolution algorithm and to the LBANC algorithm.

[REIS 82]
Reiser, M., "Performance Evaluation of Data Communication Systems", *Proc. IEEE*, vol. 70, no. 2, Feb. 1982, pp. 171–195.

A wide ranging survey paper.

[ROBE 72]
Roberts, L.G., "ALOHA Packet System with and without Slots and Capture", ARPANET Satellite System Note 8 (NIC 112290), June 1972. Also in *Computer Communications Review*, Vol. 5, April 1975.

[ROBE 88a]
Robertazzi, T.G. and Schwartz, S.C., "Best 'Ordering' for Floating Point Addition", *ACM Transactions on Mathematical Software*, Vol. 14, No. 1, March 1988, pp. 101–110.

An error analysis of various schemes for accurately adding large numbers of positive floating point numbers.

[ROBE 89]
Robertazzi, T.G., "Recursive Solution of a Class of Non-Product Form Protocol Models", *IEEE INFOCOM'89*, Ottawa, Canada, April 1989, pp. 38–46.

[ROBE 92]
Robertazzi, T.G., "Recursive Solution of Equilibrium State Probabilities for Three Tandem Queues with Limited Buffer Space", *SUNY at Stony Brook CEAS Technical Report 624*, April 6, 1992, available from T. Robertazzi.

[ROBE 93]
Robertazzi, T.G., editor, *Performance Evaluation of High Speed Switching Fabrics and Networks: ATM, Broadband ISDN and MAN Technology*, IEEE Press, Piscataway, N.J., April 1993.

[RODR]
Rodrigues, M.A., "Erasure Node: Performance Improvements for the IEEE 802.6 MAN", *Proceedings of IEEE INFOCOM'90*, San Francisco, CA, June 1990, pp. 636–643.

See also [GARR].

[ROM]

> Rom, R. and Sidi, M., *Multiple Access Protocols: Performance and Analysis*, Springer Verlag, New York, 1990.

> A unified treatment of multiple access protocol analysis.

[RUBI]

> Rubinstein, R.Y., *Simulation and the Monte Carlo Method*, Wiley, New York, 1981.

> A thorough mathematical treatment of Monte Carlo simulation.

[RYU]

> Ryu, B.K. and Elwalid, A., "The Importance of Long Range Dependence of VBR Video Traffic in ATM Traffic Engineering: Myths and Realities", *Proceedings of ACM SIGCOMM*, vol. 26, no. 4, Oct. 1996, pp. 3–14.

[SAAD]

> Saadawi, T., Ammar, M. and El Hakeem, A., *Fundamentals of Telecommunication Networks*, Wiley, New York, 1994.

> An excellent mathematical introduction to computer networks and protocols at the senior/1st year graduate level.

[SAAT]

> Saaty, T.L., *Elements of Queueing Theory*, McGraw-Hill, New York, 1961.

[SAHI]

> Sahinoglu, Z. and Tekinay, S., "On Multimedia Networks: Self-Similar Traffic and Network Performance", *IEEE Communications Magazine*, Vol. 37, No. 1, Jan. 1999, pp. 48–52.

> A concise tutorial introduction to self-similar traffic.

[SAND]

> Sanders, W.H. and Meyer, J.F., "METASAN: A Performability Evaluation Tool Based On Stochastic Activity Networks", *Proceedings of the ACM-IEEE Comp. Soc. Fall Joint Comp. Conf.*, Nov. 1986.

[SAUE 76]

> Sauer, C.H., et. al., "Hybrid Analysis/Simulation of Distributed Networks", IBM Thomas J. Watson Research Center, Yorktown Heights, NY, *Research Report* RC 6341, 1976.

[SAUE 81]

> Sauer, C. and Chandy, K.M., *Computer System Performance Modeling*, Prentice-Hall, Englewood Cliffs NJ, 1981.

> An excellent treatment of its topic.

[SAUE 83]

> Sauer, C.H. and MacNair, E.A., *Simulation of Computer Communication Systems"*, Prentice-Hall, Englewood Cliffs, N.J., 1983.

A very readable treatment which makes use of the RESQ queueing software package.

[SCHO]

Schoemaker, Ed., *Computer Networks and Simulation II and III*, North-Holland, Amsterdam, The Netherlands, 1982.

[SCHR]

Schruben, L., "Detecting Initialization Bias in Simulation Output", *Operations Research*, Vol. 30, 1982, pp. 569–590.

[SCHW 66]

Schwartz, M., Bennett, W.R. and Stein, S., *Communications Systems and Techniques*, McGraw-Hill, New York, 1966.

[SCHW 82]

Schwartz, M., "Performance Analysis of the SNA Virtual Route Pacing Control", *IEEE Transactions on Communications*, Vol. COM-30, Jan. 1982, pp. 172–184.

[SCHW 87]

Schwartz, M., *Telecommunication Networks: Protocols, Modeling and Analysis*, Addison-Wesley, Reading, Mass., 1987.

An excellent introduction to commonly used analytical techniques.

[SCHW 96]

Schwartz, M., *Broadband Integrated Networks*, Prentice-Hall, Upper Saddle River, N.J., 1996.

An excellent introduction to integrated traffic modeling and analysis.

[SCHWE]

Schweitzer, P., "Approximate Analysis of Multiclass Closed Networks of Queues", *Proc. of the International Conference on Stochastic Control and Optimization*, Amsterdam, 1979.

[SHAN]

Shanthikumar, J. and Sargent, R., "A Unifying View of Hybrid Simulation/Analytical Models and Modeling", *Operations Research*, Vol. 31, Nov.–Dec. 1983, pp. 1030–1052.

[SHAP]

Shapiro, S.D., "A Stochastic Petri Net with Applications to Modeling Occupancy Times for Concurrent Task Systems", *Networks*, Vol. 9, 1979, pp. 375–379.

[SHED]

Shedler, G.S. and Haas, P.J., "Regenerative Simulation of Stochastic Petri Nets", *International Workshop on Timed Petri Nets*, Torino, Italy, July 1985, IEEE Computer Society Press, 1986, pp. 14–21.

[STAL]

Stallings, W., *High Speed Networks: TCP/IP and ATM Design Principles*, Prentice Hall, Upper Saddle River, N.J., 1998.

Contains an excellent chapter long introduction to self-similar traffic and modeling.

[STEP]

Stephenson, M. and Molloy, M.K., "Measurements of Terminal Character Input Behavior", *Proceedings of the Computer Networking Symposium*, Washington D.C. 1986, IEEE Computer Society Press, pp. 227–235.

This paper describes the results of an experimental study of character interarrival times from asynchronous terminals attached to a time sharing system in a university environment.

[STID]

Stidham Jr., S., "A Last Word on $L = \lambda W$", *Operations Research*, Vol. 22, 1974, pp. 417–421.

[SURI 83]

Suri, R., "Robustness of Queueing Network Formulas", *Journal of the Association of Computing Machinery*, Vol. 30, No. 3, July 1983, pp. 564–594.

[SURI 84]

Suri, R., "Perturbation Analysis Gives Strongly Consistent Estimates for the M/G/1 Queue", Div. of Applied Science, Harvard University, Cambridge, MA, *Tech. Rep.*, 1984.

[SURI 87]

Suri, R., "Infinitesimal Perturbation Analysis of Discrete Event Dynamic Systems," *Journal of the ACM*, Vol. 34, July 1987, pp. 686–717.

[SZYM 86]

Szymanski, T.H., "A VLSI Comparison between Crossbar and Switch-Recursive Banyan Interconnection Networks", *Proceedings of the International Conference on Parallel Processing*, Aug. 1986, pp. 192–199.

[SZYM]

Szymanski, T. and Fang, C., "Design and Analysis of Buffered Crossbars and Banyans with Cut-Through Switching", *Proceedings of IEEE Supercomputing'90*, New York, N.Y., Nov. 1990, pp. 264–273.

[TAKA]

Takacs, L., *Introduction to the Theory of Queues*, Oxford University Press, New York, 1962.

[TANE]

Tanenbaum, A., *Computer Networks*, 2nd edition, Prentice-Hall, Englewood Cliffs N.J., 1988.

[TAQQ 85]

Taqqu, M.S., "A Bibliographical Guide to Self-Similar Processes and Long Range Dependence", in *Dependence in Probability and Statistics*, Eberlin, E. and Taqqu, M.S., editors, Birkhauser, Basel, 1985, pp. 137–165.

[TAQQ 95]

Taqqu, M.S., Teverovsky, V. and Willinger, W., "Estimators for Long Range Dependence: An Empirical Study", *Fractals*, Vol. 3, No. 4, 1995, pp. 785–798.

[TAQQ 97]

√ Taqqu, M.S., Teverovsky, V. and Willinger, W., "Is Network Traffic Self-Similar or Multifractal?", *Fractals*, World Scientific Publishing Co., Vol. 5, No. 1, 1997, pp. 63–73.

[TAQQ 97B]

Taqqu, M.S., Willinger, W. and Sherman, R., "Proof of a Fundamental Result in Self-Similar Traffic Modeling", *ACM Computer Communication Review*, Vol. 27, April 1997, pp. 5–23.

[THOM]

Thomas, J.B., *An Introduction to Statistical Communication Theory*, Wiley, New York, 1969.

A very readable introductory work.

[TSYB]

Tsybakov, B. and Georganas, N.D., "On Self-Similar Traffic in ATM Queues: Definition, Overflow Probability Bound and Cell Delay Distribution", *IEEE/ACM Transactions on Networking*, Vol. 5, No. 3, June 1997, pp. 397–409.

[TURN]

Turner, J., "Design of a Broadcast Packet Network", *IEEE INFOCOM 86*, Miami, FL, April 1986, pp. 667–675.

[VAND]

van Dijk, N.B., "A Simple Bounding Methodology for Non-Product Form Finite Capacity Queueing Systems", *Proceedings of the First International Workshop on Queueing Networks with Blocking*, North Carolina State University, May 1988, Elsevier North-Holland, H.G. Perros and T. Altiok, eds., 1989.

[VANS]

Van Slyke, R., Chou, W. and Frank, H., "Avoiding Simulation in Simulating Communication Networks", *Proceedings of the 1972 National Computer Conference*, pp. 165–169.

[VITE]

Viterbi, A.M., "Mean Delay in Synchronous Packet Networks with Priority Queueing Discipline", *IEEE Transactions on Communications*, Vol. 39, No. 4, April 1991, pp. 469–473.

[WAGN]

Wagner, D.B. and Lazowska, E.D., "Parallel Simulation of Queueing Networks: Limitations and Potentials", *Proceedings of the 1989 ACM SIGMETRICS and PERFORMANCE'89 International Conference on Mea-

surement and Modeling of Computer Systems, Berkeley, California, May
1989, pp. 146–155.

[WALR]

Walrand, J., "Quick Simulation of Queueing Networks", *Proceedings of
the 2nd International Workshop on Applied Mathematics and Perfor-
mance/Reliability Models of Comput./Commun. Systems*, Rome, Italy,
May 1987, pp. 275–286.

[WALR 88]

Walrand, J., *An Introduction to Queueing Networks*, Prentice-Hall, Englewood Cliffs N.J., 1988.

[WANG 86]

Wang, I.Y. and Robertazzi, T.G., "The Probability Flux Circulation of
Certain Communication and Computation State Models", *Proceedings of
the 1986 Information Sciences and Systems Conference*, Princeton University, Princeton N.J., March 1986, pp. 865–870.

[WANG]

Wang, I.Y. and Robertazzi, T.G., "Recursive Computation of Steady State
Probabilities of Non-Product Form Queueing Networks Associated With
Computer Network Models", *IEEE Transactions on Communications*,
Vol. 38, No. 1, Jan. 1990, pp. 115–117.

[WANG 89]

Wang, I.Y. and Robertazzi, T.G., "Service Stage Petri Net Protocols with
Product Form Solution", *Proceedings of the 1989 ACM Sigmetrics and
Performance '89 International Conference on Measurement and Modeling
of Computer Systems*, Berkeley, CA, May 1989, pg. 233. Journal version
appears in *Queueing Systems*, Vol. 7, No. 3, 1990, pp. 355–374.

[WELC]

Welch, P., "The Statistical Analysis of Simulation Results", in *The Performance Modeling Handbook*, S. Lavenberg, Ed., Academic Press, New
York, 1983.

[WHIT]

Whitt, W., "A Review of $L = \lambda W$ and Extensions", *Queueing Systems*,
Vol. 9, No. 3, Oct. 1991, pp. 235–268.

[WHIT 81]

Whitt, W., "Approximating a Point Process by a Renewal Process: The
View through the Queue, An Indirect Approach", *Management Science*,
Vol. 27, No. 6, June 1981, pp. 619–636.

[WHIT 82]

Whitt, W., "Approximating a Point Process by a Renewal Process I: Two
Basic Methods", *Operations Research*, Vol. 30, No. 1, Jan./Feb. 1982, pp.
125–147.

[WIES]

Wieselthier, J.E. and Ephremides, A., "Some Markov Chain Problems in the Evaluation of Multiple-Access Protocols", *Proceedings of the First International Conference on the Numerical Solution of Markov Chains*, sponsored by N.C. State University Computer Science Dept., N.C. State University, Raleigh, N.C., Jan. 1990, pp. 258–282.

[WILL]

Willinger, W., Taqqu, M.S., Sherman, R. and Wilson, D.V., "Self-Similarity Through High-Variability: Statistical Analysis of Ethernet LAN Traffic at the Source Level", *IEEE/ACM Transactions on Networking*, Vol. 5, No. 1, Feb. 1997, pp. 71–86.

[WOLF]

Wolff, R.W., *Stochastic Modeling and the Theory of Queues*, Prentice-Hall, 1989.

A very well written text that has strong coverage of stochastic processes associated with queues.

[WONG]

Wong, J., Moura, J. and Field J., "Hierarchical Modeling of Local Area Computer Networks", *Proceedings of the IEEE National Telecommunications Conference*, Houston, Texas, Dec. 1980. pp. 37.1.1–37.1.5.

[WONG CY]

Wong, C.Y., Dillon, T.S. and Forward, K.E., "Timed Places Petri Nets with Stochastic Representation of Place Time", *International Workshop on Timed Petri Nets*, Torino, Italy, IEEE Computer Society Press, July 1985, pp. 96–103.

[WU]

Wu, C.-l. and Feng, T.-y., *Tutorial: Interconnection Networks for Parallel and Distributed Processing*, IEEE Computer Society Press, 1984.

A large collection of previously published papers on interconnection networks.

[XION]

Xiong, Y. and Bruneel, H., "Performance of Statistical Multiplexers with Finite Number of Inputs and Train Arrivals", *Proceedings of the IEEE INFOCOM'92*, Florence, Italy, May 1992, pp. 2036–2044.

[YANG]

Yang, T., Posner, M.J.M. and Templeton, J.G.C., "A Generalized Recursive Technique for Finite Markov Processes", *Proceedings of The First International Conference on the Numerical Solution of Markov Chains*, sponsored by N.C. State University Computer Science Dept., N.C. State University, Raleigh, N.C., Jan. 1990, pp. 216–237.

[YCAR]

Ycart, B., "The Philosophers' Process: An Ergodic Reversible Nearest Particle System", *Technical Report, Statistics 91/18, Laboratoire de Math-*

ematiques Appliquees, University of Pau, France. Address: Laboratoire de Mathematiques Appliquees, U.R.A. CNRS 1204, Faculte' des Sciences, Av. de l'Universite', 64000 Pau-France.

[YEGE]

Yegenoglu, F. and Jabbari, B., "Characterization of Modeling of Aggregate Traffic for Finite Buffer Statistical Multiplexers", *Computer Networks and ISDN Systems*, Vol. 26, 1994, pp. 1169–1185.

[YOON]

Yoon, H., Lee, K.Y. and Liu, M.T., "Performance Analysis of Multibuffered Packet-Switching Networks in Multiprocessor Systems", *IEEE Transactions on Computers*, Vol. 39, No. 3, March 1990, pp. 319–327.

[ZAHO]

Zahorjan, J., Eager, D.L., and Sweillam, H.M., "Accuracy, Speed and Convergence of Approximate Mean Value Analysis", *Performance Evaluation*, North-Holland, Vol. 8, No. 4, August 1988, pp. 255–270.

A new modification of the approximate Mean Value Analysis algorithm for load independent networks is presented. The convergence of AMVA techniques is also examined.

[ZAZA]

Zazanis, M. and Suri, R., "Estimating First and Second Derivatives of Response Time for G/G/1 Queues from a Single Sample Path", Div. of Applied Science, Harvard University, *Tech. Rep.,* 1985.

[ZENI]

Zenie, A., "Colored Stochastic Petri Nets", *International Workshop on Timed Petri Nets*, July 1985, Torino, Italy, IEEE Computer Society Press, 1986, pp. 262–271.

About the Author

Thomas G. Robertazzi was born in Brooklyn NY and now lives on Long Island, NY. He received the Ph.D from Princeton University in 1981 and the B.E.E. from the Cooper Union in 1977. During 1982–83 he was an assistant professor in the electrical engineering department of Manhattan College, Riverdale N.Y. Prof. Robertazzi is presently an associate professor of electrical and computer engineering at the University at Stony Brook. During Fall 1990 he was a visiting research scientist at Columbia University's Center for Telecommunication Research.

In recent years he has taught engineering courses at Stony Brook, The Cooper Union and in industry. These courses include ones on network management and planning, networking, performance evaluation, wireless technology and communication systems.

Prof. Robertazzi's research interests involve the performance evaluation of computer and communication systems as well as network planning. He maintains a very active research program and supervises several Ph.D students. He has published extensively in the areas of parallel processor scheduling, ATM switching, queueing networks, Petri networks and multi-hop radio networks. Along with Dr. James Cheng, Prof. Robertazzi is the co-creator of divisible load models of parallel processor scheduling. Prof. Robertazzi has served as editor for books for the IEEE Communications Society and an associate editor of the journal Wireless Networks. Besides this book he has previously edited a second book on performance evaluation, co-authored a third book on scheduling and authored a fourth book on telecommunications network planning.

Since 1993 Prof. Robertazzi has also been faculty director of the Stony Brook Interdisciplinary Program in Science and Engineering. This is based in a residential undergraduate college and serves to provide an academically enriched environment to the college's residents.

Prof. Robertazzi likes to relax by reading about exploration—be it in space or on ancient seas. He enjoys music, research, seeing all of his students do well and spending time with his family. He is looking forward to the 100th anniversaries of Markov's and Erlang's work.

You can visit Prof. Robertazzi's web page at:
www.ece.sunysb.edu

Index

A

Algebraic Topology, 112–132,237–274
ARMA Process, 337
Arrival Theo., 197,202–203
Arthurs, E., 299
Asymptotic Exp., 214
Autocorrelation, 349
Autocovariance, 349

B

Bailey, N., 69
Banyan Net, 308–309
Baskett, F., 101
Batch Means, 226
Bernoulli Proc., 280
Bessel Func., 72
BCMP Theorem, 111
Binomial Dist., 284–288,300,305
Birman, A., 231
Birth-Death Process, 30
Blocking, 56,59
Blocking Prob., 58
Bodnar, B., 9
Boguslavsky, L.B., 231
Bruell, S./Balbo, G., 165,166, 167,172,173,174,175
Build. Block, 112–129,246
Burke, P., 44,47,52
Burke's Theo., 47–52,91
Burstiness, 343–348
Buzen, J., 165,170,172,176

C

Cavin III, R., 72

Chandy, K., 101,132
Chao, X., 148
Circl. Struc., 118–129
Classes, 5
Coef. of Var., 344
Combinations, 363
Confidence, Int., 225–226
Consist. Cond., 130–132
Consist. Graph, 131
Cont. Time Models, 333–338
Convolution, 362
Convolution Algo., 164–176
  Examples, 176–196
  Marginal Dist., 170–174
  State Dep., 166–167
  State Ind., 167–170
  State Space, 164–166
  Throughput, 174–175
  Utilization, 175–176
Conway, A., 231
Cooper, R., 29,44
Coyle, E., 155
Crossbar, 298–299
Crovella, M. & Bestavros, A., 353
Cum. Dist. Func., 358
Customer, 2
Cyclic Flow, 119

D

DAC, 231
Delay, 43–47,55–56,198, 204
Depart. Inst., 83–84
De Souza E Silva, E., 231
Deterministic Proc., 339

Discrete Time Models, 275–
  325,338–341
DQDB, 312–319

E
Electric Analog, 32,34
Equilibrium Anal., 34
Emer, J., 6
Ephremides, A., 279
Erlang, A., 1
Erlang B Form., 65
Erlang C Form., 63
Erramilli, A., et al., 354
Expectation, 361–362
Exponential Dist., 359–360

F
File Service, 6–8
Florin, G., 237
Fluid Model, 17
Fluid Flow Approx., 337,342–343

G
Garrett, M., 315
Gelenbe, E., 141–142
Geom. Replication, 118
Georganas, N., 231
Gidron, R., 11
Global Balance, 34
Gopinath, B., 34
Gordon, W., 101
Gross, D./Harris, C.,
  39,43,45,69,75,83,86
Grunenfelder, R., 346
Guerin, R., 13

H
Hamilton, R., 155
Hammond, J./O'Reilly, P., 75
Haviv, M., 202

Herzog, U., 132
Heyman, D., 354
Hluchyj, M., 299
Hsiao, M.-T., 221
Huang, H.-Y., 290
Hui, J., 299
Hurst Param., 350,352
Hurst Effect, 352

I
Imbedded Chain, 86
Independence, 358
Ind. Repls., 225
Index of Disp., 344–346
Insensitivity, 155
Inter-arrivals, 27–28
Interconnection, 298
Inter. Det. Proc., 340–341
Inter. Poisson Proc., 336–337
Isolated Circ., 118

J
Jackson, J., 101
Jenq, Y., 309
Johnston, D., 72
Joint Dens. Func., 360
Joint Dist. Func., 360
Jones, S., 72

K
Karol, M., 299
Kaufman, L., 34
Kelly, F., 48
Kendall, D., 75,84
Khinchin, A., 75
Kleinrock, L., 22,28,29,30,
  39,45,56,69,75,90
Kobayashi, H., 45,102
Kogan, Y., 231
Kuehn, P., 333–341
Kurose, J., 223

L
Lavenberg, S., 197,231
Lazar, A., 91,112,221,238
Leland, W., et al., 348,350–352
Length, 119
Li, S.-Q., 315,346
Lightwave, 11–13
Likelihood Ratios, 227
Lindley, D., 84
Linear Eq. Sol., 342
Little's Law, 43–47
Liu, A., 9
Local Balance, 36,53,112–114,246
Lucatoni, D., 337

M
MAN, 313
Mandelbrot, B., 349
Marcum, J., 73
Marginal Dens. Func., 360
Markov, A., 1
Markov Arrival Process, 337
Markov Chain, 29
Markov Modulation, 334–335,339–
  340
Markov Property, 28–29
McKenna, J., 206
Mean Value Analysis, 197–206
  Theory, 197–198,203–204
  Algo., 198,204
  Arrival Theo., 202–203
  Random Routing, 203–204
  Examples, 198–201,204–206
Michiel, H. & Laevens, K., 335,341–
  349,351–354
Mitra, D., 206
Molloy, M., 237,253
Moment, 361–362
Moment Gen. Func., 68–69
Morgan, S., 299
Mouftah, H., 223
Multiprocessor, 9–11,122–
  125,238–249

Muntz, R.R., 101
MVAC, 231

N
Natkin, S., 237
Neg. Customers, 141–154
Neuts, M., 337
Newell, G.F., 101
Nichols, W., 6
Normalization, 111
Norton's Equiv., 219–223

P
Packet Switch, 13–18,34–
  35,297–319
Palacios, F., 101
PANACEA, 206–218
  Asymp. Expans., 214–216
  Error Analysis, 218
  Integral Rep., 208–211
  Normal Usage, 212
  Perf. Meas., 211–212
  P.F. Solution, 207–208
  Pseudonetworks, 216–218
  Transformations, 212–214
Park, K. et al., 353
Paxson, V. and Floyd, S., 348
Peak to Mean Rate, 343–344
Performance Measures, 55–
  56
Perturbation Anal., 227
Petri, C., 237
Petri Nets
  see Stochastic P.N.
Pinedo, M., 148
plaNet, 14
Poisson Dist.
  Mean, Variance, 25–26
Poisson Process, 20–22
Poisson Approx., 288–290
P-K MVA Form., 74–83
Pollaczek, F., 74–75

Probability
  Conditional, 357
  "Density, 360
  Density Func., 359–360
  Flux, 32
  Gen. Func., 68,342
  Joint, 358
  Marginal, 358
  Measure, 357
Processor Sharing, 4–5
Product Form Solution
  Open Nets, 102–108
  Closed Nets, 108–111
Pseudorandom Numbers,
  224

Q
Q Functions, 73
QPSX, 313
Queueing Systems
  Blocking, 56–59
  Discrete Time, 275–319
  FIFO, 4
  Geom/Geom/m/N, 290–293
  Geom/Geom/1/N, 294–297
  Geom/Geom/1, 296–297
  Input, 303–307
  ISDN, 138–139
  LIFO, 4
  LIFOPR, 4
  M/M/1 State Ind., 30–42
  M/M/1 State Dep., 53–67
  M/M/1/N, 56–59
  M/M/∞, 59–61
  M/M/m, 61–63
  M/M/m/m, 64–65
  CPU Model, 65–67
  M/G/1, 74–89
  M/D/1, 82–83
  Neg. Customers, 141–154
  Notation, 3
  Output, 300–303
  Tandem, 135–138
  Window Flow, 139–141

R
Ramakrishnan, K., 206
Random Join, 20–21
Random Split, 20–21
RECAL, 231
Recursions, 132–141
Regenerative Meth., 226
Regeneration Point, 84
Reich, E., 91
Reiser, M., 197,231
Relay Seq., 120,123
Renewal Proc., 338,341
Robertazzi, T., 112,132,135,238,
  290
R/S Plot, 354

S
Sample Point, 357
Sample Space, 357
Schwartz, M., 45,102,138–
  140,167,334,343
Self-Similar Traffic
  Basics, 348–352
  Detection, 353–354
  Hurst Effect, 352
  Network Perf., 354
  Roots, 353
Semi-Markov Proc., 338
Service Discipline, 4–5
Service Times
  Exponential, 29
  General, 74
  Geometric, 290
Shift Reg. Seq., 126–129
Simulation
  Decomposition, 229–230
  Optimization, 227–228
  Parallel, 228–229
  Sensitivity, 226–228
  Stat. Nature, 224–226
  Variance Reduc., 230
Solution Methods, 341–343
Spectral Charac., 346–347
State Space Size, 35,309–310

Steady-State Anal., 34
Stochastic Petri Nets
  Alt. Bit. Model, 253–255
  Conferences, 269
  CSMA Model, 251–253
  Dining Phil. Prob., 249–251
  Multiprocessor, 238–249
  Non-Prod. Form Sol., 260
  Packages, 269
  Prod. Form Solution, 243–249
  Resource Sharing, 257–263
  Rules, 247–248
  Synchronization, 263–268
  Toroidal Lattices, 243–249
Summation Formula, 363
Switched Poisson Proc., 336,
  340

T
Temple, A., 11
TeraNet, 11–13
Thomas, J., 26
Throughput, 55,174–175,198,204
TOMP, 239
Traffic Eqs., 103,109
Transient Effects, 343

Transient M/M/1
  Analysis, 68–72
  Computation, 72–74
Type A, 133
Type B, 133

U
Utilization, 56,175–176

V
Variance, 362
Var.-Time Plot, 354

W
Walrand, J., 48
Wang., I., 112,135,269
Wieselthier, J., 279
Willinger, W. et al., 353
Woo, L., 132
Wunderlich, E., 34

Y
Ycart, B., 251